AF594989

Stem Cells and Irradiation

Stem Cells and Irradiation

Editor

Alain Chapel

MDPI • Basel • Beijing • Wuhan • Barcelona • Belgrade • Manchester • Tokyo • Cluj • Tianjin

Editor
Alain Chapel
Laboratoire de Radiobiologie
des Expositions Médicales
(LRMED)
Institut de Radioprotection et
de Sûreté Nucléaire (IRSN)
Fontenay-aux-Roses
France

Editorial Office
MDPI
St. Alban-Anlage 66
4052 Basel, Switzerland

This is a reprint of articles from the Special Issue published online in the open access journal *Cells* (ISSN 2073-4409) (available at: www.mdpi.com/journal/cells/special_issues/stem_cell_irradiation).

For citation purposes, cite each article independently as indicated on the article page online and as indicated below:

LastName, A.A.; LastName, B.B.; LastName, C.C. Article Title. *Journal Name* **Year**, *Volume Number*, Page Range.

ISBN 978-3-0365-6901-7 (Hbk)
ISBN 978-3-0365-6900-0 (PDF)

Contents

 cells MDPI

Editorial

Stem Cells and Irradiation

Alain Chapel [1,2]

[1] Service de Recherche en Radiobiologie et en Médecine Régénérative (SERAMED), Laboratoire de Radiobiologie des Expositions Médicales (LRMED), Institut de Radioprotection et de Sûreté Nucléaire (IRSN), F-92260 Fontenay-aux-Roses, France; alain.chapel@irsn.fr; Tel.: +33-1-5835-9546

[2] Centre de Recherche Saint-Antoine, CRSA, INSERM, Sorbonne Université, Saint Antoine Hospital, 75012 Paris, France

The main difficulty of radiotherapy is to destroy cancer cells without depletion of healthy tissue. Stem cells and cancers are tightly interrelated. On the one hand, radiosensitivity/radioresistance of cancer stem cells affects the radiocurability of tumors, on the other hand, radiosensitivity is responsible for the stem cell depletion of organs at risk exposed to irradiation. Efficient solid cancer destruction is limited by the preservation of organ homeostasis. For this reason, targeted irradiation is an effective cancer therapy, however, damage inflicted to normal tissues surrounding the tumor may cause severe complications. The consequences of stem cell depletion of healthy tissue irradiated are acute and chronic radiation diseases. The depletion of endogenous stem cells can be compensated by a supply of exogenous stem cells. For this reason, cell therapy is a therapeutic approach that offers a therapeutic alternative to patients who have failed conventional treatment. This domain will bring forth the solution for optimal radiocurability associated with long-term patients' quality of life.

This special issue covers research on the radiosensitivity of cancer stem cells and adult stem cells associated with tissue regenerative medicine.

Integration of the cross talk of these two types of stem cells is essential. Nagle and colleague studied the roles of organoids as model to understand relationship between normal tissue and tumor responses in radiobiological studies [1]. Understanding the radioresistance mechanisms of cancer cells is fundamental in order to be able to eradicate the tumor while preserving healthy tissue. Two research articles dealt with cancer cells. The article of Park and colleagues demonstrated that SOX2 contributed to the induction of colorectal cancer cells and is regulated by radio-induced activation of the PI3K/AKT pathway [2]. Kamble and colleagues studied the relationships between radioresistance and breast cancer stem cells mediated by Nrf2-Keap1 pathway. Nrf2-Keap1 signaling controls mesenchymal–epithelial plasticity and regulates tumor-initiating ability and promotes the radioresistance of breast cancer stem cells [3].

Citation: Chapel, A. Stem Cells and Irradiation. *Cells* **2021**, *10*, 760. https://doi.org/10.3390/cells10040760

Received: 19 March 2021
Accepted: 26 March 2021
Published: 30 March 2021

Publisher's Note: MDPI stays neutral with regard to jurisdictional claims in published maps and institutional affiliations.

The consequence of restoring the homeostasis of healthy tissue is first of all to allow tissue regeneration and then organ functionality on a permanent basis. A supply of mesenchymal stromal cells (MSCs) ensures this functionality, mainly through a trophic effect.

Two research articles dealt with MSCs in the treatment of radiological burns. Brunchukov and colleagues studied the effect of human MSCs derived from the placenta and their conditioned medium concentrate on skin-regenerative processes. The use of conditioned MSCs in severe local radiation injuries accelerates the transition of the healing process to the stage of regeneration and epithelization [4]. Cavallero and colleagues settled a method founded on an in vitro bioengineering of human skin organoids, joined with in vivo xenografting in immune-deficient mice. This model was used to understand significances of exposure epidermal stem cells to low-dose irradiation and their consequences in radio-dermatitis [5]. Thanks to its trophic effect, the treatment of healthy tissue by MSCs seems to be able to address many of the pathologies resulting from radiotherapy. Synthesis reports explore the recent progress and discuss the future perspectives about MSCs and MSC-exosomes for mitigating radiotherapy side effects [6–8].

Each pathology related to radiotherapy is complex, which is why it is necessary to apprehend all the elements of a pathology and its treatment but also to be able to understand in detail the associated mechanisms. Helissey and colleagues reviewed radiation cystitis and its treatments. The authors investigated the role of immunity with a special focus on macrophages. They concluded that MSCs seem to be an excellent therapeutic substitute for the treatment of fibrosis in chronic radiation cystitis [9].

To go further into the abovementioned issue and end on an optimistic note, it is interesting to associate the beneficial effect of irradiation with cell therapy in order to propose novel treatments for new pathologies. Tovar and colleagues tried a therapeutic approach, based on MSCs stimulated with radiation, to improve pneumonia caused by SARS-CoV-2. The activation of the immune system by the irradiated tumor to trigger the beneficial abscopal effect is decisively improving radiotherapy applications and their outcomes [10].

References

1. Nagle, P.W.; Coppes, R.P. Current and Future Perspectives of the Use of Organoids in Radiobiology. *Cells* **2020**, *9*, 2649. [CrossRef] [PubMed]
2. Park, J.H.; Kim, Y.H.; Shim, S.; Kim, A.; Jang, H.; Lee, S.J.; Park, S.; Seo, S.; Jang, W.I.; Lee, S.B.; et al. Radiation-Activated PI3K/AKT Pathway Promotes the Induction of Cancer Stem-Like Cells via the Upregulation of SOX2 in Colorectal Cancer. *Cells* **2021**, *10*, 135. [CrossRef] [PubMed]
3. Kamble, D.; Mahajan, M.; Dhat, R.; Sitasawad, S. Keap1-Nrf2 Pathway Regulates ALDH and Contributes to Radioresistance in Breast Cancer Stem Cells. *Cells* **2021**, *10*, 83. [CrossRef] [PubMed]
4. Brunchukov, V.; Astrelina, T.; Usupzhanova, D.; Rastorgueva, A.; Kobzeva, I.; Nikitina, V.; Lishchuk, S.; Dubova, E.; Pavlov, K.; Brumberg, V.; et al. Evaluation of the Effectiveness of Mesenchymal Stem Cells of the Placenta and Their Conditioned Medium in Local Radiation Injuries. *Cells* **2020**, *9*, 2558. [CrossRef] [PubMed]
5. Cavallero, S.; Neves Granito, R.; Stockholm, D.; Azzolin, P.; Martin, M.T.; Fortunel, N.O. Exposure of Human Skin Organoids to Low Genotoxic Stress Can Promote Epithelial-to-Mesenchymal Transition in Regenerating Keratinocyte Precursor Cells. *Cells* **2020**, *9*, 1912. [CrossRef]
6. Wang, K.-X.; Cui, W.-W.; Yang, X.; Tao, A.-B.; Lan, T.; Li, T.-S.; Luo, L. Mesenchymal Stem Cells for Mitigating Radiotherapy Side Effects. *Cells* **2021**, *10*, 294. [CrossRef]
7. Pu, X.; Ma, S.; Gao, Y.; Xu, T.; Chang, P.; Dong, L. Mesenchymal Stem Cell-Derived Exosomes: Biological Function and Their Therapeutic Potential in Radiation Damage. *Cells* **2021**, *10*, 42. [CrossRef] [PubMed]
8. Rezvani, M. Therapeutic Potential of Mesenchymal Stromal Cells and Extracellular Vesicles in the Treatment of Radiation Lesions—A Review. *Cells* **2021**, *10*, 427. [CrossRef]
9. Helissey, C.; Cavallero, S.; Brossard, C.; Dusaud, M.; Chargari, C.; François, S. Chronic Inflammation and Radiation-Induced Cystitis: Molecular Background and Therapeutic Perspectives. *Cells* **2021**, *10*, 21. [CrossRef] [PubMed]
10. Tovar, I.; Guerrero, R.; López-Peñalver, J.J.; Expósito, J.; Ruiz de Almodóvar, J.M. Rationale for the Use of Radiation-Activated Mesenchymal Stromal/Stem Cells in Acute Respiratory Distress Syndrome. *Cells* **2020**, *9*, 2015. [CrossRef] [PubMed]

 cells

Review

Therapeutic Potential of Mesenchymal Stromal Cells and Extracellular Vesicles in the Treatment of Radiation Lesions—A Review

Mohi Rezvani [1,2]

[1] CSO, Swiss Bioscience GmbH, Wagistrasse 27 A, Schlieren, 8952 Zurich, Switzerland; mohi@swissbioscience.com

[2] Reader Emeritus, Faculty of Clinical Medicine, University of Oxford, Kidlington, Oxfordshire OX5 2AX, UK

Abstract: Ionising radiation-induced normal tissue damage is a major concern in clinic and public health. It is the most limiting factor in radiotherapy treatment of malignant diseases. It can also cause a serious harm to populations exposed to accidental radiation exposure or nuclear warfare. With regard to the clinical use of radiation, there has been a number of modalities used in the field of radiotherapy. These includes physical modalities such modified collimators or fractionation schedules in radiotherapy. In addition, there are a number of pharmacological agents such as essential fatty acids, vasoactive drugs, enzyme inhibitors, antioxidants, and growth factors for the prevention or treatment of radiation lesions in general. However, at present, there is no standard procedure for the treatment of radiation-induced normal tissue lesions. Stem cells and their role in tissue regeneration have been known to biologists, in particular to radiobiologists, for many years. It was only recently that the potential of stem cells was studied in the treatment of radiation lesions. Stem cells, immediately after their successful isolation from a variety of animal and human tissues, demonstrated their likely application in the treatment of various diseases. This paper describes the types and origin of stem cells, their characteristics, current research, and reviews their potential in the treatment and regeneration of radiation induced normal tissue lesions. Adult stem cells, among those mesenchymal stem cells (MSCs), are the most extensively studied of stem cells. This review focuses on the effects of MSCs in the treatment of radiation lesions.

Keywords: radiation; mesenchymal; stem cell; extracellular vesicles; micro vesicles; paracrine effect; adipose tissue derived stem cells

Citation: Rezvani, M. Therapeutic Potential of Mesenchymal Stromal Cells and Extracellular Vesicles in the Treatment of Radiation Lesions—A Review. *Cells* **2021**, *10*, 427. https://doi.org/10.3390/cells10020427

Academic Editors: Alain Chapel and Joni H. Ylostalo

Received: 11 November 2020
Accepted: 13 February 2021
Published: 18 February 2021

1. Development of Radiation-Induced Lesions

It is now well accepted that the human body contains adult stem cells or in other words post-natal stem cells that are capable of differentiating into other tissues and can regenerate or repair damaged tissues. Over the last decades, stem cell hypothesis, the development of tissue deficits due to the inability of stem cells to replenish lost cells, has become a reality. Stem cells were in a way studied by radiobiologists well before it was proposed as a hypothesis. In fact, the initial theory of the development of radiation lesions' "target cell theory" was based on radiation-induced cell loss. Target cell theory introduced by Puck and Marcus [1] considers cell loss as the cardinal cause of radiation induced normal tissue damage or tumour ablation. In recent years, it has been shown that the process of development of radiation damage and the damage itself starts by molecular changes long before denudation of target cells. However, one cannot deny the fact that the ultimate lesions manifest as loss of functional cells. Most bodily tissues possess a pool of clonogenic cells that are mobilised in response to assaults such as trauma or radiation. Damage to the tissue is repaired by proliferation of clonogenic or tissue specific stem cells. Sterilisation of these clonogenic cells by radiation manifests as radiation damage. In mild cases as the damage is sensed, these clonogenic cells migrate to the site of damage, and together with local surviving clonogic cells, proliferate to repair the tissue. However, in severe cases of tissue repairs, there might not be enough surviving clonogenic cells as the site of damage or sufficient number of mobilised cells to reach the site and repair the damage. Thus, the damage gets established as a result of failure of endogenous stem cells to regenerate the damaged tissue.

In early responding tissues, such as gut, oral mucosa, or epidermis of skin, the initiation of molecular events triggered by radiation results in the loss of both clonogenic and differentiated functional cells. Loss of clonogenic cells or in other words basal stem cells results in a deficiency to replace the lost functional cells. In the event of survival of a sufficient number of proliferating tissue specific stem cells in the irradiated region or its vicinity, complete healing is observed. However, in severe cases, where the radiation causes sterilisation of the tissue-specific clonogeic cells, denudation of the tissue will follow. Deficiency of the stem cells to produce new cells to replace the lost cells and resulting imbalance brings about the erosion of the epithelial layer.

In late responding tissues, such as late dermal reaction of skin or central nervous system damage, the involvement of stem cells are also established. However, the pattern of development of lesions in late responding tissues is more complex, as the response of slow turnover tissues (such as neural tissue) differ from the response of rapid turnover tissues (such as epithelial tissue). In late responding tissues, the overall tissue response is dependent on more than one cell type and their response to irradiation. The complex process of late radiation damage is initiated by a cascade of molecular events from injured cells that result in eventual denudation of functional differentiated cells. The response develops as the consequence of damage to both slow and rapid turnover tissues. For example, a rapid unset of radiation-induced apoptosis has been reported as early as 3–6 h in dentate gyrus after irradiation of rat brain [2,3]. These authors also reported a higher number of apoptosis than the number of proliferating cells and concluded that non-proliferating cells as well as proliferating cells in the subgranular zone of rat brain were sensitive to radiation and cell number, in this region, was significantly lower than age-matched controls 120 days after irradiation. This in part can be the cause of radiation-induced cognitive deficit. A dose-dependent reduction in the number of subepidermal cells in irradiated rat brain and the inability of surviving stem cells in regenerating the subepiderma, that manifest a clear deficit at 180 days after irradiation, was reported [2]. Deficiency of stem cells to regenerate the lost tissue results in the development of scaring or fibrosis as a final lesion. Therefore, the replacement of stem cells by donor stem cells, possibly before establishment of the lesion, may prevent the development or shorten the duration/severity of the lesions in both early and late responding tissues.

2. Treatment of Radiation Lesions with Stem Cells

Radiation lesions is amenable to treatment by methods that result in repairing or regeneration of the damaged tissue. In fact, stem cell transplantation in medical practice is not new and have been used for decades in bone marrow transplantation [4].

Stem cell treatment of radiation damage is based on the assumption that the transplanted cells integrate with the damaged host tissue to replace the damaged/lost cells or stimulate the host cells to prevent the damage or regenerate the damaged tissue. The later will obviously be more efficient before establishment of the radiation damage. Transplanting the stem cells before the full establishment of radiation lesion can prevent the development of radiation damage or shorten the duration of the manifestation of the lesion.

Bone marrow transplantation has been successfully used in the treatment of leukaemia, lymphoma, and certain types of anaemia procedures. Initial efforts in this field were directed towards transplantation of pre-differentiated stem cells and a good example of this is bone marrow transplantation that started as early as 1951 with the work of Lorenz [5] who found that infusion of the spleen or marrow cells could protect the irradiated mice. Bone marrow transplantation is based on allogenic use of stem cells. Whole marrow or stem cells of the marrow are extracted from a donor and transplanted to the host to reconstruct the haemopoietic tissues of cancer patients. The patient, prior to bone-marrow transplantation, is myeloablated by radiation or chemotherapy. The process of bone marrow transplantation is reviewed by [4].

Later, non-tissue specific or naive stem cells were transplanted on the basis of the opinion that the niche, or local microenvironment, consisting of surrounding cells, will define the fate of the transplanted cells and direct the administered stem cells to lodge into target tissue and differentiate into the required cells to restore structural and functional deficits.

In this article, a number of papers indicating the application of stem cells in the treatment of radiation-induced lesions are reviewed. It is also argued that the beneficial effect of transplanted stem cells in irradiated bodies is not necessarily due to the lodging of the transplanted stem cells in the irradiated tissue to replace the lost/damaged cells. It is suggested that perhaps the result is by paracrine effect; i.e., transplanted stem cells secrete bioactive substances that are capable of stimulating the host cells to reproduce and repair the damaged tissue. This means that the transplanted stem cells, besides integrating in the structure of damaged tissues, secrete biologically active factors, mainly in the form of extracellular vesicles, such as exosomes and microvesicles, that stimulate and mobilise the endogenous stem cells to repair the damage. Recently, it was shown by many researchers including ourselves, that the effect of stem cells is exerted in a paracrine fashion [6–8]. Transplanted stem cells, by integration with the host tissue, mobilisation of endogenous stem cells, or a combination of both mechanisms, result in functional and structural improvements of injured tissues. For a review on extracellular vesicles, see [9,10].

3. Types of Stem Cells

Stem cells are undifferentiated cells that are capable of dividing to produce more stem cells and/or differentiate specialised cells. Stem cells are classified by their potentiality into three main types; multipotent, pluripotent, and totipotent. Totipotent stem cells can generate an entire individual. Pluripotency is the ability of certain cells to differentiate into the three embryonic layers (ectoderm, mesoderm, and endoderm). Multipotency is the ability of stem cells to differentiate into one or two embryonic layers such as mesoderm and endoderm. In contrast, adult stem cells are multipotent cells.The stem cells currently used in medical applications or studied in research can be divided into three main types.

(1) Embryonic stem cells (ES): these are pluripotent cells located at the inner cell mass of blastocysts. Embryonic stem cells are usually harvested around four days after fertilisation when the embryo is in its blastula phase [11]. Embryonic stem cells can be differentiated into any one of the three germ layers; endoderm, mesoderm, or ectoderm.

(2) Induced pluripotent stem cells (iPCs): these cells, as indicated by their name are pluripotent that are generated from mature somatic cells, like skin or blood cells, by introduction of transcription factors for encoding certain genes. This is in fact back reprogramming of mature cells to embryonic stem cell state. The classic mixture of transcriptions factors to produce iPSCs consist of Oct3/4, Sox2, Klf4, and c-Myc [12].

(3) Adult stem cells: This is another group of stem cells that are multipotent. Adult stem cells or adult progenitor cells are tissue-specific stem cells are available almost in all body tissues [13] such as epidermal stem cells of skin, stem cells of human hair follicles, cardiac stem cells of heart, neural stem cells of the brain, hepatic stem cells, intestinal stem cells, dental pulp stem cells, ovarian epithelial stem cells, mammary stem cells, testicular stem cells, and satellite cells/myogenic stem cells of the skeletal muscle. Hemopoietic stem cells and mesenchymal stem cells are other groups of adult stem cells. Hemopoietic stem cells are derived from blood vessels and bone marrow. Mesenchymal stromal cells (MSCs) are another type of multipoint adult cells [14–16] found in bone marrow, adipose tissue [17,18], and almost all postnatal tissues [19]. MSCs are non-hematopoietic stem cell-like cells first identified by Friedenstein [20,21] and their characteristics are described [22]. In bone marrow, MSCs have a supportive role for hematopoietic stem progenitor cells (HSPCs) that is also involved

in the maintenance of marrow microenvironment by secreting bioactive factors [23]. MSCs of adipose tissue are termed Adipose Tissue-derived Stem cells (ADSCs), which, like other MSCs are spindle-shaped plastic adherent cells, capable of differentiating to other cells [24,25]. Another source of MSCs (UC-MSCs) is umbilical cord blood [26] or Wharton jelly of umbilical cord [27,28]. UC-MSCs like other MSCs differentiate into three germ layers and contribute to tissue repair and regeneration [29].

ES and IPS cells have the advantage of indefinite renewal and the ability to differentiate into all cell types. This property gives them a role in replacing damaged cells by direct differentiation. On the other hand, adult stem cells are limited in their proliferation. Adult stem cells can either differentiate to replace specialized cells but in a limited number of cases. This is the case, for example, with MSCs that differentiate into osteoblasts. On the other hand, when adult stem cells come to repairing tissue from which they did not originate, they preferentially act by trophic effect, such as MSCs to allow intestinal regeneration.

The Mesenchymal and Tissue Stem Cell Committee of the International Society for Cellular Therapy [30] states three conditions as the minimal criteria for definition of human MSC. (1) MSC must be plastic-adherent, (2) express CD105, CD73, and CD90, and lack expression of CD45, CD34, CD14 or CD11b, CD79alpha or CD19 and HLA-DR surface molecules, and (3) differentiate into osteoblasts, adipocytes, and chondroblasts in vitro. MSCs has been shown to differentiate into endodermal lineage such as hepatocytes [31], cardiomyocytes [32], and ectodermal lineage neurons [33].

MSCs are the most extensively studied adult stem cells and BM-MSCs are the first to be transplanted and used in regenerative medicine, including treatment of radiation lesions. Alternatively, ADSCs appear to be a better kind of MSCs [34]. Furthermore, ADSCs can be obtained by lipoaspiration, which is much less invasive than obtaining BM-MSCs by bone marrow aspiration. ADSCs exhibit intermediate radiation sensitivity [35] and it appears that irradiation of human ADSCs with low-level laser changes their morphology and enhances their proliferation and therapeutic potential [36]. The potential of mesenchymal stem cell therapy in the treatment of radiation-induced lesions has been reviewed [37].

4. Homing of Transplanted Stem cells

MSCs, for regenerative purposes, can be transplanted directly into the site of damage or introduced systemically. In the latter, it is assumed that homing of the transplanted cells is regulated by the local microenvironment and they are directed to the site of injury by cues from damaged tissues of the host through a series of signals. Furthermore, the transplanted cells secrete diverse trophic factors and immunomodulatory substances that contribute to the process of regeneration by stimulating the endogenic stem cells. In majority of the studies of the distribution of transplanted cells in irradiated animals, it has been shown that the transplanted cells home to the radiation-damaged tissues. MSCs intravenously transplanted to rats with myocardial lesions home to the infarct region of the heart, while in uninjured control animals, the transplanted cells migrated to the bone marrow [38]. In the treatment of radiation-induced multi-organ failure in non-human primates, transplanted MSCs home to injured tissues [39]. Human MSCs were systemically transplanted into total body or abdominal irradiated NOD/SCID mice [40,41]. It was reported that the transplanted cells home to the irradiated organs and were found three months post irradiation. These observations support the hypothesis that transplanted stem cells migrate to radiation-induced injury sites in irradiated animals. However, this does not seem to be specific to radiation lesions as migration of transplanted stem cells to non-radiation damaged tissues has been reported too. In an acute nontransmural myocardial infarct model [42], it was shown that transplanted MSCs mainly home to the infarct myocardial region observed 24 h after intravenous transplantation that lasted for 7 days after transplantation. However, these authors observed some migration to non-target organs as well but the main concentration was in the infarct region.

Homing factors are crucial in the delivery of stem cells to damaged tissues. Some homing factors have been identified. For example stromal cell-derived factor-1 (SDF-1) is known to allow the targeting of hematopoietic stem cells to the marrow when it needs to be recolonized by hematopoietic stem cells. The secretion of SDF-1 similarly allows the homing of MSCs that express the C-X-C Motif Chemochine Receptor-4 (CXCR4) molecule, which is the receptor for the SDF-1 molecule. Another chemokine, Monocyte Chemotactic Protein-1 (MCP-1), was found to be a key regulator for stem cell recruitment to the myocardium in or cochlear tissue.

5. Stem Cell Treatment of Radiation Lesions

Interest in the application of mesenchymal cells as therapeutics has increased recently. A few early stage clinical trials have also been reported [43–46] but in general one can say that treatment with MSCs is still in an experimental phase and larger clinical trials are needed before its clinical use. Safety of MSCs in clinical trials have been reviewed and adverse effects listed [47]. The safety of MSCs for the treatment of radiation lesions has also been reported [48].

Like other cells, irradiation of MSCs induces senescence and/or apoptosis [49]. This has been shown in MSCs isolated from irradiated human skin, where colony formation, proliferation, and differentiation capacity are reduced [50].

MSCs have been shown not to give rise to tumours [51] as they are non-tumourgenic [52].

6. Studies on Hematopoietic System

Although interest in stem cell treatment increased over the last two decades, stem cell transplantation started more than half a century ago with bone marrow transplantation by Lorenz et al. [5] followed by Barnes et al. [53]. These authors demonstrated that transplantation of bone marrow cells could protect mice against ionising radiation. This was the pioneering process of bone marrow transplantation that developed as a routine clinical procedure, where whole marrow or marrow cells extracted from bone marrow are transplanted into myeloablated host in the treatment of both malignant and non-malignant diseases such as leukaemia, lymphoma, and certain types of anaemia [54].

The effect of transplantation of bone marrow-derived mononuclear cells in non-human primates were studied by Bertho et al. [55]. These authors demonstrated that cell transplantation 24 h after 8 Gy total body irradiation shortened the period and severity of pancytopenia. Acute radiation syndrome (ARS), besides multi-organ failure, causes pancytopenia too. The efficacy of transplantation of human UC-MSCs to combat the effects of ARS was also studied [56]. However, in this study, UC-MSCs were modified to to express human extracellular superoxide dismutase. The regenerative potential of MSCs combined with the antioxidant effect of human extracellular superoxide dismutase was intended to produce a rapid and effective strategy for the treatment of radiation accident victims.

The protective effects of allogenic stem cell transplantation against acute radiation syndrome was demonstrated by transplantation of human umbilical cord-derived MSCs in mice [57].

7. Studies on Nervous System

Study of the regeneration of nervous system after irradiation was first started by transplantation of oligodendrocyte progenitor cells [58,59]. The results showed significant remyelation of radiation-induced demyelinated rat spinal cord.

Later, regenerative properties of transplantation of two types of neural stem cell were studied in a rat model of radiation myelopathy [60]. Twelve millimetre of rats' spinal cord was irradiated with 22 Gy gamma rays. This was ED_{100} in six months in this model of radiation myelopathy. Neuroepithelial stem cells were obtained from the hipocampal proliferative analogue on embryonic day 14 from an H-2kb-tsA58 transgenic mouse. It was believed that both cell types were multipoint stem cells because they were 90% nestin positive in culture and they had been shown to differentiate into neurons, oligodendrocytes, and astrocytes. Stem cells were transplanted, intradurally, three months after spinal cord irradiation. While control animals developed front leg paralysis within 183 days after irradiation, 30% of animals in stem cell transplanted group stayed paralysis free until day 200.

Wei et al. [61] using rat cervical spinal cord irradiation model irradiated 20 mm of cervical spinal cord of rats and injected one million UC-MSCs through the tail vein at 90 days after injection followed by three weekly injections. These authors demonstrated that multiple injections of stem cells significantly improved neuron survival and locomotor recovery at 180 days post irradiation.

In a rat model of cranial irradiation [62], human embryonic stem cells were transplanted to the hippocampus of athymic nude rats two days after 10 Gy cranial irradiation. This resulted in a significant cognitive improvement four months after irradiation compared to the controls that did not receive stem cell transplantation. The same authors observed similar results in the same model after transplantation of human neural stem cells [63]. These authors reported equivalent cognitive restoration with both types of stem cell transplantations [64]. Efficacy of stem cell therapy in amelioration of radiation-induced brain damage is reviewed by Chu et al. [65].

8. Studies on the Gut

Semont et al. [66] studied the regenerative effects of transplantation of human BM-MSCs in NOD/SCID mice. Transplantation was by infusion and the results were assessed by functional and histological assessment of the jejunum. The results demonstrated both structural and functional improvements by MSC transplantation.

The effect of autologous bone marrow derived stem cell treatment was studied in a pig model of irradiation proctitis, developed by 4MV photons [67]. It was demonstrated that repeated administration of mesenchymal stem cells resulted in reduction of collagen deposition and radiation-induced fibrosis. Reduction in expression of inflammatory cytokines both systemically and in rectal mucosa were also observed.

In a rat model of colorectal cancer, transplantation of allogenic MSCs significantly improved normal tissue damage induced by radiotherapy [68]. This study also demonstrated that MSC transplantation increased the tumour-free survival of the animals. The number of tumour free animals was higher than expected while the incidence and size of the tumours were reduced.

In our own laboratory (unpublished work), the effect of transplantation by ip injection of human ADSCs on gut was studied in rats. In this study, four cm of rats' distal colon were irradiated with 11 Gy 250 kV X-rays while the rest of the animal was shielded. Twenty four hours after irradiation, the animals were grouped into six groups and treated. Group 1: unirradiated controls received only one ml PBS injection, Group 2–6 received radiation followed by one ml saline injection (radiation only- Group 2), two million ADSCs suspended in one ml PBS (Group 3), two million ADSCs lysate in one ml saline (Group 4). One ml conditioned media collected from 2 million ADSC cultures (Group 5) injected ip and finally conditioned media administered three times 24 h, 72 h, and 120 hrs after irradiation (Group 6). The results were assessed by counting the number of crypts per circumference by light microscopy nine days after irradiation. As expected, radiation only reduced the number of crypts significantly compared with unirradiated control group. Injection of 2 million intact ADSCs, lysate, or a single dose of conditioned media increased the number of crypts almost equally. However, the best result was obtained by three consecutive injections of conditioned media. Comparable results obtained from injection of intact MSCs or the lysate of the equivalent number of cell indicates the possibility of a paracrine effect. This was also confirmed that the outcome of conditioned media injection that usually contains mi-RNA, a number of proteins, and biologically active lipids was more effective than the intact stem cells injections.

The possibility of the paracrine effect was indicated in a similar study [69] where the effectiveness of secretions of human UC-MSC to prevent radiation-induced intestinal injury was investigated in BALB/C mice after 10 Gy cobalt irradiation. In this study, UC-MSCs were expanded under hypoxic conditions. Multiple injections of the hypoxic conditioned media was delivered to the animals after irradiation for seven days. This treatment improved the structure of the intestine, decreased diarrhoea, and increased the survival rate.

Paracrine effect of stem cell transplantation was also shown in a study by Chen et al. [70] where conditioned media obtained from rat bone marrow MSCs were injected into rats just before irradiation. The results indicated that the conditioned media injection increased the expression of anti-inflammatory cytokines and reduced the expression of inflammatory cytokines.

In a recent study [71], total body irradiated mouse, at a dose of 7 Gy (^{60}Co γ-rays), received intravenous injections of one million human placenta-derived stem cells for 10 days after irradiation and compared with another group of animals that received radiation only. Ten days after irradiation, radiation-induced small intestinal damage was compared with that of a control group. It was shown that stem cell transplantation significantly improved ($p < 0.01$) the outcome of radiation enteropathy or lethal radiation syndrome. It was also shown that stem cell transplantation exerted inhibitory actions on inflammatory cytokines and assisted the re-establishment of epithelial homeostasis.

In a rat model of colonic anastomosis performed by irradiation [72], it was shown that transplantation of rat ADSCs promoted anastomotic healing of the irradiated colon through enhanced vessel formation and reduced inflammation. In this study, the ADSC injections were delivered several times before and after the surgical procedure.

Sémont et al. [66,73] described the effects of MSCs as a consequence of their ability to improve the renewal capability of the small intestine epithelium. They also suggested that MSC treatment favours the re-establishment of cellular homeostasis by both increasing endogenous proliferation processes and inhibiting radiation-induce apoptosis of the small intestine epithelial cells.

MSC treatment decreased the interactions between mast cells and nerve fibers and reversed mechanical visceral hypersensitivity [74]. These authors suggest that the mechanism of effect is that the MSCs release cytokines and growth factors, such as interleukin (IL)-11, human hepatocyte growth factor, fibroblast growth factor-2, and insulin-like growth factors. Each of these factors have been described earlier as facilitating intestinal mucosa repair, either through enhancement of cell proliferation or inhibition of epithelial cell apoptosis [66,69,73,74].

9. Studies on the Liver

Prevention of radiation-induced liver damage was the subject of study well before the establishment of mesenchymal cells as stem cells. In an earlier work [75], lethally irradiated mice were treated with syngeneic fetal liver cells that resulted in longer survival.

Later, the effects of BM-MSC transplantation on irradiated liver was studied in NOD/SCID mice [76]. In this study, animals received 10.5 Gy of ^{60}Co gamma rays, followed by intravenous delivery of 5 million human BM-MSCs five hours after irradiation. This study demonstrated that MSC transplantation reduced radiation-induced apoptosis and significantly reduced the transaminase values (AST and ALT) compared with irradiated but not transplanted animals.

In a study of the effects of hepatic irradiation on transplanted BM-MSCs in cirrhotic rats and the underlying mechanism by which mesenchymal stem cells (MSCs) relieve liver fibrosis [77], the BM-MSCs from male rats were injected via portal vein into two groups of thioacetamide-induced cirrhotic rats. The right hemiliver of one cirrhotic rat group was irradiated (15 Gy) four days before transplantation. It was shown that the transplantation of MSCs alleviated liver fibrosis and reduced expression of transforming growth factor-β1, Smad2, and collagen type I. In addition, hepatic irradiation promoted homing and repopulation of MSCs and enhanced the effect of BM-MSCs in improving thioacetamide-induced liver fibrosis in rats. The authors concluded that BM-MSCs may function by inhibiting transforming growth factor-β-Smad signaling pathway in the liver.

10. Studies on the Lung

Mice exposed to thoracic irradiation were injected intravenously on days 0 and 14 after irradiation with genetically modified MSCs, expressing soluble transforming growth factor-b, MSCs conditioned media (MSC-CM). Sixty weeks after irradiation, all animals in the control group that had received only PBS injection after irradiation died. The survival rate of MSC and MSC-CM groups was 40% and 80%, respectively. The thickness of alveolar septa, malondialdehyde in lung homogenates, and plasma TGF-β1 levels significantly decreased in mice treated with either MSCs or MSC-CM, indicating the protective effects of MSC transplantation or MSC-CM injection, which reflects the paracrine effect of MSCs [78].

Improvements in acute radiation-induced lung injury has been demonstrated by Jiang et al. [79]. These authors injected rat ADSCs through the tail vein to right lung irradiated rats two hours after irradiation with 15 Gy X-rays. ADSC transplantation resulted in increased serum levels of anti-inflammatory cytokine IL-10 and reduced serum levels of the pro-inflammatory cytokines TNF-alpha, IL-1, and IL-6.

Human umbilical cord stem cells were transplanted 24 h before or 24 h after lung irradiation in rats [80]. The results demonstrated alleviation of radiation pneumonitis in both groups in comparison with the controls. Transplantation of umbilical cord MSCs have also been shown to be beneficial in the prevention of radiation-induced lung fibrosis [81,82]. However, these authors have shown that modification of stem cells to produce manganese superoxide dismutase significantly enhances the modulatory effect of MSC transplantation. Furthermore, MSC transplantation has been shown to reduce the incidence of lung metastasis in mice [83], beside lowering radiation-induced lung injury.

Feasibility and mode of action of mesenchymal stem cell therapy in amelioration of radiation-induced lung injury have been reported by Xu [84].

11. Studies on the Skin

Francois et al. [85] irradiated the skin of the hind leg of NOD/SCID mouse with 30 Gy single dose of Cobalt-60 gamma rays. Human BM-MSCs was transplanted by intravenous injection 24 h after irradiation. In stem cell transplanted animals, partial healing of the skin lesions was observed two weeks earlier; at six weeks after irradiation. Complete healing of epithelium was observed at eight weeks after irradiation in this group. While in control animals that had received radiation only, only partial healing of the skin lesions were observed at eight weeks.

BM-MSCs were injected into the skin of mini-pigs irradiated with large dose of 50 Gy of ^{60}Co gamma rays [86]. Autologous BM-MSCs were injected intradermally 4–14 weeks after irradiation, 2–3 times a week. Each injection contained 99–128 million autologous cells. Minipigs were followed up for over 30 weeks and it was shown that the treatment lead to local accumulation of lymphocytes at the dermis/subcutis border, improved vascularization, and reduction of inflammatory reactions. In another study of acute cutaneous radiation syndrome [87], skin of mini-pigs were irradiated with 50 Gy of ^{60}Co gamma rays. At day 76 post irradiation, inflammatory cytokines IL-1α and IL-6 (specific markers of M1 macrophage) and IL-10 and TGF-β (specific markers of M2 macrophages) were assessed. Treatment with autologous ADSCs resulted in increased M2 macrophage markers associated with CD68+/CD206+ cells, indicating that MSC treatment directed the inflammatory response to proregenerapive outcome.

ADSC treatment of irradiated wounds on rats resulted in accelerated healing of wounds in rats [88]. Three-cm diameter of rats dorsal skin was irradiated with 50 Gy of 6MeV electrons. Three weeks after irradiation, rats received one million MSCs in PBS, compared with those that received only PBS. At week six after irradiation, wounds on ADSC treated rats were significantly smaller than controls. Histological examination of the wounds also indicated re-epithelialsation and neoangiogenesis in MSC-treated wounds.

This was supported by the reported beneficiary effects of ADSC injection on healing of irradiated wounds in nude mice [89]. The dorsal skin of nude mice were irradiated non-lethally and wounds were created by skin biopsy punch. Wounds were injected with ADSCs and compared with vehicle injected wounds.

Beneficial effects of cell therapy was demonstrated after transplantation of bone marrow mononuclear cells (BMMNC) on irradiated wounds [90]. Skin wounds were created by skin biopsy punch after cobalt-60 irradiation. It was demonstrated that cell therapy resulted in increased vascular density and improved matrix remodelling.

With respect to the clinical effects of MSC transplantation on treating radiation-induced lesions, a 27-year-old Chilean radiation accident victim was treated by skin allograft after excision of the irradiated tissue [91]. The graft did not last very long and got infected. The patient was treated with skin allograft again but this time with addition of autologous BM-MSCs. A second dose of stem cells was delivered nine days after that, resulting in complete healing and wound closure at 75 days after first MSC transplantation.

A number of clinical studies that are not stem cell transplantation per se but can be attributed to the existence of ADSCs in fat have been reported. These include the treatment of radiation induced normal tissue lesions by autologous fat grafting. A 67-year-old cancer patient who developed a chronic non-healing ulcer in her leg after surgery and radiotherapy of a squamous cell carcinoma was treated with fat infiltrated around and under the ulcer area. The ulcer fully healed two months after treatment [92]. Rigoti et al. [93] treated 20 patients suffering from radiation-induced normal tissue lesions as side effects of radiotherapy with autologous fat grafting that resulted in improvements in all cases. Fat grafting was successfully used in rectifying aesthetic defects caused radiotherapy in head and neck cancer patients [94]. Breast irradiated patients do not respond favourably to allogenic reconstruction [95]. However, favourable outcomes and formation of new subcutaneous tissue have been reported after fat grafting in mastectomy patients who had received breast irradiation [96,97].

12. Studies on the Salivary Gland and Oral Mucosa

In a mouse model, the ability of ADSCs to minimize and/or repair single dose radiation-induced oral mucositis was demonstrated after 18 Gy single-dose of orthovoltage X-ray [98]. It was shown that intraperitoneal transplantation of 5 doses of 2.5 million freshly cultured syngenic ADSCs significantly and reproducibly reduced the duration of radiation-induced oral mucositis from 5.6 ± 0.3 days to 1.6 ± 0.3 days. The therapeutic benefits were shown to be significantly dependent on dose, frequency, and the start of cell transplantation.

Effects of BM-MSCs on irradiated salivary gland was assessed by mobilisation of autologus BM-MSC by administration of granulocyte stimulating factor (G-CSF) [99]. It was shown that the mobilised MSCs promoted regeneration of irradiated salivary glands and increased gland weight, number of ancinar cells, and salivary flow rate.

In another study [100], it was shown that the local transplantation of human ADSCs resulted in tissue remodelling with a greater number of salivary epithelial cells in a rat model of salivary gland irradiation. This indicated that local transplantation of ADSCs alleviated radiation-induced cell death. It was also shown that when an injectable porcine small intestinal submucosa matrix was used as a cell delivery carrier, the anti-apoptotic and anti-oxidative effects of ADSCs and salivary protein synthesis were enhanced.

Protective and regenerative effects of ADSCs on radiation induced salivary gland was also studied in rats [101]. These authors reported statistically significant improvements in the salivary gland of rats treated with ADMSc, 48 h after irradiation. The efficacy of stem cell transplantation and mobilisation in the treatment of radiation-induced xerostomia was discussed and reviewed [102,103].

Clinically, in a randomised placebo-controlled phase 1/2 trial [48], 30 patients were studied. In this study ADSCs or placebo were transplanted in submandibular glands of patients who had had previously received radiotherapy for oropharyngeal squamous cell carcinoma. No adverse events were detected from ADSC transplantation, indicating its safety. Unstimulated whole salivary flow rates in the transplanted group significantly increased compared to the placebo-arm. The xerostomia symptom scores significantly decreased and salivary gland function improved in the ADSC group.

13. Discussion

In a living body, cell loss and regeneration takes place continually as a natural process. Tissues function takes place as a result of a continued cell loss and replacement with new cells. Cells are lost due to ageing, wear and tear, or other insults such as radiation, and are replaced by new cells produced by indigenous stem cells or tissue-specific progenitor cells that differentiate into functional cells. Target cell theory of radiation damage [1] was developed exactly on this basis. According to this theory, cell loss is the cardinal cause of development of radiation lesions or tumour eradication by irradiation. In fact, radiation disturbs the usually continuous process of cell loss and cell replacement. The cells killed or damaged by radiation fail to produce sufficient progenies to replace the lost cells, therefore, the number of lost cells exceed the number of cells produced. When the deficit goes beyond a critical level where the number of progenies become so low that it cannot produce sufficient differentiated cells to maintain the tissue function, radiation lesion manifests. In early responding tissues, the latency period of the development of radiation lesion corresponds to the turnover time of the cells. For example, radiation mucositis, and radiation-induced moist desquamation of the skin are considered as a result of sterilisation of epithelia and their latency period corresponds to the turnover time of the target cells. However, late radiation damage cannot be described by turnover time of a certain cell type; however, it develops as a result of loss of a number of cells and subsequent events. For example, in the development of radiation-induced late dermal damage or late submucosal damage, loss or damage to endothelial cells play an important role. Loss of endothelial cells and damage to the vasculature impair the circulation and loss of parenchyma ensues. This is also true of radiation damage in two central nervous tissues, where late radiation damage manifests as demyelination of axons and necrosis. Some believe that the reproductive death of glial cells is the cardinal cause and demyelination and necrosis develop as a consequence of gradual loss of these cells [104]. However, some authors [105,106] consider vascular damage and lack of sufficient blood supply as the cardinal cause of the development of radiation-induced demyelination and necrosis of nervous tissue. Whatever the cause, both schools of thought agree that the demyelination and subsequent necrosis of nervous tissue is initiated by cell death, reproductive sterilisation of vascular or glial cells. The severity and duration of radiation-induced lesions are dose-dependent. This implies that the more cell loss, the more severe and long lasting the lesion. Besides radiation dose, radiation quality is another determinant factor on the degree of cell loss and consequently lesion development. However, treatment of radiation lesions, particularly treatment with stem cells, is in its infancy and there is not much data to be discussed.

Not all radiation lesions are fatal. Radiation lesions heal after sublethal doses; when surviving cells in the irradiated region regenerate or healthy cells from the margin of the irradiated region migrate to the irradiated area and revive the damaged tissue. However, when the cell loss is extensive or the number of surviving/migrating cells is not sufficient, the lesions remain unhealed. On this basis, replacement of lost cells by stem cell transplantation was a plausible attempt to modify radiation induced tissue damage.

Regeneration of irradiated salivary gland by mobilizing endogenous stem cells [99] supports the idea that there is always a number of stem cells in the damaged tissue and whole body, and their stimulation and mobilisation either by secretory factors from other stem cells or by cytokines could rescue damaged tissue. Protection of salivary glands from radiation-induced apoptosis and preservation of acinar structure and function were attributed to the activation of FGFR-PI3K signalling via actions of ADSC-secreted factors, including FGF10 [107].

The effectiveness of cell transplantation in amelioration of radiation lesions is supported by the works reviewed in this paper. Radiation lesions that develop due to lack or insufficiency of functional cells are modified by the transplantation of exogenous cells [90,108–111]. Besides, amelioration of radiation-induced lesions and subcutaneous tissue formation in patients who received fat grafting after mastectomy of breast-irradiated patients can be attributed to the stem cell component of the fat graft [96,97].

However, it is not certain that the beneficial effect of stem cell transplantation is the result of direct integration of transplanted cells in the damaged tissue or the result of stimulation of the surviving endogenous cells by the transplanted cells—the paracrine effect. Some authors, while reporting the beneficial effects of stem cell transplantation, fail to demonstrate the integration of the transplanted donor cells in host tissue or demonstrate a very low level of engraftment that cannot justify the significant functional improvements as a result of transplantation. In the study of the effect of stem cell transplantation on amelioration of radiation-induced salivary gland damage by mobilisation of endogenous bone marrow stem cells [99], significant improvements were seen in the gland weight and salivary flow but transdifferentiation of stimulated bone marrow cells in the salivary glands were not observed. Stem cell transplantation showed therapeutic effects on irradiated lung tissue but the number of transplanted cells in irradiated lungs were so low that they could not justify the observed improvements [78]. Neural stem cells transplanted intradurally in spinal cord irradiated rats resulted in 30% reduction in the development of radiation myelopathy [60] but these authors failed to demonstrate the transdifferentiation of the transplanted cells in the irradiated spinal cords of engrafted rats. Similarly, despite improvements in irradiated liver tissue by exogenous cell transplantation, the transplanted cells were not found in the liver of the irradiated animals [76].

These findings suggest that the beneficial effects of stem cell transplantation are not necessarily due to the replacement of damaged cells by healthy transplanted cells or their trans differentiation into functional cells. It is probable that the paracrine effect also plays a role [112–114]. It is to say that the transplanted cells secrete some bioactive factors that stimulate endogenous stem cells. Bioactive factors secreted by MSCs are both immunomodulatory and trophic. Secretion of angiogenic and antiapoptotic factors by transplanted human ADSCs have been reported [115]. VEGF secretion were increased manyfold when the ADSCs were cultured under hypoxic conditions. In fact, paracrine effect was reported as early as 1971 by Little [116] who reported the repair of potentially lethal radiation damage by a conditioned medium of cultured mammalian cells. Later, it was shown that the growth of cultured endothelial cells was enhanced and endothelial apoptosis was reduced by the addition of conditioned media obtained from ADSCs grown under hypoxic conditions [115]. Regeneration of radiation damaged tissues by transplanted MSCs has been attributed to the indirect effect of stem cell transplantation due to the secretion of cytokines and growth factors [76]. Tissue regeneration, acceleration of angiogenesis, and growth of nerves have been reported after transplantation of ADSCs in mice [117]. The beneficial effects of ADSC transplantation were attributed to the secretion of neurotrophic genes and extracellular matrix proteins required for nerve growth and myelination. MSCs, besides trophic effects, exert immunomodulatory effects too [109] that inhibit the surveillance ability of lymphocytes. This prevents the immunogenicity and allows allogenic transplantation of MSCs. A total of 73 proteins secreted by human ADSCs have been reported that includes factors such as heat shock proteins, macrophage inflammatory proteins, proteases, protease inhibitors, cycloskelethal components, extracellular matrix components, metabolic enzymes, anti-inflammatory proteinsVEGF, IGF-1, EGF, EGF, and many others [118,119]. Besides, RNA-containing microparticles are also involved in the paracrine effect. Microparticles or microvesicles consist of extracellular vesicles (EVs) that are released by almost all bodily cells, including stem cells. Evs are referred to a heterogenous population of membrane-coated small vesicles with diameter of 30–1000 nm. Exosomes constitute the microvesicles of diameter less than 200 nanometer. EVs consist of a bilipid membrane and a cargo consisting of various proteins and miRNA. Intracellular communication of cells is facilitated by secreted microvesicles [120–123]. Microvesicles released by stimulation of MSCs show therapeutic characteristics against ischemia-repurfusion induced acute and chronic kidney injury [124]. The same authors also demonstrated that inactivating RNA by pretreatment of microvesicles by RNase abrogated its therapeutic effect. This indicates the importance of the RNA component of microvesicles in exerting its therapeutic effect.

Evidence is mounting in support of paracrine effect of stem cells; in recent years particularly, EVs derived from stem cells have been the focus of extensive research efforts in the fields of regenerative medicine and radiation. The beneficial effects of MSC-secreted microvesicles have been demonstrated in vitro and in vivo treatment of many lesions [113,125–133]. Inhibition of tumour growth by MSC-derived microvesicles have also been demonstrated [68,128]. It has also been shown that platelet-derived microvesicles facilitate the homing of transplanted bone marrow stem cells in irradiated mice [134]. EVs extracted from human MSCs were injected into nude mice by three consecutive applications after a lethal whole body irradiation that resulted in 85% reduction in mortality [135]. Recently, the efficacy of MSC-derived EVs in amelioration of radiation-induced hematopoietic syndrome was reported [136]. Exosomes derived from mesenchymal stem cells have been used for conditioning macrophages to be used in the treatment of acute radiation syndrome [137]. It appears that, besides proteins and bioactive lipids, the RNA content of the cargo of EVs is the major component of the action of the beneficiary effects of EVs. This mode of action have been shown to be responsible for the amelioration of radiation-induced lung injury by mesenchymal cell-derived EVs [138]. The mode of action and potential of EVs in the treatment of radiation lesions are reviewed by Forsberg et al. [139]. EVs have also been indicated in mediating radiation-induced bystander effects [140].

14. Conclusions

The results of publications reviewed in this article indicate the beneficial effects of stem cell transplantation in the treatment of radiation lesions and tumour inhibition. Transplantation of intact stem cells or EVs derived from stem cells exert beneficial effects. However, it must be noted that radiation dose can play a major role in defining the results of the stem cell transplantation. The main principal of paracrine effect is based on the fact that paracrine factors excreted by transplanted stem cells stimulate endogenous stem cells to regenerate damaged tissue. After mild radiation doses, the donor cells partially contribute in regeneration of the damaged tissue and partially stimulate the endogenous stem cells to repair the damaged tissues. However, if a substantial large radiation dose is delivered to an organ, depleting almost all of the endogenous stem cells within the irradiated volume, the regeneration will be dependent almost entirely on the direct effect of the transplanted stem cells. This is to say that after a substantially large radiation dose, the paracrine effect will not be sufficiently effective and a substantially large stem cell dose will be required. Finally, it must be borne in mind that the conclusions made in this article are on the basis of limited experimental results published during recent years. Further research on the efficacy of stem cell transplantation and microvesicles secreted by activated stem cells, in amelioration of radiation lesions, is required.

Funding: This review received no external funding.

Institutional Review Board Statement: Not applicable.

Informed Consent Statement: Not applicable.

Data Availability Statement: Not applicable.

Conflicts of Interest: The author declares no conflict of interest.

References

1. Puck, T.T.; Marcus, P.I. Action of x-rays on mammalian cells. *J. Exp. Med.* **1956**, *103*, 653–666. [CrossRef]
2. Tada, E.; Yang, C.; Gobbel, G.T.; Lamborn, K.R.; Fike, J.R. Long-term impairment of subependymal repopulation following damage by ionizing irradiation. *Exp. Neurol.* **1999**, *160*, 66–77. [CrossRef]
3. Tada, E.; Parent, J.M.; Lowenstein, D.H.; Fike, J.R. X-irradiation causes a prolonged reduction in cell proliferation in the dentate gyrus of adult rats. *Neuroscience* **2000**, *99*, 33–41. [CrossRef]

4. Thomas, E.D. A history of haemopoietic cell transplantation. *Br. J. Haematol.* **1999**, *105*, 330–339. [CrossRef]
5. Lorenz, E.; Uphoff, D.; Reid, T.R.; Shelton, E. Modification of irradiation injury in mice and guinea pigs by bone marrow injections. *J. Natl. Cancer Inst.* **1951**, *12*, 197–201. [CrossRef]
6. Collino, F.; Deregibus, M.C.; Bruno, S.; Sterpone, L.; Aghemo, G.; Viltono, L.; Tetta, C.; Camussi, G. Microvesicles derived from adult human bone marrow and tissue specific mesenchymal stem cells shuttle selected pattern of miRNAs. *PLoS ONE* **2010**, *5*, e11803. [CrossRef] [PubMed]
7. Camussi, G.; Deregibus, M.C.; Cantaluppi, V. Role of stem-cell-derived microvesicles in the paracrine action of stem cells. *Biochem. Soc. Trans.* **2013**, *41*, 283–287. [CrossRef] [PubMed]
8. Anger, F.; Camara, M.; Ellinger, E.; Germer, C.T.; Schlegel, N.; Otto, C.; Klein, I. Human Mesenchymal Stromal Cell-Derived Extracellular Vesicles Improve Liver Regeneration After Ischemia Reperfusion Injury in Mice. *Stem Cells Dev.* **2019**, *28*, 1451–1462. [CrossRef] [PubMed]
9. National Toxicology Program. Final Report on Carcinogens Background Document for Formaldehyde. *Rep. Carcinog. Backgr. Doc.* **2010**, i-512.
10. Zha, Q.B.; Yao, Y.F.; Ren, Z.J.; Li, X.J.; Tang, J.H. Extracellular vesicles: An overview of biogenesis, function, and role in breast cancer. *Tumour Biol.* **2017**, *39*. [CrossRef]
11. Thomson, J.A.; Itskovitz-Eldor, J.; Shapiro, S.S.; Waknitz, M.A.; Swiergiel, J.J.; Marshall, V.S.; Jones, J.M. Embryonic stem cell lines derived from human blastocysts. *Science* **1998**, *282*, 1145–1147. [CrossRef] [PubMed]
12. Yamanaka, S. Strategies and new developments in the generation of patient-specific pluripotent stem cells. *Cell Stem Cell* **2007**, *1*, 39–49. [CrossRef] [PubMed]
13. Rogers, E.H.; Hunt, J.A.; Pekovic-Vaughan, V. Adult stem cell maintenance and tissue regeneration around the clock: Do impaired stem cell clocks drive age-associated tissue degeneration? *Biogerontology* **2018**, *19*, 497–517. [CrossRef] [PubMed]
14. Caplan, A.I. Mesenchymal stem cells. *J. Orthop. Res. Off. Publ. Orthop. Res. Soc.* **1991**, *9*, 641–650. [CrossRef]
15. Pittenger, M.F.; Mackay, A.M.; Beck, S.C.; Jaiswal, R.K.; Douglas, R.; Mosca, J.D.; Moorman, M.A.; Simonetti, D.W.; Craig, S.; Marshak, D.R. Multilineage potential of adult human mesenchymal stem cells. *Science* **1999**, *284*, 143–147. [CrossRef]
16. Keating, A. Mesenchymal stromal cells. *Curr. Opin. Hematol.* **2006**, *13*, 419–425. [CrossRef]
17. Zuk, P.A.; Zhu, M.I.N.; Mizuno, H.; Huang, J.; Futrell, J.W.; Katz, A.J.; Benhaim, P.; Lorenz, H.P.; Hedrick, M.H. Multilineage cells from human adipose tissue: Implications for cell-based therapies. *Tissue Eng.* **2001**, *7*, 211–228. [CrossRef]
18. Zuk, P.A., Zhu, M.; Ashjian, P.; De Ugarte, D.A.; Huang, J.I.; Mizuno, H.; Alfonso, Z.C.; Fraser, J.K.; Benhaim, P.; Hedrick, M.H. Human adipose tissue is a source of multipotent stem cells. *Mol. Biol. Cell.* **2002**, *13*, 4279–4295. [CrossRef]
19. Hennrick, K.T.; Keeton, A.G.; Nanua, S.; Kijek, T.G.; Goldsmith, A.M.; Sajjan, U.S.; Bentley, J.K.; Lama, V.N.; Moore, B.B.; Schumacher, R.E.; et al. Lung cells from neonates show a mesenchymal stem cell phenotype. *Am. J. Respir. Crit. Care Med.* **2007**, *175*, 1158–1164. [CrossRef] [PubMed]
20. Friedenstein, A.J.; Deriglasova, U.F.; Kulagina, N.N.; Panasuk, A.F.; Rudakowa, S.F.; Luriá, E.A.; Ruadkow, I.A. Precursors for fibroblasts in different populations of hematopoietic cells as detected by the in vitro colony assay method. *Exp. Hematol.* **1974**, *2*, 83–92.
21. Friedenstein, A.J. Precursor cells of mechanocytes. *Int. Rev. Cytol.* **1976**, *47*, 327–359.
22. Li, Z.; Hu, X.; Zhong, J.F. Mesenchymal Stem Cells: Characteristics, Function, and Application. *Stem Cells Int.* **2019**, *2019*, 8106818. [CrossRef]
23. Crippa, S.; Bernardo, M.E. Mesenchymal Stromal Cells: Role in the BM Niche and in the Support of Hematopoietic Stem Cell Transplantation. *Hemasphere* **2018**, *2*, e151. [CrossRef] [PubMed]
24. Casteilla, L.; Planat-Benard, V.; Laharrague, P.; Cousin, B. Adipose-derived stromal cells: Their identity and uses in clinical trials, an update. *World J. Stem Cells* **2011**, *3*, 25–33. [CrossRef] [PubMed]
25. Baer, P.C.; Koch, B.; Hickmann, E.; Schubert, R.; Cinatl, J.J.; Hauser, I.A.; Geiger, H. Isolation, Characterization, Differentiation and Immunomodulatory Capacity of Mesenchymal Stromal/Stem Cells from Human Perirenal Adipose Tissue. *Cells* **2019**, *8*, 1346. [CrossRef]
26. Amati, E.; Sella, S.; Perbellini, O.; Alghisi, A.; Bernardi, M.; Chieregato, K.; Lievore, C.; Peserico, D.; Rigno, M.; Zilio, A.; et al. Generation of mesenchymal stromal cells from cord blood: Evaluation of in vitro quality parameters prior to clinical use. *Stem Cell Res. Ther.* **2017**, *8*, 1–15. [CrossRef]
27. Wang, H.S.; Hung, S.C.; Peng, S.T.; Huang, C.C.; Wei, H.M.; Guo, Y.J.; Fu, Y.S.; Lai, M.C.; Chen, C.C. Mesenchymal stem cells in the Wharton's jelly of the human umbilical cord. *Stem Cells* **2004**, *22*, 1330–1337. [CrossRef]
28. Friedman, R.; Betancur, M.; Boissel, L.; Tuncer, H.; Cetrulo, C.; Klingemann, H. Umbilical cord mesenchymal stem cells: Adjuvants for human cell transplantation. *Biol. Blood Marrow Transplant.* **2007**, *13*, 1477–1486. [CrossRef]
29. Nagamura-Inoue, T.; He, H. Umbilical cord-derived mesenchymal stem cells: Their advantages and potential clinical utility. *World J. Stem Cells* **2014**, *6*, 195–202. [CrossRef]
30. Dominici, M.; Le Blanc, K.; Mueller, I.; Slaper-Cortenbach, I.; Marini, F.; Krause, D.; Deans, R.J.; Keating, A.; Prockop, D.J.; Horwitz, E.M.; et al. Minimal criteria for defining multipotent mesenchymal stromal cells. The International Society for Cellular Therapy position statement. *Cytotherapy* **2006**, *8*, 315–317. [CrossRef]
31. Lange, C.; Bassler, P.; Lioznov, M.V.; Bruns, H.; Kluth, D.; Zander, A.R.; Fiegel, H.C. Liver-specific gene expression in mesenchymal stem cells is induced by liver cells. *World J. Gastroenterol.* **2005**, *11*, 4497–4504. [CrossRef]

32. Toma, C.; Pittenger, M.F.; Cahill, K.S.; Byrne, B.J.; Kessler, P.D. Human mesenchymal stem cells differentiate to a cardiomyocyte phenotype in the adult murine heart. *Circulation* **2002**, *105*, 93–98. [CrossRef]
33. Takeda, Y.S.; Xu, Q. Neuronal Differentiation of Human Mesenchymal Stem Cells Using Exosomes Derived from Differentiating Neuronal Cells. *PLoS ONE* **2015**, *10*, e0135111. [CrossRef]
34. Zhu, Y.; Liu, T.; Song, K.; Fan, X.; Ma, X.; Cui, Z. Adipose-derived stem cell: A better stem cell than BMSC. *Cell Biochem. Funct.* **2008**, *26*, 664–675. [CrossRef] [PubMed]
35. Baasse, A.; Machoy, F.; Juerss, D.; Baake, J.; Stang, F.; Reimer, T.; Krapohl, B.D.; Hildebrandt, G. Radiation Sensitivity of Adipose-Derived Stem Cells Isolated from Breast Tissue. *Int. J. Mol. Sci.* **2018**, *19*, 1988. [CrossRef]
36. Nurković, J.; Zaletel, I.; Nurković, S.; Hajrović, Š.; Mustafić, F.; Isma, J.; Škevin, A.J.; Grbović, V.; Filipović, M.K.; Dolićanin, Z. Combined effects of electromagnetic field and low-level laser increase proliferation and alter the morphology of human adipose tissue-derived mesenchymal stem cells. *Lasers Med. Sci.* **2016**, *32*, 151–160. [CrossRef]
37. Nicolay, N.H.; Lopez Perez, R.; Debus, J.; Huber, P.E. Mesenchymal stem cells—A new hope for radiotherapy-induced tissue damage? *Cancer Lett.* **2015**, *366*, 133–140. [CrossRef]
38. Saito, T.; Kuang, J.Q.; Bittira, B.; Al-Khaldi, A.; Chiu, R.C. Xenotransplant cardiac chimera: Immune tolerance of adult stem cells. *Ann. Thorac. Surg.* **2002**, *74*, 19–24. [CrossRef]
39. Chapel, A.; Bertho, J.M.; Bensidhoum, M.; Fouillard, L.; Young, R.G.; Frick, J.; Demarquay, C.; Cuvelier, F.; Mathieu, E.; Trompier, F.; et al. Mesenchymal stem cells home to injured tissues when co-infused with hematopoietic cells to treat a radiation-induced multi-organ failure syndrome. *J. Gene Med.* **2003**, *5*, 1028–1038. [CrossRef]
40. François, S.; Bensidhoum, M.; Mouiseddine, M.; Mazurier, C.; Allenet, B.; Semont, A.; Frick, J.; Saché, A.; Bouchet, S.; Thierry, D.; et al. Local Irradiation Not Only Induces Homing of Human Mesenchymal Stem Cells at Exposed Sites but Promotes Their Widespread Engraftment to Multiple Organs: A Study of Their Quantitative Distribution After Irradiation Damage. *Stem Cells* **2006**, *24*, 1020–1029. [CrossRef]
41. Mouiseddine, M.; Francois, S.; Semont, A.; Sache, A.; Allenet, B.; Mathieu, N.; Frick, J.; Thierry, D.; Chapel, A. Human mesenchymal stem cells home specifically to radiation-injured tissues in a non-obese diabetes/severe combined immunodeficiency mouse model. *Br. J. Radiol.* **2007**, *80*. [CrossRef]
42. Kraitchman, D.L.; Tatsumi, M.; Gilson, W.D.; Ishimori, T.; Kedziorek, D.; Walczak, P.; Segars, W.P.; Chen, H.H.; Fritzges, D.; Izbudak, I.; et al. Dynamic imaging of allogeneic mesenchymal stem cells trafficking to myocardial infarction. *Circulation* **2005**, *112*, 1451–1461. [CrossRef]
43. Golpanian, S.; Difede, D.L.; Khan, A.; Schulman, I.H.; Landin, A.M.; Tompkins, B.A.; Heldman, A.W.; Miki, R.; Goldstein, B.J.; Mushtaq, M.; et al. Allogeneic Human Mesenchymal Stem Cell Infusions for Aging Frailty. *J. Gerontol. Ser. A Boil. Sci. Med. Sci.* **2017**, *72*, 1505–1512. [CrossRef]
44. Pang, Y.; Xiao, H.W.; Zhang, H.; Liu, Z.H.; Li, L.; Gao, Y.; Li, H.B.; Jiang, Z.J.; Tan, H.; Lin, J.R.; et al. Allogeneic Bone Marrow-Derived Mesenchymal Stromal Cells Expanded In Vitro for Treatment of Aplastic Anemia: A Multicenter Phase II Trial. *Stem Cells Transl. Med.* **2017**, *6*, 1569–1575. [CrossRef]
45. Tompkins, B.A.; DiFede, D.L.; Khan, A.; Landin, A.M.; Schulman, I.H.; Pujol, M.V.; Heldman, A.W.; Miki, R.; Goldschmidt-Clermont, P.J.; Goldstein, B.J.; et al. Allogeneic Mesenchymal Stem Cells Ameliorate Aging Frailty: A Phase II Randomized, Double-Blind, Placebo-Controlled Clinical Trial. *J. Gerontol. A Biol. Sci. Med. Sci.* **2017**, *72*, 1513–1522. [CrossRef]
46. Jo, C.H.; Chai, J.W.; Jeong, E.C.; Oh, S.; Kim, P.S.; Yoon, J.Y.; Yoon, K.S. Intratendinous Injection of Autologous Adipose Tissue-Derived Mesenchymal Stem Cells for the Treatment of Rotator Cuff Disease: A First-In-Human Trial. *Stem Cells* **2018**, *36*, 1441–1450. [CrossRef]
47. Toyserkani, N.M.; Jørgensen, M.G.; Tabatabaeifar, S.; Jensen, C.H.; Sheikh, S.P.; Sørensen, J.A. Concise Review: A Safety Assessment of Adipose-Derived Cell Therapy in Clinical Trials: A Systematic Review of Reported Adverse Events. *Stem Cells Transl. Med.* **2017**, *6*, 1786–1794. [CrossRef] [PubMed]
48. Grønhøj, C.; Jensen, D.H.; Vester-Glowinski, P.; Jensen, S.B.; Bardow, A.; Oliveri, R.S.; Fog, L.M.; Specht, L.; Thomsen, C.; Darkner, S.; et al. Safety and Efficacy of Mesenchymal Stem Cells for Radiation-Induced Xerostomia: A Randomized, Placebo-Controlled Phase 1/2 Trial (MESRIX). *Int. J. Radiat. Oncol.* **2018**, *101*, 581–592. [CrossRef] [PubMed]
49. Alessio, N.; Del Gaudio, S.; Capasso, S.; Di Bernardo, G.; Cappabianca, S.; Cipollaro, M.; Peluso, G.; Galderisi, U. Low dose radiation induced senescence of human mesenchymal stromal cells and impaired the autophagy process. *Oncotarget* **2015**, *6*, 8155–8166. [CrossRef] [PubMed]
50. Johnson, M.B.; Niknam-Bienia, S.; Soundararajan, V.; Pang, B.; Jung, E.; Gardner, D.J.; Xu, X.; Park, S.Y.; Wang, C.; Chen, X.; et al. Mesenchymal Stromal Cells Isolated from Irradiated Human Skin Have Diminished Capacity for Proliferation, Differentiation, Colony Formation, and Paracrine Stimulation. *Stem Cells Transl. Med.* **2019**, *8*, 925–934. [CrossRef]
51. Ra, J.C.; Shin, I.S.; Kim, S.H.; Kang, S.K.; Kang, B.C.; Lee, H.Y.; Kim, Y.J.; Jo, J.Y.; Yoon, E.J.; Choi, H.J.; et al. Safety of Intravenous Infusion of Human Adipose Tissue-Derived Mesenchymal Stem Cells in Animals and Humans. *Stem Cells Dev.* **2011**, *20*, 1297–1308. [CrossRef]
52. Ogura, F.; Wakao, S.; Kuroda, Y.; Tsuchiyama, K.; Bagheri, M.; Heneidi, S.; Chazenbalk, G.; Aiba, S.; Dezawa, M. Human adipose tissue possesses a unique population of pluripotent stem cells with nontumorigenic and low telomerase activities: Potential implications in regenerative medicine. *Stem Cells Dev.* **2014**, *23*, 717–728.

53. Barnes, D.W.; Corp, M.J.; Loutit, J.F.; Neal, F.E. Treatment of murine leukaemia with X rays and homologous bone marrow; preliminary communication. *Br. Med. J.* **1956**, *2*, 626–627. [CrossRef] [PubMed]
54. Friedrichs, B.; Tichelli, A.; Bacigalupo, A.; Russell, N.H.; Ruutu, T.; Shapira, M.Y.; Beksac, M.; Hasenclever, D.; Socié, G.; Schmitz, N. Long-term outcome and late effects in patients transplanted with mobilised blood or bone marrow: A randomised trial. *Lancet Oncol.* **2010**, *11*, 331–338. [CrossRef]
55. Bertho, J.-M.; Frick, J.; Demarquay, C.; Lauby, A.; Mathieu, E.; Dudoignon, N.; Jacquet, N.; Trompier, F.; Chapel, A.; Joubert, C.; et al. Reinjection of Ex Vivo–Expanded Primate Bone Marrow Mononuclear Cells Strongly Reduces Radiation-Induced Aplasia. *J. Hematotherapy* **2002**, *11*, 549–564. [CrossRef] [PubMed]
56. Gan, J.; Meng, F.; Zhou, X.; Li, C.; He, Y.; Zeng, X.; Jiang, X.; Liu, J.; Zeng, G.; Tang, Y.; et al. Hematopoietic recovery of acute radiation syndrome by human superoxide dismutase-expressing umbilical cord mesenchymal stromal cells. *Cytotherapy* **2015**, *17*, 403–417. [CrossRef] [PubMed]
57. Bandekar, M.; Maurya, D.K.; Sharma, D.; Checker, R.; Gota, V.; Mishra, N.; Sandur, S.K. Xenogeneic transplantation of human WJ-MSCs rescues mice from acute radiation syndrome via Nrf-2-dependent regeneration of damaged tissues. *Arab. Archaeol. Epigr.* **2020**, *20*, 2044–2057. [CrossRef]
58. Groves, A.K.; Barnett, S.C.; Franklin, R.J.; Crang, A.J.; Mayer, M.; Blakemore, W.F.; Noble, M. Repair of demyelinated lesions by transplantation of purified O-2A progenitor cells. *Nature* **1993**, *362*, 453–455. [CrossRef]
59. Franklin, R.J.; Bayley, S.A.; Milner, R.; Ffrench-Constant, C.; Blakemore, W.F. Differentiation of the O-2A progenitor cell line CG-4 into oligodendrocytes and astrocytes following transplantation into glia-deficient areas of CNS white matter. *Glia* **1995**, *13*, 39–44. [CrossRef]
60. Rezvani, M.; Birds, D.A.; Hodges, H.; Hopewell, J.W.; Milledew, K.; Wilkinson, J.H. Modification of radiation myelopathy by the transplantation of neural stem cells in the rat. *Radiat. Res.* **2001**, *156*, 408–412. [CrossRef]
61. Wei, L.; Zhang, J.; Xiao, X.B.; Mai, H.X.; Zheng, K.; Sun, W.L.; Wang, L.; Liang, F.; Yang, Z.L.; Liu, Y.; et al. Multiple injections of human umbilical cord-derived mesenchymal stromal cells through the tail vein improve microcirculation and the microenvironment in a rat model of radiation myelopathy. *J. Transl. Med.* **2014**, *12*, 246. [CrossRef] [PubMed]
62. Acharya, M.M.; Christie, L.A.; Lan, M.L.; Donovan, P.J.; Cotman, C.W.; Fike, J.R.; Limoli, C.L. Rescue of radiation-induced cognitive impairment through cranial transplantation of human embryonic stem cells. *Proc. Natl. Acad. Sci. USA* **2009**, *106*, 19150–19155. [CrossRef]
63. Acharya, M.M.; Christie, L.A.; Lan, M.L.; Giedzinski, E.; Fike, J.R.; Rosi, S.; Limoli, C.L. Human Neural Stem Cell Transplantation Ameliorates Radiation-Induced Cognitive Dysfunction. *Cancer Res.* **2011**, *71*, 4834–4845. [CrossRef]
64. Acharya, M.M.; Christie, L.A.; Lan, M.L.; Limoli, C.L. Comparing the functional consequences of human stem cell transplantation in the irradiated rat brain. *Cell Transplant.* **2013**, *22*, 55–64. [CrossRef]
65. Chu, C.; Gao, Y.; Lan, X.; Lin, J.; Thomas, A.M.; Li, S. Stem-Cell Therapy as a Potential Strategy for Radiation-Induced Brain Injury. *Stem Cell Rev. Rep.* **2020**, *16*, 639–649. [CrossRef] [PubMed]
66. Sémont, A.; Mouiseddine, M.; François, A.; Demarquay, C.; Mathieu, N.; Chapel, A.; Saché, A.; Thierry, D.; Laloi, P.; Gourmelon, P. Mesenchymal stem cells improve small intestinal integrity through regulation of endogenous epithelial cell homeostasis. *Cell Death Differ.* **2009**, *17*, 952–961. [CrossRef] [PubMed]
67. Linard, C.; Busson, E.; Holler, V.; Strup-Perrot, C.; Lacave-Lapalun, J.V.; Lhomme, B.; Prat, M.; Devauchelle, P.; Sabourin, J.C.; Simon, J.M.; et al. Repeated autologous bone marrow-derived mesenchymal stem cell injections improve radiation-induced proctitis in pigs. *Stem Cells Transl. Med.* **2013**, *2*, 916–927. [CrossRef]
68. François, S.; Usunier, B.; Forgue-Lafitte, M.E.; L'Homme, B.; Benderitter, M.; Douay, L.; Gorin, N.C.; Larsen, A.K.; Chapel, A. Mesenchymal Stem Cell Administration Attenuates Colon Cancer Progression by Modulating the Immune Component within the Colorectal Tumor Microenvironment. *Stem Cells Transl. Med.* **2019**, *8*, 285–300. [CrossRef]
69. Gao, Z.; Zhang, Q.; Han, Y.; Cheng, X.; Lu, Y.; Fan, L.; Wu, Z. Mesenchymal stromal cell-conditioned medium prevents radiation-induced small intestine injury in mice. *Cytotherapy* **2012**, *14*, 267–273. [CrossRef]
70. Chen, Y.X.; Zeng, Z.C.; Sun, J.; Zeng, H.Y.; Huang, Y.; Zhang, Z.Y. Mesenchymal stem cell-conditioned medium prevents radiation-induced liver injury by inhibiting inflammation and protecting sinusoidal endothelial cells. *J. Radiat Res.* **2015**, *56*, 700–708. [CrossRef]
71. Han, Y.-M.; Park, J.-M.; Choi, Y.S.; Jin, H.; Lee, Y.-S.; Han, N.-Y.; Lee, H.; Hahm, K.B. The efficacy of human placenta-derived mesenchymal stem cells on radiation enteropathy along with proteomic biomarkers predicting a favorable response. *Stem Cell Res. Ther.* **2017**, *8*, 1–15. [CrossRef]
72. Van de Putte, D.; Demarquay, C.; Van Daele, E.; Moussa, L.; Vanhove, C.; Benderitter, M.; Ceelen, W.; Pattyn, P.; Mathieu, N. Adipose-Derived Mesenchymal Stromal Cells Improve the Healing of Colonic Anastomoses Following High Dose of Irradiation Through Anti-Inflammatory and Angiogenic Processes. *Cell Transplant.* **2017**, *26*, 1919–1930. [CrossRef] [PubMed]
73. Semont, A.; Demarquay, C.; Bessout, R.; Durand, C.; Benderitter, M.; Mathieu, N. Mesenchymal stem cell therapy stimulates endogenous host progenitor cells to improve colonic epithelial regeneration. *PLoS ONE* **2013**, *8*, e70170. [CrossRef]
74. Durand, C.; Pezet, S.; Eutamène, H.; Demarquay, C.; Mathieu, N.; Moussa, L.; Daudin, R.; Holler, V.; Sabourin, J.C.; Milliat, F.; et al. Persistent visceral allodynia in rats exposed to colorectal irradiation is reversed by mesenchymal stromal cell treatment. *Pain* **2015**, *156*, 1465–1476. [CrossRef] [PubMed]

75. Tulunay, O.; Good, R.A.; Yunis, E.J. Protection of lethally irradiated mice with allogeneic fetal liver cells: Influence of irradiation dose on immunologic reconstitution. *Proc. Natl. Acad. Sci. USA* **1975**, *72*, 4100–4104. [CrossRef]
76. Mouiseddine, M.; Francois, S.; Souidi, M.; Chapel, A. Intravenous human mesenchymal stem cells transplantation in NOD/SCID mice preserve liver integrity of irradiation damage. *Methods Mol. Biol.* **2012**, *826*, 179–188. [PubMed]
77. Shao, C.-H.; Chen, S.-L.; Dong, T.-F.; Chai, H.; Yu, Y.; Deng, L.; Wang, Y.; Cheng, F. Transplantation of bone marrow–derived mesenchymal stem cells after regional hepatic irradiation ameliorates thioacetamide-induced liver fibrosis in rats. *J. Surg. Res.* **2014**, *186*, 408–416. [CrossRef]
78. Xue, J.; Li, X.; Lu, Y.; Gan, L.; Zhou, L.; Wang, Y.; Lan, J.; Liu, S.; Sun, L.; Jia, L.; et al. Gene-modified Mesenchymal Stem Cells Protect Against Radiation-induced Lung Injury. *Mol. Ther.* **2013**, *21*, 456–465. [CrossRef]
79. Jiang, X.; Jiang, X.; Qu, C.; Chang, P.; Zhang, C.; Qu, Y.; Liu, Y. Intravenous delivery of adipose-derived mesenchymal stromal cells attenuates acute radiation-induced lung injury in rats. *Cytotherapy* **2015**, *17*, 560–570. [CrossRef]
80. Wang, R.; Zhu, C.Z.; Qiao, P.; Liu, J.; Zhao, Q.; Wang, K.J.; Zhao, T.B. Experimental treatment of radiation pneumonitis with human umbilical cord mesenchymal stem cells. *Asian Pac. J. Trop. Med.* **2014**, *7*, 262–266. [CrossRef]
81. Chen, H.-X.; Xiang, H.; Xu, W.-H.; Li, M.; Yuan, J.; Liu, J.; Sun, W.-J.; Zhang, R.; Li, J.; Ren, Z.-Q.; et al. Manganese Superoxide Dismutase Gene–Modified Mesenchymal Stem Cells Attenuate Acute Radiation-Induced Lung Injury. *Hum. Gene Ther.* **2017**, *28*, 523–532. [CrossRef] [PubMed]
82. Wei, L.; Zhang, J.; Yang, Z.L.; You, H. Extracellular superoxide dismutase increased the therapeutic potential of human mesenchymal stromal cells in radiation pulmonary fibrosis. *Cytotherapy* **2017**, *19*, 586–602. [CrossRef]
83. Klein, D.; Schmetter, A.; Imsak, R.; Wirsdörfer, F.; Unger, K.; Jastrow, H.; Stuschke, M.; Jendrossek, V. Therapy with Multipotent Mesenchymal Stromal Cells Protects Lungs from Radiation-Induced Injury and Reduces the Risk of Lung Metastasis. *Antioxid. Redox Signal.* **2016**, *24*, 53–69. [CrossRef]
84. Xu, T.; Zhang, Y.; Chang, P.; Gong, S.; Shao, L.; Dong, L. Mesenchymal stem cell-based therapy for radiation-induced lung injury. *Stem Cell Res. Ther.* **2018**, *9*, 18. [CrossRef]
85. François, S.; Mouiseddine, M.; Mathieu, N.; Semont, A.; Monti, P.; Dudoignon, N.; Saché, A.; Boutarfa, A.; Thierry, D.; Gourmelon, P.; et al. Human mesenchymal stem cells favour healing of the cutaneous radiation syndrome in a xenogenic transplant model. *Ann. Hematol.* **2006**, *86*, 1–8. [CrossRef]
86. Agay, D.; Scherthan, H.; Forcheron, F.; Grenier, N.; Hérodin, F.; Meineke, V.; Drouet, M. Multipotent mesenchymal stem cell grafting to treat cutaneous radiation syndrome: Development of a new minipig model. *Exp. Hematol.* **2010**, *38*, 945–956. [CrossRef]
87. Riccobono, D.; Nikovics, K.; François, S.; Favier, A.L.; Jullien, N.; Schrock, G.; Scherthan, H.; Drouet, M. First Insights into the M2 Inflammatory Response after Adipose-Tissue-Derived Stem Cell Injections in Radiation-Injured Muscles. *Health Phys.* **2018**, *115*, 37–48. [CrossRef]
88. Huang, S.-P.; Huang, C.-H.; Shyu, J.-F.; Lee, H.-S.; Chen, S.-G.; Chan, J.Y.-H.; Huang, S.-M. Promotion of wound healing using adipose-derived stem cells in radiation ulcer of a rat model. *J. Biomed. Sci.* **2013**, *20*, 51. [CrossRef] [PubMed]
89. Wu, S.-H.; Shirado, T.; Mashiko, T.; Feng, J.; Asahi, R.; Kanayama, K.; Mori, M.; Chi, D.; Sunaga, A.; Sarukawa, S.; et al. Therapeutic Effects of Human Adipose-Derived Products on Impaired Wound Healing in Irradiated Tissue. *Plast. Reconstr. Surg.* **2018**, *142*, 383–391. [CrossRef] [PubMed]
90. Holler, V.; Buard, V.; Roque, T.; Squiban, C.; Benderitter, M.; Flamant, S.; Tamarat, R. Early and Late Protective Effect of Bone Marrow Mononuclear Cell Transplantation on Radiation-Induced Vascular Dysfunction and Skin Lesions. *Cell Transplant.* **2019**, *28*, 116–128. [CrossRef] [PubMed]
91. Bertho, J.M.; Bey, E.; Bottolier-Depois, J.F.; Revel, T.D.; Gourmelon, P. *A New Therapeutic Approach of Radiation Burns by Mesenchymal Stem Cell Transplantation*; Partial-Body Radiation Diagnostic Biomarkers and Medical Management of Radiation Injury Workshop AFFRI: Bethesda, MD, USA, 2008.
92. Mohan, A.; Singh, S. Use of fat transfer to treat a chronic, non-healing, post-radiation ulcer: A case study. *J. Wound Care* **2017**, *26*, 272–273. [CrossRef]
93. Rigotti, G.; Marchi, A.; Galie, M.; Baroni, G.; Benati, D.; Krampera, M.; Pasini, A.; Sbarbati, A.; Rubin, J.P.; Marra, K.G. Clinical Treatment of Radiotherapy Tissue Damage by Lipoaspirate Transplant: A Healing Process Mediated by Adipose-Derived Adult Stem Cells. *Plast. Reconstr. Surg.* **2007**, *119*, 1409–1422. [CrossRef]
94. Phulpin, B.; Gangloff, P.; Tran, N.; Bravetti, P.; Merlin, J.L.; Dolivet, G. Rehabilitation of irradiated head and neck tissues by autologous fat transplantation. *Plast. Reconstr. Surg.* **2009**, *123*, 1187–1197. [CrossRef]
95. Cordeiro, P.G.; Pusic, A.L.; Disa, J.J.; McCormick, B.; VanZee, K. Irradiation after immediate tissue expander/implant breast reconstruction: Outcomes, complications, aesthetic results, and satisfaction among 156 patients. *Plast. Reconstr. Surg.* **2004**, *113*, 877–881. [CrossRef]
96. Serra-Renom, J.M.; Munoz-Olmo, J.L.; Serra-Mestre, J.M. Fat grafting in postmastectomy breast reconstruction with expanders and prostheses in patients who have received radiotherapy: Formation of new subcutaneous tissue. *Plast. Reconstr. Surg.* **2010**, *125*, 12–18. [CrossRef]
97. Salgarello, M.; Visconti, G.; Barone-Adesi, L. Fat grafting and breast reconstruction with implant: Another option for irradiated breast cancer patients. *Plast. Reconstr. Surg.* **2012**, *129*, 317–329. [CrossRef] [PubMed]
98. Maria, O.M.; Kumala, S.; Heravi, M.; Syme, A.; Eliopoulos, N.; Muanza, T. Adipose mesenchymal stromal cells response to ionizing radiation. *Cytotherapy* **2016**, *18*, 384–401. [CrossRef]

99. Lombaert, I.M.; Wierenga, P.K.; Kok, T.; Kampinga, H.H.; deHaan, G.; Coppes, R.P. Mobilization of bone marrow stem cells by granulocyte colony-stimulating factor ameliorates radiation-induced damage to salivary glands. *Clin. Cancer Res.* **2006**, *12*, 1804–1812. [CrossRef] [PubMed]
100. Choi, J.S.; An, H.Y.; Shin, H.S.; Kim, Y.M.; Lim, J.Y. Enhanced tissue remodelling efficacy of adipose-derived mesenchymal stem cells using injectable matrices in radiation-damaged salivary gland model. *J. Tissue Eng. Regen Med.* **2018**, *12*, e695–e706. [CrossRef]
101. Saylam, G.; Bayır, Ö.; Gültekin, S.S.; Pınarlı, F.A.; Han, Ü.; Korkmaz, M.H.; Sancaktar, M.E.; Tatar, İ.; Sargon, M.F.; Tatar, E.Ç. Protective/restorative Role of the Adipose Tissue-derived Mesenchymal Stem Cells on the Radioiodine-induced Salivary Gland Damage in Rats. *Radiol Oncol.* **2017**, *51*, 307–316. [CrossRef] [PubMed]
102. Nevens, D.; Nuyts, S. The role of stem cells in the prevention and treatment of radiation-induced xerostomia in patients with head and neck cancer. *Cancer Med.* **2016**, *5*, 1147–1153. [CrossRef] [PubMed]
103. Isern, J.C.; Rich, R.S.; Tarruella, J.R.; Bustamante, J.Z. Radiation-induced Xerostomia in a Patient With Head and Neck Cancer Treated With Injection of Autologous Expanded Adipose Tissue-derived Mesenchymal Stem Cells: A New Therapeutic Opportunity. *Acta Otorrinolaringol.* **2021**, *72*, 63–65. [CrossRef]
104. Burns, R.J.; Jones, A.N.; Robertson, J.S. Pathology of radiation myelopathy. *J. Neurol. Neurosurg. Psychiatry* **1972**, *35*, 888–898. [CrossRef]
105. Delattre, J.Y.; Rosenblum, M.K.; Thaler, H.T.; Mandell, L.; Shapiro, W.R.; Posner, J.B. A model of radiation myelopathy in the rat. Pathology, regional capillary permeability changes and treatment with dexamethasone. *Brain* **1988**, *111*, 1319–1336. [CrossRef] [PubMed]
106. Powers, B.E.; Beck, E.R.; Gillette, E.L.; Gould, D.H.; LeCouter, R.A. Pathology of radiation injury to the canine spinal cord. *Int. J. Radiat. Oncol. Biol. Phys.* **1992**, *23*, 539–549. [CrossRef]
107. Shin, H.S.; Lee, S.; Kim, Y.M.; Lim, J.Y. Hypoxia-Activated Adipose Mesenchymal Stem Cells Prevents Irradiation-Induced Salivary Hypofunction by Enhanced Paracrine Effect Through Fibroblast Growth Factor 10. *Stem Cells* **2018**, *36*, 1020–1032. [CrossRef]
108. Herodin, F.; Mayol, J.F.; Mourcin, F.; Drouet, M. Which place for stem cell therapy in the treatment of acute radiation syndrome? *Folia Histochemica et Cytobiologica* **2005**, *43*, 223–227.
109. Caplan, A.I. Why are MSCs therapeutic? New data: New insight. *J. Pathol.* **2009**, *217*, 318–324. [CrossRef] [PubMed]
110. Coppes, R.P.; Stokman, M.A. Stem cells and the repair of radiation-induced salivary gland damage. *Oral Dis.* **2011**, *17*, 143–153. [CrossRef]
111. Benderitter, M.; Caviggioli, F.; Chapel, A.; Coppes, R.P.; Guha, C.; Klinger, M.; Malard, O.; Stewart, F.; Tamarat, R.; Van Luijk, P.; et al. Stem cell therapies for the treatment of radiation-induced normal tissue side effects. *Antioxid. Redox Signal.* **2014**, *21*, 338–355. [CrossRef]
112. Janowska-Wieczorek, A.; Majka, M.; Ratajczak, J.; Ratajczak, M.Z. Autocrine/paracrine mechanisms in human hematopoiesis. *Stem Cells* **2001**, *19*, 99–107. [CrossRef]
113. Zhang, S.; Chen, L.; Liu, T.; Zhang, B.; Xiang, D.; Wang, Z.; Wang, Y. Human Umbilical Cord Matrix Stem Cells Efficiently Rescue Acute Liver Failure Through Paracrine Effects Rather than Hepatic Differentiation. *Tissue Eng. Part A* **2012**, *18*, 1352–1364. [CrossRef]
114. Pierro, M.; Ionescu, L.; Montemurro, T.; Vadivel, A.; Weissmann, G.; Oudit, G.; Emery, D.; Bodiga, S.; Eaton, F.; Péault, B.; et al. Short-term, long-term and paracrine effect of human umbilical cord-derived stem cells in lung injury prevention and repair in experimental bronchopulmonary dysplasia. *Thorax* **2013**, *68*, 475–484. [CrossRef]
115. Rehman, J.; Traktuev, D.; Li, J.; Merfeld-Clauss, S.; Temm-Grove, C.J.; Bovenkerk, J.E.; Pell, C.L.; Johnstone, B.H.; Considine, R.V.; March, K.L. Secretion of angiogenic and antiapoptotic factors by human adipose stromal cells. *Circulation* **2004**, *109*, 1292–1298. [CrossRef]
116. Little, J.B. Repair of potentially-lethal radiation damage in mammalian cells: Enhancement by conditioned medium from stationary cultures. *Int. J. Radiat. Biol. Relat. Stud. Phys. Chem. Med.* **1971**, *20*, 87–92. [CrossRef] [PubMed]
117. Lopatina, T.; Kalinina, N.; Karagyaur, M.; Stambolsky, D.; Rubina, K.; Revischin, A.; Pavlova, G.; Parfyonova, Y.; Tkachuk, V. Adipose-Derived Stem Cells Stimulate Regeneration of Peripheral Nerves: BDNF Secreted by These Cells Promotes Nerve Healing and Axon Growth De Novo. *PLoS ONE* **2011**, *6*, e17899. [CrossRef] [PubMed]
118. Chen, L.; Tredget, E.E.; Wu, P.Y.; Wu, Y. Paracrine factors of mesenchymal stem cells recruit macrophages and endothelial lineage cells and enhance wound healing. *PLoS ONE* **2008**, *3*, e1886. [CrossRef] [PubMed]
119. Chiellini, C.; Cochet, O.; Negroni, L.; Samson, M.; Poggi, M.; Ailhaud, G.; Alessi, M.-C.; Dani, C.; Amri, E.-Z. Characterization of human mesenchymal stem cell secretome at early steps of adipocyte and osteoblast differentiation. *BMC Mol. Biol.* **2008**, *9*, 26. [CrossRef]
120. Distler, J.H.W.; Huber, L.C.; Hueber, A.J.; Reich, C.F.; Gay, S.; Distler, O.; Pisetsky, D.S. The release of microparticles by apoptotic cells and their effects on macrophages. *Apoptosis* **2005**, *10*, 731–741. [CrossRef] [PubMed]
121. Distler, J.H.W.; Pisetsky, D.S.; Huber, L.C.; Kalden, J.R.; Gay, S.; Distler, O. Microparticles as regulators of inflammation: Novel players of cellular crosstalk in the rheumatic diseases. *Arthritis Rheum.* **2005**, *52*, 3337–3348. [CrossRef]
122. Distler, J.H.; Huber, L.C.; Gay, S.; Distler, O.; Pisetsky, D.S. Microparticles as mediators of cellular cross-talk in inflammatory disease. *Autoimmunity* **2006**, *39*, 683–990. [CrossRef]

123. Chen, T.S.; Lai, R.C.; Lee, M.M.; Choo, A.B.; Lee, C.N.; Lim, S.K. Mesenchymal stem cell secretes microparticles enriched in pre-microRNAs. *Nucleic Acids Res.* **2010**, *38*, 215–224. [CrossRef] [PubMed]
124. Gatti, S.; Bruno, S.; Deregibus, M.C.; Sordi, A.; Cantaluppi, V.; Tetta, C.; Camussi, G. Microvesicles derived from human adult mesenchymal stem cells protect against ischaemia-reperfusion-induced acute and chronic kidney injury. *Nephrol. Dial. Transplant.* **2011**, *26*, 1474–1483. [CrossRef] [PubMed]
125. Parekkadan, B.; Van Poll, D.; Suganuma, K.; Carter, E.A.; Berthiaume, F.; Tilles, A.W.; Yarmush, M.L. Mesenchymal stem cell-derived molecules reverse fulminant hepatic failure. *PLoS ONE* **2007**, *2*, e941. [CrossRef]
126. Van Poll, D.; Parekkadan, B.; Cho, C.H.; Berthiaume, F.; Nahmias, Y.; Tilles, A.W.; Yarmush, M.L. Mesenchymal stem cell-derived molecules directly modulate hepatocellular death and regeneration in vitro and in vivo. *Hepatology* **2008**, *47*, 1634–1643. [CrossRef] [PubMed]
127. Bruno, S.; Grange, C.; Deregibus, M.C.; Calogero, R.A.; Saviozzi, S.; Collino, F.; Morando, L.; Busca, A.; Falda, M.; Bussolati, B.; et al. Mesenchymal Stem Cell-Derived Microvesicles Protect Against Acute Tubular Injury. *J. Am. Soc. Nephrol.* **2009**, *20*, 1053–1067. [CrossRef] [PubMed]
128. Bruno, S.; Collino, F.; Deregibus, M.C.; Grange, C.; Tetta, C.; Camussi, G. Microvesicles derived from human bone marrow mesenchymal stem cells inhibit tumor growth. *Stem Cells Dev.* **2012**, e33115. [CrossRef] [PubMed]
129. Li, T.; Yan, Y.; Wang, B.; Qian, H.; Zhang, X.; Shen, L.; Wang, M.; Zhou, Y.; Zhu, W.; Li, W.; et al. Exosomes Derived from Human Umbilical Cord Mesenchymal Stem Cells Alleviate Liver Fibrosis. *Stem Cells Dev.* **2013**, *22*, 845–854. [CrossRef] [PubMed]
130. Chamberlain, C.S.; Kink, J.A.; Wildenauer, L.A.; McCaughey, M.; Henry, K.; Spiker, A.M.; Halanski, M.A.; Hematti, P.; Vanderby, R. Exosome-educated macrophages and exosomes differentially improve ligament healing. *Stem Cells* **2020**, *39*, 55–61.
131. Ren, Z.; Qi, Y.; Sun, S.; Tao, Y.; Shi, R. Mesenchymal Stem Cell-Derived Exosomes: Hope for Spinal Cord Injury Repair. *Stem Cells Dev.* **2020**, *29*, 1467–1478. [CrossRef] [PubMed]
132. You, J.; Zhou, O.; Liu, J.; Zou, W.; Zhang, L.; Tian, D.; Dai, J.; Luo, Z.; Liu, E.; Fu, Z.; et al. Human Umbilical Cord Mesenchymal Stem Cell-Derived Small Extracellular Vesicles Alleviate Lung Injury in Rat Model of Bronchopulmonary Dysplasia by Affecting Cell Survival and Angiogenesis. *Stem Cells Dev.* **2020**, *29*, 1520–1532. [CrossRef]
133. You, M.-J.; Bang, M.; Park, H.-S.; Yang, B.; Jang, K.B.; Yoo, J.; Hwang, D.-Y.; Kim, M.; Kim, B.; Lee, S.-H.; et al. Human umbilical cord-derived mesenchymal stem cells alleviate schizophrenia-relevant behaviors in amphetamine-sensitized mice by inhibiting neuroinflammation. *Transl. Psychiatry* **2020**, *10*, 1–15. [CrossRef] [PubMed]
134. Janowska-Wieczorek, A.; Majka, M.; Kijowski, J.; Baj-Krzyworzeka, M.; Reca, R.; Turner, A.R.; Ratajczak, J.; Emerson, S.G.; Kowalska, M.A.; Ratajczak, M.Z. Platelet-derived microparticles bind to hematopoietic stem/progenitor cells and enhance their engraftment. *Blood* **2001**, *98*, 3143–3149. [CrossRef]
135. Accarie, A.; L'Homme, B.; Benadjaoud, M.A.; Lim, S.K.; Guha, C.; Benderitter, M.; Tamarat, R.; Sémont, A. Extracellular vesicles derived from mesenchymal stromal cells mitigate intestinal toxicity in a mouse model of acute radiation syndrome. *Stem Cell Res. Ther.* **2020**, *11*, 371. [CrossRef] [PubMed]
136. Cavallero, S.; Riccobono, D.; Drouet, M.; François, S. MSC-Derived Extracellular Vesicles: New Emergency Treatment to Limit the Development of Radiation-Induced Hematopoietic Syndrome? *Health Phys.* **2020**, *119*, 21–36. [CrossRef]
137. Kink, J.A.; Forsberg, M.H.; Reshetylo, S.; Besharat, S.; Childs, C.J.; Pederson, J.D.; Gendron-Fitzpatrick, A.; Graham, M.; Bates, P.D.; Schmuck, E.G.; et al. Macrophages Educated with Exosomes from Primed Mesenchymal Stem Cells Treat Acute Radiation Syndrome by Promoting Hematopoietic Recovery. *Biol. Blood Marrow Transplant.* **2019**, *25*, 2124–2133. [CrossRef] [PubMed]
138. Lei, X.; He, N.; Zhu, L.; Zhou, M.; Zhang, K.; Wang, C.; Huang, H.; Chen, S.; Li, Y.; Liu, Q.; et al. Mesenchymal Stem Cell-Derived Extracellular Vesicles Attenuate Radiation-Induced Lung Injury via miRNA-214-3p. *Antioxid. Redox Signal.* **2020**, in press. [CrossRef]
139. Forsberg, M.H.; Kink, J.A.; Hematti, P.; Capitini, C.M. Mesenchymal Stromal Cells and Exosomes: Progress and Challenges. *Front. Cell Dev. Biol.* **2020**, *8*, 665. [CrossRef]
140. Szatmári, T.; Kis, D.; Bogdándi, E.N.; Benedek, A.; Bright, S.; Bowler, D.; Persa, E.; Kis, E.; Balogh, A.; Naszályi, L.N.; et al. Extracellular Vesicles Mediate Radiation-Induced Systemic Bystander Signals in the Bone Marrow and Spleen. *Front. Immunol.* **2017**, *8*, 347. [CrossRef]

 MDPI

Review

Mesenchymal Stem Cells for Mitigating Radiotherapy Side Effects

Kai-Xuan Wang [1], Wen-Wen Cui [1], Xu Yang [1], Ai-Bin Tao [2], Ting Lan [1], Tao-Sheng Li [3,*] and Lan Luo [1,3,*]

1 School of Medical Technology, Xuzhou Key Laboratory of Laboratory Diagnostics, Xuzhou Medical University, Xuzhou 221004, China; 301910411466@stu.xzhmu.edu.cn (K.-X.W.); 301910411469@stu.xzhmu.edu.cn (W.-W.C.); 300104120621@stu.xzhmu.edu.cn (X.Y.); tinglan@xzhmu.edu.cn (T.L.)
2 Division of Cardiology, The Affiliated People's Hospital of Jiangsu University, Zhenjiang 212132, China; taoab@jskfhn.org.cn
3 Department of Stem Cell Biology, Atomic Bomb Disease Institute, Nagasaki University, Nagasaki 852-8523, Japan
* Correspondence: litaoshe@nagasaki-u.ac.jp (T.-S.L.); luolan@xzhmu.edu.cn (L.L.); Tel.: +81-95-819-7099 (T.-S.L.); +86-15852233024 (L.L.); Fax: +81-95-819-7100 (T.-S.L.)

Abstract: Radiation therapy for cancers also damages healthy cells and causes side effects. Depending on the dosage and exposure region, radiotherapy may induce severe and irreversible injuries to various tissues or organs, especially the skin, intestine, brain, lung, liver, and heart. Therefore, promising treatment strategies to mitigate radiation injury is in pressing need. Recently, stem cell-based therapy generates great attention in clinical care. Among these, mesenchymal stem cells are extensively applied because it is easy to access and capable of mesodermal differentiation, immunomodulation, and paracrine secretion. Here, we summarize the current attempts and discuss the future perspectives about mesenchymal stem cells (MSCs) for mitigating radiotherapy side effects

Keywords: radiation-induced injury; radiotherapy; mesenchymal stem cells

Citation: Wang, K.-X.; Cui, W.-W.; Yang, X.; Tao, A.-B.; Lan, T.; Li, T.-S.; Luo, L. Mesenchymal Stem Cells for Mitigating Radiotherapy Side Effects. *Cells* **2021**, *10*, 294. https://doi.org/10.3390/cells10020294

Academic Editors: Alexander Ljubimov and Alain Chapel

Received: 24 November 2020
Accepted: 29 January 2021
Published: 1 February 2021

1. Introduction

Malignant tumors are one of the most aggressive diseases and have high mortality. Currently, there are no efficient methods capable of eradicating cancers clinically. As a conventional cancer treatment modality, radiotherapy (RT) can kill cancer cells and improve patient survival rates. Unfortunately, cancer patients also have to risk radiotoxicity to healthy tissues around the tumor. Clinical studies have revealed skin, intestinal, brain, pulmonary, hepatic, and cardiovascular injuries in cancer patients who received RT [1–5]. Although developments in RT devices and techniques (e.g., intensity-modulated RT, IMRT; image-guided RT, IGRT.) have significantly decreased radiation dose, exposure volume, and area, radiation injury is still unavoidable [6–9]. There is no evidence showing the existing dose threshold that would not damage the cell [10]. Emerging epidemiological data have consistently confirmed that low-dose radiation could also cause tissue damage [11,12]. Thus, when optimizing the RT technique to reduce the risk of radiation exposure, more effort should be made to seek satisfactory treatment for radiation-induced tissue injury.

In recent decades, stem cells have become a hot topic of research in regenerative medicine, bioengineering, and other clinical settings. Among the various stem cell types, mesenchymal stem cells (MSCs) are the most frequently studied. Thousands of publications are issued, and more than 490 clinical trials utilizing MSCs have been carried out or ongoing [13]. The reasons might be that MSCs are easy to access due to their abundant resources, including bone marrow, adipose tissue, umbilical cord, and placental tissue. Additionally, MSCs possess stable genomes, great self-renewal ability, mesodermal differentiation capacity, and immunomodulatory and paracrine secretome [14]. Indeed, MSCs reveal the tremendous therapeutic potential in various diseases such as cancer, diabetes mellitus, autoimmune disease, liver injury, and cardiovascular disease [15–19]. Thus, scientists attempt to investigate whether MSCs therapy could also mitigate radiation injury. Here, we will first introduce the underlying mechanisms of radiation injury and the features of MSCs briefly. Then, we focus on the recent progress on MSCs therapy in treating radiation injury. Last, we discuss the challenges and future perspectives of the MSCs therapy.

2. Pathophysiological Mechanisms of Radiation Injury

RT utilizes high doses of radioactive energy, known as ionizing radiation (IR), to kill cancer cells. Notably, IR also injuries the healthy cells around the tumor, causing various complications. However, the pathophysiological mechanisms of radiation injury remain mostly unclear. IR induces increased production of reactive oxygen species (ROS), referred to as oxidative stress, injuring cell components such as DNA, proteins, organelles, etc. [20]. The damages to DNA mainly comprise single- and double-stranded breaks and base lesions [21]. Incorrect DNA repair would give rise to mutagenesis or chromosomal instability resulting in cell apoptosis and carcinogenesis [22]. Excessive ROS activates unfolded protein response in the endoplasmic reticulum (ER), which further elicits Ca^{2+} release from ER, causing ER stress [23]. If the ER stress was uncontrolled, the unfolded protein response pathways trigger downstream signals such as c-Jun N-terminal kinase and Bcl-2 protein family members, initiating cell apoptosis or autophagy [24]. The enhanced ROS and imbalanced Ca^{2+} in the cytoplasm cause mitochondrial membrane permeabilization [25], leading to Bax's activation and the release of cytochrome c, promoting apoptosis development [26]. Moreover, mutated mitochondrial DNA, impaired PPAR-α pathways, and dysregulated ROS production induce mitochondrial dysfunction [26]. The proper functionality of cellular components is closely connected with the cell fate. Thus, clarifying the alterations of intercellular and intracellular signal cascades is beneficial for understanding the radiation injury.

Inflammatory responses, endothelial cell injuries, and fibrosis are vital radiation injury features [27–29]. At the acute phase after IR, inflammatory cytokines (tumor necrosis factor, TNF; interleukin-1, IL-1; IL-6; IL-8), chemokines (C-C motif chemokine ligand, CCL; C-C motif chemokine, CXC), and adhesion molecules (intercellular cell adhesion molecule, vascular cell adhesion molecule, E-selectin) are secreted, inducing vasodilation and vascular permeability [30]. Subsequently, coagulation cascade signals are triggered, and endothelial basement membrane is degraded, enabling clearance of damaged tissue and repairing initiation. This acute response may sustain from minutes to several days after IR [29]. Notably, chronic inflammation and oxidative stress would induce fibrosis at the later phase of diseases [31]. The transforming growth factor-β1 (TGF-β1)/Smad signaling has been recognized as the primary player that mediates myofibroblasts proliferation and regulates extracellular matrix and collagens deposition [32]. IR also upregulates the connecting tissue growth factor levels that can enhance the binding of TGF-β1 with its receptor (Smad2, Smad3), promoting fibroblast trans-differentiation [33]. By dissociating TGF-β from its complex, the enhanced ROS promotes TGF-β1/Smad signaling, which further modulates ROS generation via upregulating NADPH oxidase 4 transcriptional activity [34]. Moreover, myofibroblasts are also found to originate from the process named epithelial or endothelial to mesenchymal transition [35]. Other profibrotic cytokines, such as CCL3, CCL2, IL-1, and IL-6, are also essential for fibrosis progress. Elevated IL-6 levels post IR is correlated with radiation toxicity in breast cancer patients and the degree of fibrosis in the irradiated lung [36,37]. Fibrosis formation is usually a chronic but ongoing progressing process, and it lacks sensitive tools allowing for early detection.

Apart from these mechanisms, telomere erosion, miRNAs alterations, epigenetic regulations, and stem cell damage are also engaged in the pathophysiological development of radiation injury [38–41]. Moreover, these underlying mechanisms interconnect with each other and vary depending on the tissue/cell types, IR patterns (types, doses, and dose rates), and patient-related factors (individual comorbidities and risk factors, such as body mass index, smoking, and genetic predisposition). Thus, determining factors that promote radiation injury progression from asymptomatic remains challenging.

3. Characteristics of MSCs

Currently, there is no absolute definition of MSCs. To facilitate the development of MSCs-based study, the International Society for Cellular Therapy proposes several minimal criteria identifying MSCs [42–44]. Firstly, surface CD antigens are the most primary and necessary verification method. MSCs positively express stro-1, CD44, CD73, CD90, and CD105. Different from hematopoietic stem cells, MSCs lack CD34, CD45, CD14 (or CD11b), CD79α (or CD19), and HLA-DR. Secondly, MSCs are considered to be plastic-adherent when cultured under standard conditions. Lastly, MSCs must possess the capability of differentiating into osteoblasts, adipocytes, and chondroblasts. This report largely standardizes the definition of MSCs and instructs investigators to estimate the authenticity of their cells.

MSCs can be obtained from multiple tissues (bone marrow, adipose tissue, peripheral blood, umbilical cord, and placenta), providing researchers with great convenience and increasing its clinical application popularity [45]. MSCs derived from differed tissues show distinct characteristics, including proliferation and differentiation potential, paracrine effect, immunophenotypes, and immunomodulatory capacity [46,47]. For example, umbilical cord blood-derived MSCs (UC-MSCs) show more significant proliferation and slower senescence compared with that from bone marrow (BM-MSCs) and adipose tissue (AT-MSCs) [48]. However, BM- and AT-MSCs are capable of tri-lineage differentiation (osteogenic, adipogenic, and chondrogenic) under respective culture conditions, while placenta- and UC-MSCs only differentiate into two cell lineage [46]. Additionally, discrepant paracrine activity reflected by the expression of various cytokines and growth factors was observed in UC- and AT-MSCs [49]. All these differences may influence the function of MSCs from multiple sources. A comprehensive understanding of these features would promote a more efficient clinical application of MSCs.

In most MSCs-based therapy studies, immunomodulation is regarded as the leading factor of the therapeutic property. MSCs can interact with immune system cells (T cell, B cell, natural killer cells, etc.) and regulate immune response depending on direct cell-cell contact and various immunomodulated factors [50]. High inflammation levels would stimulate MSCs to release anti-inflammatory cytokines, inhibiting overactivated inflammation and immune responses. The involved molecules include inducible nitric oxide synthase (iNOs), TGF-β, IL-10, prostaglandin E2 (PGE2), and hepatocyte growth factor (HGF) [51]. T cells would be deactivated by inducing apoptosis or suppressing proliferation [52]. On the contrary, the silent immune system would induce the pro-inflammatory phenotype of MSCs to ensure basic self-defense against the external pathogen. Such plastic immunomodulation function protect tissue against pathogen invasion or self-attack, making MSCs a popular object in the study of tissue repair and regeneration [53].

4. Current Attempts of MSCs for Mitigating Radiation Injury

Considerable progress in medications has dramatically reduced the mortality and morbidity of cancer patients. The increased number of cancer survivors enables clinicians to realize the side effects of related treatments such as RT. To date, it has gained remarkable improvements in achieving high-precision RT. For instance, breast cancer patients receiving IMRT exhibited significantly lower occurrence, severity, and persistent of radiodermatitis than those receiving conventional RT [8]. A significant reduction in gastrointestinal toxicity was observed in IMRT than conventional two-dimensional RT (IMRT vs. RT: 33% vs. 77%) [6]. Moreover, the combination of IGRT and IMRT (IG-IMRT) showed more significant superiority than conventional three-dimensional conformal RT in the treatments of rectal cancer and hepatocellular carcinoma [7,9]. With IG-IMRT, hepatocellular carcinoma patients showed longer median survival (IG-IMRT vs. RT: 44.7 vs. 24.0 months) [7,9]. Although modern RT doses have been minimized and are precise, radiation complications still typically occur acutely or chronically. Here, we mainly discuss the latest advances in MSCs therapy application mitigating radiation injury involving the skin, intestine, brain, lung, liver, and heart.

4.1. MSCs in Radiation-Induced Skin Injury

Radiation-induced skin injury or radiodermatitis is the most common side effect in people exposed to IR. Up to 95% of cancer patients undergoing RT experienced radiodermatitis [54]. Among the manifestation of radiodermatitis, erythema is the most apparent and mild symptom (incidence with more than 90%), followed by moist desquamation (incidence of 30%) [55]. These varying severity levels are associated with direct radiation injuries and consequent inflammations affecting different skin structures, including epidermis, dermis, and vasculature (well described in [56,57]). The release of cytokines and chemokines by recruited immune cells activates dermal fibroblasts, causing chronic dermatitis and skin fibrosis [58]. Regular treatment of radiodermatitis comprises self-care (daily hygiene habits, loose clothing, avoiding tobacco and alcohol, adequate water intake, etc.) and prophylactic topical corticosteroids [59]. Such therapies are usually based on hearsay or physician preferences lacking powered studies to demonstrate their efficiency [60,61]. The occurrence of radiodermatitis has destroyed patients' physical appearance and beauty, and also delayed wound healing [29]. Thus, novel therapeutic validating by a more systematic and rigorous design is urgently needed.

It has demonstrated that bone marrow-derived cells such as MSCs, endothelial progenitors, and myelomonocytic cells are recruited to the injured sites by chemotactic signals SDF-1 and CXCR4 participating in the healing process [62]. The intravenous injection of MSCs significantly accelerates the wound healing rate [63]. Increased survival of BM-MSCs ameliorates injury induced by IR combined with traumatic tissue injury [64]. Thus, scientists have attempted to mitigate radiodermatitis using exogenous administration of MSCs. For instance, Moghaddam et al. intradermally transplanted AT-MSCs (2×10^6) to guinea pigs receiving 60 Gy abdominal radiation. These irradiated guinea pigs showed alleviated skin damage, and the combination of low-intensity ultrasound enhanced the curative effect of AT-MSCs [65]. However, the exact mechanism underlying the therapeutic potential of MSCs for radiodermatitis is unclear. Anti-inflammation and anti-fibrosis may be the main ways for MSCs to inhibit radiation injury [66,67]. Inflammation-related cytokines (IL1β and IL10) were regulated by BM-MSCs (5×10^5) in radiation mice models with a 35 Gy dose [67]. Similarly, BM-MSCs injection (2×10^6) via tail vein efficiently reduced 45 Gy radiation-induced rats' skin fibrosis reflected by decreased TGF-β1 [66]. Notably, the MSCs conditioned medium (CM) could also accelerate wound healing after pipetting onto the irradiated rats' skin wound [68]. This result indicated that paracrine factors from MSCs play a critical role in repairing radiodermatitis by mitigating the injury site's inflammatory microenvironment. Apart from animal studies, limited clinical trials were also carried out. A case report analyzed the treatment potential of cadaveric MSCs on a necrotic ulcer in a patient receiving 50–60 Gy dose RT for right leg angioma [69]. Two years after the treatment, clinicians observed a reduced ulcer size and improved the skin quality, confirming the MSC therapy's efficiency. Thus, MSCs or their secretiome could be novel therapeutics for mitigating the radiodermatitis.

4.2. MSCs in Radiation-Induced Intestinal Injury

Radiation-induced intestinal injury (RIII) or radiation enteropathy develops in RT-treated patients with abdominal or pelvic tumors. About 60–80% of patients have nausea, abdominal pain, and diarrhea within 2–3 weeks of RT [70]. Such symptoms usually disappear within 1–3 months of completing therapy. However, a few patients may experience delayed RIII, including disorders in intestine motility and nutrient absorption. Some severe chronic RIII may progress to intestinal obstruction or perforation and fistulae formation. The pathological changes in acute RIII involve inflammation reaction and consequent crypt cell death [71,72]. On the other hand, chronic RIII is more complex and is characterized by mucosa atrophy, intestinal wall fibrosis, and microvascular sclerosis [70]. Numerous preclinical studies utilizing natural products [73], peptides [72], and small molecules [74] to alleviate RIII have been carried out. However, researcher have not yet reached a consensus on the clinical application. Amifostine, a free-radical scavenger, is the earliest drug proved by the FDA to mitigate radiation therapy-related injury [75]. Nevertheless, the narrow treatment time window and lingering concerns of amifostine hinder its clinical uses [76]. Moreover, the US FDA has approved Neupogen and Neulasta in 2015 and leukine in 2018 for acute radiation syndrome [77]. Thus, novel therapeutic strategies are eagerly needed, especially drugs specific for each radiation-induced organ injury.

MSCs were initially found to migrate and settle in the injured intestine after RT [78]. Lately, studies revealed that the transplanted MSCs can reverse the disrupted intestinal function by RT [79,80]. Such benefits were attributed to the MSCs secretome-mediated intestinal regeneration via inflammation inhibition, neovascularization, and epithelial homeostasis maintenance [81]. Additionally, there exist specific stem cells in the intestinal crypt responsible for intestinal repair and regeneration [82]. BM-MSCs (1×10^6) transplantation via tail vein injection was found to increase $Lgr5^+$ intestinal stem cell populations, thus facilitating the repair of radiation-induced intestinal injury via activated Wnt/β-catenin signaling [83]. Based on the excellent paracrine effect, MSCs-CM were also applied to preclinical experiments of RIII. Repeated injection of AT-MSCs-CM (abundant angiogenic factors such as IL-8, angiogenin, HGF, and vascular endothelial growth factor) promoted intra-villi microvascular recovery in the irradiated intestine via activating the PI3K/AKT signal pathway [84]. Nevertheless, MSCs cultured under normal conditions only secrete slight cytokines that may possess unsatisfactory therapeutic potential. Given this, Chen et al. pretreated BM-MSCs with pro-inflammatory factors (TNF-α, IL-1β, nitric oxide) and found an enhanced paracrine effect of MSCs, primarily represented by the secretion of IGF [85]. The pretreated BM-MSCs-CM exhibited a more significant therapeutic efficacy in modulating inflammatory responses and mediating epithelial regeneration [85]. Moreover, other modifications such as carrying foreign genes (HGF, CXCL12) or cytokines (R-Spondin1) and engineered MSCs (hydrogel loaded) have also been tested for their capacity in alleviating RIII [86–88]. Preclinical studies have shown the therapeutic potential of MSCs (modified or not) in treating radiation injury. MSCs was also tested for clinical treatment of RIII, in which reduced intestinal inflammation and hemorrhage were exhibited after systematic usage of MSCs [89]. However, a detailed treatment strategy remains unknown.

4.3. MSCs in Radiation-Induced Brain Injury

Radiation-induced brain injury (RIBI) is mainly presented as cognitive dysfunction in patients experiencing head and neck RT [90]. The degree of tissue injury is unequal based on different periods (acute, early delayed, late delayed) [91]. Acute response is sporadic under current RT techniques. Early RIBI involves angioedema and manifested clinically as headache and drowsiness [92]. Acute and early RIBI are generally recovered within 1 to 6 months. However, late RIBI often represents severe irreversible lesions such as vascular injury and demyelination, leading to ultimate white matter necrosis and brain atrophy [93,94]. Apart from the vascular endothelial cells, neurons and glial cells are also susceptible to IR [95]. In all, RIBI is intractable due to the complex dynamic process [91]. Early epidemiological data showed 11% of morbidity of severe dementia in cancer patients receiving whole brain radiation [96]. In fact, sensitive neurocognitive tests suggested that 90% of irradiated patients had neurological impairment [97]. With regard to the treatment of RIBI, anti-inflammatory drugs have been applied to counteract RIBI, such as eicosapentaenoic acid and fenofibrate [98,99]. Moreover, traditional Chinese medicines are also beneficial for neuroprotection against radiation [100]. In preclinical studies, intrahippocampal transplantation of human neural stem cells restored neural plasticity of irradiated rats by improving the expression of activity-regulated cytoskeletal [101]. At present, MSCs-based cell transplantation and secretome administration are also considered as therapeutic strategies preclinically. UC-MSCs (1×10^6) transplantation via caudal vein infusion showed anti-inflammatory and anti-apoptotic effects on mice with RIBI [102,103]. The RT-triggered inflammation was inhibited, reflected by the decreased IL-1, TNF-α, and the increased IL-10 [102]. On the other hand, the downregulation of pro-apoptotic proteins (p53, Bax) and the upregulation of anti-apoptotic Bcl-2 confirmed apoptosis reduction. This anti-apoptotic benefit was further enhanced through the combined administration of UC-MSCs and nimodipine [103]. MSCs-mediated regulation of both inflammation and apoptosis rescued neurons and astrocytes from necrosis. Additionally, microglia were activated during RIII and initiated inflammation reaction by cytokine and chemokine secretion [104]. Intensive inflammation further accelerated microglia pyroptosis related to the increased expression of NLRP3 inflammasome and caspase-1 [105]. Human trophoblast-derived MSCs (1×10^5) transplantations via brain cortex are able to reverse the microglia pyroptosis, promoting tissue repair [105]. Others also identified that the intranasally administered human MSCs (5×10^5) restored neurological function by reducing inflammation and oxidative stress via declined damage-induced c-AMP response element-binding signals [106]. Unfortunately, only a few researches on applying MSCs therapy in RIBI have been reported so far. The finding that MSCs are also homed to gliomas would encourage more efforts to be devoted to this area [107].

4.4. MSCs in Radiation-Induced Lung Injury

Thoracic tumors patients receiving RT tend to suffer from radiation-induced lung injury (RILI) with a mortality of approximately 15% [108]. The RILI is a complex dynamic process, including early pneumonitis and delayed pulmonary fibrosis [109]. The common pathological changes of RILI include epithelial and endothelial cell injuries, inflammatory responses, resulting in the dysfunction of the blood-air barrier and vascular permeability [109]. Moreover, the alveolar macrophages are also stimulated to secrete abundant cytokines (TGF-β1, TNF-α, IL-1β, IL-6, and IL-12) that further participate in the inflammatory process [110]. TGF-β1 is an essential factor that mediates alveolar epithelial cells undergo EMT, a typical feature of fibrosis [111]. The occurrence of a vicious cycle of inflammation would promote delayed pulmonary fibrosis. Once the fibrosis is formed, it is difficult to reverse and leads to a poor prognosis. Apart from the amifostine, steroids, growth factors (IL-7, IL-11, etc.), antioxidants, and signaling inhibitors have been used to treat RILI, yielding unsatisfactory effects [108]. Thus, clinicians ask for novel and more effective therapeutic approaches.

The potential of treatment with MSCs to mitigate RILI has been evaluated and its underlying mechanisms have been explored. A preclinical study showed that BM-MSCs injected into irradiated mice via tail vein could differentiate into lung epithelial and endothelial cells [112]. They also observed an upregulated IL-10 and downregulated TNF-α and TGF-β in RILI mice [112]. Because excessive inflammation and irreversible fibrosis are the leading causes of RILI, the MSCs-mediated anti-inflammation and anti-fibrosis effects may play a vital role in lung tissue repair and regeneration. Consistently, Hao et al. found that intratracheal transplantation of human UC-MSCs (1×10^6/kg) inhibited canine pulmonary inflammation and fibrosis in beagle dogs induced by radiation through reducing IL-1, TGF-β, and hyaluronic acid [113]. Dong et al. first identified two anti-fibrotic factors, HGF and PGE2, that exhibited increased expression in irradiated rat lung tissue after administration of AT-MSCs [114]. Additionally, radiation-induced lung endothelial dysfunction could be alleviated by MSCs-CM [115]. This perhaps further suggested that the paracrine effect rather than differentiation plays a dominant role in the MSCs therapy. In fact, paracrine-depended secretome and vesicles derived from MSCs have also shown a significant efficacy on RILI [116]. Notably, growing evidence showed that gene-modified MSCs may possess more tremendous therapeutic potential than unmodified MSCs in RILI. For example, human UC-MSCs modified with CXCR4 showed a significant anti-fibrotic effect in irradiated mice [117]. This mainly depended on more accurate homing and colonization that was critical for enhancing targeted therapy of MSCs. Liu et al. injected

UC-MSCs expressing decorin (an inhibitor of TGF-β and fibrogenesis) into irradiated mice and observed improved lung inflammation and fibrosis [118]. Additionally, manganese superoxide dismutase (ROS scavenger) modified MSCs also exerted a therapeutic effect on RILI reflected by decreased lung cell apoptosis [119]. In fact, gene-modified MSCs overexpress certain soluble factors, which can protect tissues from radiation injury. The combination of natural MSCs properties and overexpressed beneficial factors consolidates the therapeutic effect of MSCs. Despite abundant preclinical evidence of the beneficial effect of MSCs on RILI, relevant clinical data are incredibly lacking. A report involving 11 patients with RILI confirmed autologous MSCs administration safety, but the actual efficacy could not be assessed [120].

4.5. MSCs in Radiation-Induced Hepatic Injury

Radiation-induced hepatic injury (RIHI) presents two different clinical types (classic and non-classic RIHI) reflected by distinct characteristics [121]. Both of them occurred in 36% of patients receiving reirradiation for hepatocellular carcinoma [122]. Classic RIHI is recognized by hepatomegaly, anicteric ascites, and increased abdominal circumference [123]. Patients with classic RIHI show upregulated alkaline phosphatase but normal transaminase and bilirubin levels [124]. The veno-occlusive disease, an essential manifestation of classic RIHI, is described as a complete blockage of the central vein by erythrocytes attached to a dense network of reticulin and collagen fibers [125]. Non-classic RIHI represents an impaired liver function in those patients with chronic hepatic injury, such as viral hepatitis and cirrhosis. Jaundice or significantly elevated serum transaminases levels (five times higher than the standard value) could be used to confirm non-classic RIHI [126]. Transaminases are an important biomarker for assessing the hepatic injury. After irradiation, human or rat MSCs perfusion significantly reduced serum transaminase activity, indicating recovered liver function [127,128]. The mechanism might be apoptosis inhibition due to decreased ROS production and increased secretion of anti-inflammatory IL-10 [127]. In another study, the combined intravenous administration of BM-MSCs (1×10^6) and nigella sativa oil present a similar protective effect on the liver [128]. In addition to inherent medicinal value, nigella sativa oil could enhance MSCs homing in injured liver sites. However, Moubarak et al. found that intravenous MSCs were not grafted to the liver but to the intestine following abdominal irradiation. Improved intestinal damage indirectly corrects liver abnormality via enterohepatic recirculation [129]. Meanwhile, the paracrine mechanism played a more critical role and dominated the protection of MSCs against RIHI without liver engraftment. With increased recognition of the paracrine effect, MSCs-CM was also used to examine paracrine factors' repair capability to RIHI [130]. In vitro administration of MSCs-CM for culturing sinusoidal endothelial cells increased cell viability and blocked apoptosis. In vivo injection of MSCs-CM into irradiated rat reversed radiation-induced hepatic histopathological changes. Critical nutritional factors responsible for the regeneration potential were unclear, but the mechanism may be related to phosphorylation activation of AKT and ERK. Among all beneficial growth factors secreted from MSCs, hepatocyte growth factor possesses multiple tissue repair abilities, especially liver regeneration. Gene-modified AT-MSCs over-expressing HGF downregulated profibrotic proteins (α-SMA and fibronectin) and showed greater anti-fibrotic potential on the irradiated liver in comparison to unmodified MSCs [131]. Unfortunately, there are still no relevant clinical report to date.

4.6. MSCs in Radiation-Induced Heart Injury

Apart from the lung, thoracic irradiation also induces heart injury, namely, radiation-induced heart disease (RIHD). RIHDs, such as myocardial, coronary artery, pericardial, valvular, and conduction system diseases, have been observed in breast cancer and Hodgkin's lymphoma patients [132,133]. These manifestations had a 50% cumulative incidence during 40 years of follow-up in an epidemiological study [132]. RIHD often involves vascular endothelial dysfunction [134], hypertrophy [135], and fibrosis [136]. The underlying mechanisms of RIHD remain mostly indistinct, but the roles of DNA damage, inflammation, oxidative stress, and epigenetic regulation in RIHD have been well illustrated. For the treatment of RIHD, conventional statins and angiotensin-converting enzyme inhibitors are still the first-chosen drugs clinically. With increasing interest in MSCs regeneration therapy, scientists are paying attention to the application of MSCs in RIHD. Vascular injury is the most common feature of RIHD. BM-MSCs (1×10^6/kg) transplantation via tail vein can attenuate radiation-induced artery inflammation and oxidative stress [137]. The repair effect was attributed to the modulation of a series of cytokines and the differentiation potential of MSCs into endothelial cells facilitating vascular regeneration [138]. Additionally, vascular injury is usually accompanied by myocardial fibrosis and cardiac remodeling. Encouragingly, in a RIHD rat model, BM-MSCs (1.5×10^6) transplantation via caudal vein improved myocardial fibrosis and inflammation, which were related to DNA repair and downregulated PPAR-α, PPAR-γ, TGF-β, IL-6, and IL-8 [139]. As mentioned above, MSCs-CM is beneficial to radiation injury repair owing to the paracrine effect. Chen et al. assessed the therapeutic effect of human UC-MSCs-CM on radiation-induced myocardial fibrosis. They found that irradiated human cardiac fibroblasts cultured with UC-MSCs-CM showed greater viability [140]. Inhibited NF-κB activity decreased expression of several pro-fibrotic cytokines, including TGF-β1, IL-6, and IL-8, followed by mitigated collagen deposition and fibrosis [140]. Meanwhile, changes in oxidation markers (malondialdehyde) and antioxidant enzyme levels reflected reduced oxidative stress [140]. However, specific nutritional factors released by MSCs and involved in myocardial protection from IR were not clarified [140]. Thus far, there are few MSCs therapy attempts to manage RIHDs, and abundant evidence is lacking for proving its efficacy. The data on myocardial regeneration suggest that the MSCs therapy is potentially therapeutic to treat RIHD.

5. Challenges and Future Perspectives of MSCs Therapy

Although the MSCs have powerful tissue repair capacity due to their paracrine and immunomodulation activity, huge barriers hinder their clinical application. Here, we will focus on safety and efficacy, the two most concerning aspects.

Currently, the relationship between MSCs and tumor has been attracting increased attention. The tumor consists of many types of cells involving a complex pathological environment. Cancer stem cell (CSC) is a kind of multipotent stem cell with great self-renew and differentiation capability in the tumor tissue. Like normal stem cells in the body, CSC is also indispensable for supporting tumor progression, inducing tumorigenesis, maintaining tumor growth, and promoting metastasis [141]. The tumor involves a chronic inflammatory process that recruits endogenous or exogenous MSCs [142,143]. Homed MSCs promote angiogenesis [144] and interact with CSC enhancing the growth [145] and chemoresistance [146] of CSC. The tumor exploits MSCs' unique immunosuppression nature, allowing malignant cells to escape recognition and clearance by the immune system [147–149]. It is reported that once exposed to the tumor microenvironment, MSCs would be reprogrammed and become "allies" of tumor cells, accelerating tumor progress, and invading surrounding normal tissue [149–151]. Interestingly, Chen et al. found the engulfment of stromal cells by cancer cells in human breast tumors, and these engulfing breast cancer cells exhibited gene features of MSCs [152]. However, contradictory outcomes about the cancer-promoting effect of MSCs were presented in other studies [153]. For example, several groups found that co-cultured MSCs inhibited melanoma growth by inducing cell apoptosis [154,155]. Colorectal cancer progression could also be attenuated through the intravenous injection of BM-MSCs (1×10^7) [156]. The bidirectional effects of MSCs on tumor development motivate scientists to ascertain more precise mechanisms underlying MSCs and tumor tissue interaction. Unfortunately, it seems that the pro-tumorigenic effect is dominant due to more substantial preclinical evidence. Therefore, MSCs-based therapy must be performed with great caution in clinics, especially with regards to radiation injury patients with malignancy history.

On the premise that security can be guaranteed, investigators need to seek appropriate protocols by which MSCs therapy remedy would maximize radiation repair efficiency. Many questions need to be discussed, for example, how do we determine the selection of the MSC population considering heterogeneity? In addition, the most effective delivery dose and pattern are required to ensure a high retention rate and therapy efficacy. Indeed, different organizational origins give rise to MSCs heterogeneity reflected by diversities of proliferation and differentiation capability, paracrine potential, and immunomodulatory effect [46–48]. Despite the minimal criteria mentioned above, it is difficult to sort out homogeneous MSCs. Apart from shared surface CD antigens, there are no additional markers to identify each type of tissue-derived MSCs [157]. Such heterogeneity can lead to the deviation of actual results from expectation and become a significant obstacle to selecting MSCs for clinical usage [158]. Because of the heterogeneity, each MSC population may have distinct therapeutic effects on the same tissue injury. It is necessary to search for the most potent MSC population for radiation injury of a specific tissue. On the other hand, different laboratories have their respective protocols of MSCs isolation, culture, and expansion procedures, causing MSCs heterogeneity and the following difference in quality. Therefore, MSCs management system should be standardized as much as possible. This can reduce heterogeneity caused by different treating conditions and increased comparability among different research results, thus providing valuable clinical guidance of MSCs application. Apart from heterogeneity, the effective dose range and cell delivery route must be emphasized and discussed. A dose gradient experiment of MSCs therapy in radiation injury models should be carried out to find both safe and efficient dose range [159]. In a study of radiation-induced artery injury, a high dose of BM-MSCs (1×10^7/kg) showed greater therapeutic potential in irradiated mice than a low dose of BM-MSCs (1×10^6/kg) [137]. Additionally, different injection patterns, including whole-body infusion via a vein or local interventional injection, will affect the homing of MSCs to injured sites [160]. Thus far, our understandings of the therapeutic effect of MSCs in mitigating radiation injury and the underlying mechanism are basically from preclinical trials. The transition of MSCs administration from animal to clinical studies still requires lots of effort.

6. Conclusions

RT is an indispensable part of clinical cancer treatment, and more than 50% of cancer patients received RT [161]. Though the radiation doses and related radiotoxicity have been remarkably reduced due to modern RT techniques, radiation injury in normal tissue is still a thorny problem affecting patients' life quality and even survival rate. MSCs have abundant resources, excellent regenerative potential, immunomodulatory features, showing therapeutic potential in mitigating radiation injury in preclinical studies. Moreover, chemical, physical, or pharmaceutical preconditioning greatly enhanced the therapeutic potency of MSCs [162]. The overexpression of desired factors (antioxidation, differentiation, immunomodulation, angiogenesis, anti-apoptotic, and regeneration) targeting the specific disease model represents a novel approach in precision medicine. Because the local harsh environment and death signals cause MSCs to be rarely retained in the transplanted sites, MSCs-secretome or a combination with tissue engineering are emerging as a new trend. Notably, radiation-induced skin and intestine injury are easy to be aware of. Radiotoxicity that developed months or years after RT is challenging to be diagnosed or predicted early. In order to reduce or prevent radiotoxicity, more advanced radiotherapy technologies, such as IMRT and IGRT, need to be created. On the other hand, the application of MSCs as regenerative/repair agents when symptoms are presented or as preventive medicine directly after RT also needs careful consideration. The combination of prevention and regeneration/repair is the key to protect radiotherapy patients. Though there are many obstacles in the clinical application of MSCs, there is already a clinical trial evaluating the efficacy of MSC injections for the treatment of chronic radiotherapy-induced complications (PRISME, NCT02814864). We expect a promising future of MSCs therapy in mitigating radiation injury.

Author Contributions: K.-X.W., wrote the manuscript; T.-S.L., and L.L., revised and proved the final manuscript; W.-W.C., X.Y., A.-B.T., and T.L., made suggestions and revisions in the manuscript. All authors have read and agreed to the published version of the manuscript.

Funding: This study was supported by the National Natural Science Foundation of China (grant no. 81802086, 81802063, 81971817), the Natural Science Foundation of Jiangsu Province (grant no. BK20180995), the Specialized Research Fund for Senior Personnel Program (grant no. D2019028, no. D2017022), and the Young Science and Technology Innovation Team (grant no. TD202005) of Xuzhou Medical University, and a Grant-in-Aid from the Ministry of Education, Science, Sports, Culture and Technology, Japan.

Conflicts of Interest: The authors confirm that this article content has no conflict of interest.

References

1. Larsen, A.; Reitan, J.; Aase, S.; Hauer-Jensen, M. Long-term prognosis in patients with severe late radiation enteropathy: A prospective cohort study. *World J. Gastroenterol.* **2007**, *13*, 3610–3613. [CrossRef] [PubMed]
2. Welzel, G.; Fleckenstein, K.; Schaefer, J.; Hermann, B.; Kraus-Tiefenbacher, U.; Mai, S.; Wenz, F. Memory function before and after whole brain radiotherapy in patients with and without brain metastases. *Int. J. Radiat. Oncol. Biol. Phys.* **2008**, *72*, 1311–1318. [CrossRef]
3. Bracci, S.; Valeriani, M.; Agolli, L.; De Sanctis, V.; Maurizi Enrici, R.; Osti, M. Renin-angiotensin system inhibitors might help to reduce the development of symptomatic radiation pneumonitis after stereotactic body radiotherapy for lung cancer. *Clin. Lung Cancer* **2016**, *17*, 189–197. [CrossRef] [PubMed]
4. Khozouz, R.; Huq, S.; Perry, M. Radiation-induced liver disease. *J. Clin. Oncol.* **2008**, *26*, 4844–4845. [CrossRef]
5. Van Nimwegen, F.; Schaapveld, M.; Cutter, D.; Janus, C.; Krol, A.; Hauptmann, M.; Kooijman, K.; Roesink, J.; van der Maazen, R.; Darby, S.; et al. Radiation dose-response relationship for risk of coronary heart disease in survivors of Hodgkin Lymphoma. *J. Clin. Oncol.* **2016**, *34*, 235–243. [CrossRef]
6. Desai, N.; Stein, N.; LaQuaglia, M.; Alektiar, K.; Kushner, B.; Modak, S.; Magnan, H.; Goodman, K.; Wolden, S. Reduced toxicity with intensity modulated radiation therapy (IMRT) for desmoplastic small round cell tumor (DSRCT): An update on the whole abdominopelvic radiation therapy (WAP-RT) experience. *Int. J. Radiat. Oncol. Biol. Phys.* **2013**, *85*, e67–e72. [CrossRef] [PubMed]
7. Huang, C.; Huang, M.; Tsai, H.; Huang, C.; Ma, C.; Lin, C.; Huang, C.; Wang, J. A retrospective comparison of outcome and toxicity of preoperative image-guided intensity-modulated radiotherapy versus conventional pelvic radiotherapy for locally advanced rectal carcinoma. *J. Radiat. Res.* **2017**, *58*, 247–259. [CrossRef] [PubMed]
8. Lee, H.; Hou, M.; Chuang, H.; Huang, M.; Tsuei, L.; Chen, F.; Ou-Yang, F.; Huang, C. Intensity modulated radiotherapy with simultaneous integrated boost vs. conventional radiotherapy with sequential boost for breast cancer—A preliminary result. *The Breast* **2015**, *24*, 656–660. [CrossRef] [PubMed]
9. Jiang, T.; Zeng, Z.; Yang, P.; Hu, Y. Exploration of superior modality: Safety and efficacy of hypofractioned image-guided intensity modulated radiation therapy in patients with unresectable but confined intrahepatic hepatocellular carcinoma. *Can. J. Gastroenterol. Hepatol.* **2017**, *2017*, 6267981. [CrossRef]
10. Darby, S.; Ewertz, M.; McGale, P.; Bennet, A.; Blom-Goldman, U.; Brønnum, D.; Correa, C.; Cutter, D.; Gagliardi, G.; Gigante, B.; et al. Risk of ischemic heart disease in women after radiotherapy for breast cancer. *N. Engl. J. Med.* **2013**, *368*, 987–998. [CrossRef]
11. Shimizu, Y.; Kodama, K.; Nishi, N.; Kasagi, F.; Suyama, A.; Soda, M.; Grant, E.; Sugiyama, H.; Sakata, R.; Moriwaki, H.; et al. Radiation exposure and circulatory disease risk: Hiroshima and Nagasaki atomic bomb survivor data, 1950–2003. *BMJ* **2010**, *340*, b5349. [CrossRef] [PubMed]
12. Gillies, M.; Richardson, D.; Cardis, E.; Daniels, R.; O'Hagan, J.; Haylock, R.; Laurier, D.; Leuraud, K.; Moissonnier, M.; Schubauer-Berigan, M.; et al. Mortality from circulatory diseases and other non-cancer outcomes among nuclear workers in France, the United Kingdom and the United States (INWORKS). *Radiat. Res.* **2017**, *188*, 276–290. [CrossRef] [PubMed]
13. Squillaro, T.; Peluso, G.; Galderisi, U. Clinical trials with mesenchymal stem cells: An update. *Cell Transplant.* **2016**, *25*, 829–848. [CrossRef] [PubMed]
14. Naji, A.; Eitoku, M.; Favier, B.; Deschaseaux, F.; Rouas-Freiss, N.; Suganuma, N. Biological functions of mesenchymal stem cells and clinical implications. *Cell. Mol. Life Sci.* **2019**, *76*, 3323–3348. [CrossRef]
15. Mohr, A.; Zwacka, R. The future of mesenchymal stem cell-based therapeutic approaches for cancer—From cells to ghosts. *Cancer Lett.* **2018**, *414*, 239–249. [CrossRef] [PubMed]
16. Sun, Y.; Shi, H.; Yin, S.; Ji, C.; Zhang, X.; Zhang, B.; Wu, P.; Shi, Y.; Mao, F.; Yan, Y.; et al. Human mesenchymal stem cell derived exosomes alleviate type 2 Diabetes Mellitus by reversing peripheral insulin resistance and relieving β-cell destruction. *ACS Nano* **2018**, *12*, 7613–7628. [CrossRef]
17. Cipriani, P.; Carubbi, F.; Liakouli, V.; Marrelli, A.; Perricone, C.; Perricone, R.; Alesse, E.; Giacomelli, R. Stem cells in autoimmune diseases: Implications for pathogenesis and future trends in therapy. *Autoimmun. Rev.* **2013**, *12*, 709–716. [CrossRef]
18. Driscoll, J.; Patel, T. The mesenchymal stem cell secretome as an acellular regenerative therapy for liver disease. *J. Gastroenterol.* **2019**, *54*, 763–773. [CrossRef]

19. Yun, C.; Lee, S. Enhancement of functionality and therapeutic efficacy of cell-based therapy using mesenchymal stem cells for cardiovascular disease. *Int. J. Mol. Sci.* **2019**, *20*, 982. [CrossRef]
20. Wei, J.; Wang, B.; Wang, H.; Meng, L.; Zhao, Q.; Li, X.; Xin, Y.; Jiang, X. Radiation-Induced Normal Tissue Damage: Oxidative Stress and Epigenetic Mechanisms. *Oxidative Med. Cell. Longev.* **2019**, *2019*, 3010342. [CrossRef]
21. Santivasi, W.; Xia, F. Ionizing radiation-induced DNA damage, response, and repair. *Antioxid. Redox Signal.* **2014**, *21*, 251–259. [CrossRef] [PubMed]
22. Suzuki, K.; Ojima, M.; Kodama, S.; Watanabe, M. Radiation-induced DNA damage and delayed induced genomic instability. *Oncogene* **2003**, *22*, 6988–6993. [CrossRef] [PubMed]
23. Oakes, S.; Papa, F. The role of endoplasmic reticulum stress in human pathology. *Annu. Rev. Pathol.* **2015**, *10*, 173–194. [CrossRef] [PubMed]
24. Zheng, Z.; Shang, Y.; Tao, J.; Zhang, J.; Sha, B. Endoplasmic reticulum stress signaling pathways: Activation and diseases. *Curr. Protein Pept. Sci.* **2019**, *20*, 935–943. [CrossRef]
25. Javadov, S.; Jang, S.; Parodi-Rullán, R.; Khuchua, Z.; Kuznetsov, A. Mitochondrial permeability transition in cardiac ischemia-reperfusion: Whether cyclophilin D is a viable target for cardioprotection? *Cell. Mol. Life Sci.* **2017**, *74*, 2795–2813. [CrossRef]
26. Livingston, K.; Schlaak, R.; Puckett, L.; Bergom, C. The role of mitochondrial dysfunction in radiation-induced heart disease: From bench to bedside. *Front. Cardiovasc. Med.* **2020**, *7*, 20. [CrossRef]
27. Ejaz, A.; Greenberger, J.; Rubin, P. Understanding the mechanism of radiation induced fibrosis and therapy options. *Pharmacol. Ther.* **2019**, *204*, 107399. [CrossRef]
28. Baselet, B.; Sonveaux, P.; Baatout, S.; Aerts, A. Pathological effects of ionizing radiation: Endothelial activation and dysfunction. *Cell. Mol. Life Sci.* **2019**, *76*, 699–728. [CrossRef]
29. Kiang, J.; Olabisi, A. Radiation: A poly-traumatic hit leading to multi-organ injury. *Cell Biosci.* **2019**, *9*, 25. [CrossRef]
30. Schaue, D.; Micewicz, E.; Ratikan, J.; Xie, M.; Cheng, G.; McBride, W. Radiation and inflammation. *Semin. Radiat. Oncol.* **2015**, *25*, 4–10. [CrossRef]
31. Liu, R.; Desai, L. Reciprocal regulation of TGF-β and reactive oxygen species: A perverse cycle for fibrosis. *Redox Biol.* **2015**, *6*, 565–577. [CrossRef] [PubMed]
32. Hu, H.; Chen, D.; Wang, Y.; Feng, Y.; Cao, G.; Vaziri, N.; Zhao, Y. New insights into TGF-β/Smad signaling in tissue fibrosis. *Chem. Biol. Interact.* **2018**, *292*, 76–83. [CrossRef] [PubMed]
33. Tsai, C.; Wu, S., Kau, H.; Wei, Y. Essential role of connective tissue growth factor (CTGF) in transforming growth factor-β1 (TGF-β1)-induced myofibroblast transdifferentiation from Graves' orbital fibroblasts. *Sci. Rep.* **2018**, *8*, 7276. [CrossRef] [PubMed]
34. Cheng, Q.; Li, C.; Yang, C.; Zhong, Y.; Wu, D.; Shi, L.; Chen, L.; Li, Y.; Li, L. Methyl ferulic acid attenuates liver fibrosis and hepatic stellate cell activation through the TGF-β1/Smad and NOX4/ROS pathways. *Chem. Biol. Interact.* **2019**, *299*, 131–139. [CrossRef] [PubMed]
35. Wynn, T. Cellular and molecular mechanisms of fibrosis. *J. Pathol.* **2008**, *214*, 199–210. [CrossRef]
36. Marconi, R.; Serafini, A.; Giovanetti, A.; Bartoleschi, C.; Pardini, M.; Bossi, G.; Strigari, L. Cytokine modulation in breast cancer patients undergoing radiotherapy: A revision of the most recent studies. *Int. J. Mol. Sci.* **2019**, *20*, 382. [CrossRef]
37. Saito-Fujita, T.; Iwakawa, M.; Nakamura, E.; Nakawatari, M.; Fujita, H.; Moritake, T.; Imai, T. Attenuated lung fibrosis in interleukin 6 knock-out mice after C-ion irradiation to lung. *J. Radiat. Res.* **2011**, *52*, 270–277. [CrossRef]
38. Sishc, B.; Nelson, C.; McKenna, M.; Battaglia, C.; Herndon, A.; Idate, R.; Liber, H.; Bailey, S. Telomeres and telomerase in the radiation response: Implications for instability, reprograming, and carcinogenesis. *Front. Oncol.* **2015**, *5*, 257. [CrossRef]
39. Małachowska, B.; Tomasik, B.; Stawiski, K.; Kulkarni, S.; Guha, C.; Chowdhury, D.; Fendler, W. Circulating microRNAs as biomarkers of radiation exposure: A systematic review and meta-analysis. *Int. J. Radiat. Oncol. Biol. Phys.* **2020**, *106*, 390–402. [CrossRef]
40. Weigel, C.; Schmezer, P.; Plass, C.; Popanda, O. Epigenetics in radiation-induced fibrosis. *Oncogene* **2015**, *34*, 2145–2155. [CrossRef]
41. Squillaro, T.; Galano, G.; De Rosa, R.; Peluso, G.; Galderisi, U. Concise review: The effect of low-dose ionizing radiation on stem cell biology: A contribution to radiation risk. *Stem Cells.* **2018**, *36*, 1146–1153. [CrossRef] [PubMed]
42. Dominici, M.; Le Blanc, K.; Mueller, I.; Slaper-Cortenbach, I.; Marini, F.; Krause, D.; Deans, R.; Keating, A.; Prockop, D.; Horwitz, E. Minimal criteria for defining multipotent mesenchymal stromal cells. The International Society for Cellular Therapy position statement. *Cytotherapy* **2006**, *8*, 315–317. [CrossRef] [PubMed]
43. Kiang, J. Adult mesenchymal stem cells and radiation injury. *Health Phys.* **2016**, *111*, 198–203. [CrossRef] [PubMed]
44. Kolf, C.; Cho, E.; Tuan, R. Mesenchymal stromal cells. Biology of adult mesenchymal stem cells: Regulation of niche, self-renewal and differentiation. *Arthritis Res. Ther.* **2007**, *9*, 204. [CrossRef]
45. Hass, R.; Kasper, C.; Böhm, S.; Jacobs, R. Different populations and sources of human mesenchymal stem cells (MSC): A comparison of adult and neonatal tissue-derived MSC. *Cell Commun. Signal.* **2011**, *9*, 12. [CrossRef]
46. Heo, J.; Choi, Y.; Kim, H.; Kim, H. Comparison of molecular profiles of human mesenchymal stem cells derived from bone marrow, umbilical cord blood, placenta and adipose tissue. *Int. J. Mol. Med.* **2016**, *37*, 115–125. [CrossRef]
47. Strioga, M.; Viswanathan, S.; Darinskas, A.; Slaby, O.; Michalek, J. Same or not the same? Comparison of adipose tissue-derived versus bone marrow-derived mesenchymal stem and stromal cells. *Stem Cells Dev.* **2012**, *21*, 2724–2752. [CrossRef]

48. Jin, H.; Bae, Y.; Kim, M.; Kwon, S.; Jeon, H.; Choi, S.; Kim, S.; Yang, Y.; Oh, W.; Chang, J. Comparative analysis of human mesenchymal stem cells from bone marrow, adipose tissue, and umbilical cord blood as sources of cell therapy. *Int. J. Mol. Sci.* **2013**, *14*, 17986–18001. [CrossRef]
49. Dabrowski, F.; Burdzinska, A.; Kulesza, A.; Sladowska, A.; Zolocinska, A.; Gala, K.; Paczek, L.; Wielgos, M. Comparison of the paracrine activity of mesenchymal stem cells derived from human umbilical cord, amniotic membrane and adipose tissue. *J. Obstet. Gynaecol. Res.* **2017**, *43*, 1758–1768. [CrossRef]
50. Li, N.; Hua, J. Interactions between mesenchymal stem cells and the immune system. *Cell. Mol. Life Sci.* **2017**, *74*, 2345–2360. [CrossRef]
51. Jiang, W.; Xu, J. Immune modulation by mesenchymal stem cells. *Cell Prolif.* **2020**, *53*, e12712. [CrossRef] [PubMed]
52. Meisel, R.; Zibert, A.; Laryea, M.; Göbel, U.; Däubener, W.; Dilloo, D. Human bone marrow stromal cells inhibit allogeneic T-cell responses by indoleamine 2,3-dioxygenase-mediated tryptophan degradation. *Blood* **2004**, *103*, 4619–4621. [CrossRef] [PubMed]
53. Wang, Y.; Chen, X.; Cao, W.; Shi, Y. Plasticity of mesenchymal stem cells in immunomodulation: Pathological and therapeutic implications. *Nat. Immunol.* **2014**, *15*, 1009–1016. [CrossRef] [PubMed]
54. Ryan, J. Ionizing radiation: The good, the bad, and the ugly. *J. Investig. Dermatol.* **2012**, *132*, 985–993. [CrossRef]
55. Chan, R.; Larsen, E.; Chan, P. Re-examining the evidence in radiation dermatitis management literature: An overview and a critical appraisal of systematic reviews. *Int. J. Radiat. Oncol. Biol. Phys.* **2012**, *84*, e357–e362. [CrossRef]
56. Chan, R.; Webster, J.; Chung, B.; Marquart, L.; Ahmed, M.; Garantziotis, S. Prevention and treatment of acute radiation-induced skin reactions: A systematic review and meta-analysis of randomized controlled trials. *BMC Cancer* **2014**, *14*, 53. [CrossRef]
57. Wong, R.; Bensadoun, R.; Boers-Doets, C.; Bryce, J.; Chan, A.; Epstein, J.; Eaby-Sandy, B.; Lacouture, M. Clinical practice guidelines for the prevention and treatment of acute and late radiation reactions from the MASCC Skin Toxicity Study Group. *Support. Care Cancer* **2013**, *21*, 2933–2948. [CrossRef]
58. Müller, K.; Meineke, V. Radiation-induced mast cell mediators differentially modulate chemokine release from dermal fibroblasts. *J. Dermatol. Sci.* **2011**, *61*, 199–205. [CrossRef]
59. Bostock, S.; Bryan, J. Radiotherapy-induced skin reactions: Assessment and management. *Br. J. Nurs.* **2016**, *25*, S18, S20–S24. [CrossRef]
60. McQuestion, M. Evidence-based skin care management in radiation therapy: Clinical update. *Semin. Oncol. Nurs.* **2011**, *27*, e1–e17. [CrossRef]
61. Salvo, N.; Barnes, E.; van Draanen, J.; Stacey, E.; Mitera, G.; Breen, D.; Giotis, A.; Czarnota, G.; Pang, J.; De Angelis, C. Prophylaxis and management of acute radiation-induced skin reactions: A systematic review of the literature. *Curr. Oncol.* **2010**, *17*, 94–112. [CrossRef] [PubMed]
62. Kim, J.; Kolozsvary, A.; Jenrow, K.; Brown, S. Mechanisms of radiation-induced skin injury and implications for future clinical trials. *Int. J. Radiat. Biol.* **2013**, *89*, 311–318. [CrossRef] [PubMed]
63. Gorbunov, N.; McDaniel, D.; Zhai, M.; Liao, P.; Garrison, B.; Kiang, J. Autophagy and mitochondrial remodelling in mouse mesenchymal stromal cells challenged with *Staphylococcus epidermidis*. *J. Cell. Mol. Med.* **2015**, *19*, 1133–1150. [CrossRef] [PubMed]
64. Kiang, J.; Zhai, M.; Liao, P.; Elliott, T.; Gorbunov, N. Ghrelin therapy improves survival after whole-body ionizing irradiation or combined with burn or wound: Amelioration of leukocytopenia, thrombocytopenia, splenomegaly, and bone marrow injury. *Oxidative Med. Cell. Longev.* **2014**, *2014*, 215858. [CrossRef] [PubMed]
65. Hormozi Moghaddam, Z.; Mokhtari-Dizaji, M.; Nilforoshzadeh, M.; Bakhshandeh, M.; Ghaffari Khaligh, S. Low-intensity ultrasound combined with allogenic adipose-derived mesenchymal stem cells (AdMSCs) in radiation-induced skin injury treatment. *Sci. Rep.* **2020**, *10*, 20006. [CrossRef] [PubMed]
66. Zheng, K.; Wu, W.; Yang, S.; Huang, L.; Chen, J.; Gong, C.; Fu, Z.; Zhang, L.; Tan, J. Bone marrow mesenchymal stem cell implantation for the treatment of radioactivity-induced acute skin damage in rats. *Mol. Med. Rep.* **2015**, *12*, 7065–7071. [CrossRef]
67. Horton, J.; Hudak, K.; Chung, E.; White, A.; Scroggins, B.; Burkeen, J.; Citrin, D. Mesenchymal stem cells inhibit cutaneous radiation-induced fibrosis by suppressing chronic inflammation. *Stem Cells.* **2013**, *31*, 2231–2241. [CrossRef]
68. Sun, J.; Zhang, Y.; Song, X.; Zhu, J.; Zhu, Q. The healing effects of conditioned medium derived from mesenchymal stem cells on radiation-induced skin wounds in rats. *Cell Transplant.* **2019**, *28*, 105–115. [CrossRef]
69. Portas, M.; Mansilla, E.; Drago, H.; Dubner, D.; Radl, A.; Coppola, A.; Di Giorgio, M. Use of human cadaveric mesenchymal stem cells for cell therapy of a chronic radiation-induced skin lesion: A case report. *Radiat. Prot. Dosim.* **2016**, *171*, 99–106. [CrossRef]
70. Hauer-Jensen, M.; Denham, J.; Andreyev, H. Radiation enteropathy—Pathogenesis, treatment and prevention. *Nat. Rev. Gastroenterol. Hepatol.* **2014**, *11*, 470–479. [CrossRef]
71. Potten, C.; Merritt, A.; Hickman, J.; Hall, P.; Faranda, A. Characterization of radiation-induced apoptosis in the small intestine and its biological implications. *Int. J. Radiat. Biol.* **1994**, *65*, 71–78. [CrossRef] [PubMed]
72. Kiang, J.; Smith, J.; Cannon, G.; Anderson, M.; Ho, C.; Zhai, M.; Cui, W.; Xiao, M. Ghrelin, a novel therapy, corrects cytokine and NF-κB-AKT-MAPK network and mitigates intestinal injury induced by combined radiation and skin-wound trauma. *Cell Biosci.* **2020**, *10*, 63. [CrossRef] [PubMed]
73. Zhang, H.; Yan, H.; Zhou, X.; Wang, H.; Yang, Y.; Zhang, J.; Wang, H. The protective effects of Resveratrol against radiation-induced intestinal injury. *BMC Complement. Altern. Med.* **2017**, *17*, 410. [CrossRef] [PubMed]
74. Leavitt, R.; Limoli, C.; Baulch, J. miRNA-based therapeutic potential of stem cell-derived extracellular vesicles: A safe cell-free treatment to ameliorate radiation-induced brain injury. *Int. J. Radiat. Biol.* **2019**, *95*, 427–435. [CrossRef] [PubMed]

75. Kouvaris, J.; Kouloulias, V.; Vlahos, L. Amifostine: The first selective-target and broad-spectrum radioprotector. *Oncologist* **2007**, *12*, 738–747. [CrossRef] [PubMed]
76. Nicolatou-Galitis, O.; Sarri, T.; Bowen, J.; Di Palma, M.; Kouloulias, V.; Niscola, P.; Riesenbeck, D.; Stokman, M.; Tissing, W.; Yeoh, E.; et al. Systematic review of amifostine for the management of oral mucositis in cancer patients. *Support. Care Cancer* **2013**, *21*, 357–364. [CrossRef]
77. Singh, V.; Seed, T. An update on sargramostim for treatment of acute radiation syndrome. *Drugs Today* **2018**, *54*, 679–693. [CrossRef]
78. Chapel, A.; Bertho, J.; Bensidhoum, M.; Fouillard, L.; Young, R.; Frick, J.; Demarquay, C.; Cuvelier, F.; Mathieu, E.; Trompier, F.; et al. Mesenchymal stem cells home to injured tissues when co-infused with hematopoietic cells to treat a radiation-induced multi-organ failure syndrome. *J. Gene Med.* **2003**, *5*, 1028–1038. [CrossRef]
79. Sémont, A.; François, S.; Mouiseddine, M.; François, A.; Saché, A.; Frick, J.; Thierry, D.; Chapel, A. Mesenchymal stem cells increase self-renewal of small intestinal epithelium and accelerate structural recovery after radiation injury. *Adv. Exp. Med. Biol.* **2006**, *585*, 19–30. [CrossRef]
80. Chang, P.; Qu, Y.; Wang, J.; Dong, L. The potential of mesenchymal stem cells in the management of radiation enteropathy. *Cell Death Dis.* **2015**, *6*, e1840. [CrossRef]
81. Chang, P.; Qu, Y.; Liu, Y.; Cui, S.; Zhu, D.; Wang, H.; Jin, X. Multi-therapeutic effects of human adipose-derived mesenchymal stem cells on radiation-induced intestinal injury. *Cell Death Dis.* **2013**, *4*, e685. [CrossRef] [PubMed]
82. Booth, C.; Potten, C. Gut instincts: Thoughts on intestinal epithelial stem cells. *J. Clin. Investig.* **2000**, *105*, 1493–1499. [CrossRef] [PubMed]
83. Gong, W.; Guo, M.; Han, Z.; Wang, Y.; Yang, P.; Xu, C.; Wang, Q.; Du, L.; Li, Q.; Zhao, H.; et al. Mesenchymal stem cells stimulate intestinal stem cells to repair radiation-induced intestinal injury. *Cell Death Dis.* **2016**, *7*, e2387. [CrossRef] [PubMed]
84. Chang, P.; Zhang, B.; Cui, S.; Qu, C.; Shao, L.; Xu, T.; Qu, Y.; Dong, L.; Wang, J. MSC-derived cytokines repair radiation-induced intra-villi microvascular injury. *Oncotarget* **2017**, *8*, 87821–87836. [CrossRef]
85. Chen, H.; Min, X.; Wang, Q.; Leung, F.; Shi, L.; Zhou, Y.; Yu, T.; Wang, C.; An, G.; Sha, W.; et al. Pre-activation of mesenchymal stem cells with TNF-α, IL-1β and nitric oxide enhances its paracrine effects on radiation-induced intestinal injury. *Sci. Rep.* **2015**, *5*, 8718. [CrossRef]
86. Wang, H.; Sun, R.; Li, Y.; Yang, Y.; Xiao, F.; Zhang, Y.; Wang, S.; Sun, H.; Zhang, Q.; Wu, C.; et al. HGF gene modification in mesenchymal stem cells reduces radiation-induced intestinal injury by modulating immunity. *PLoS ONE* **2015**, *10*, e0124420. [CrossRef]
87. Chang, P.; Zhang, B.; Shao, L.; Song, W.; Shi, W.; Wang, L.; Xu, T.; Li, D.; Gao, X.; Qu, Y.; et al. Mesenchymal stem cells over-expressing cxcl12 enhance the radioresistance of the small intestine. *Cell Death Dis.* **2018**, *9*, 154. [CrossRef]
88. Moussa, L.; Pattappa, G.; Doix, B.; Benselama, S.; Demarquay, C.; Benderitter, M.; Sémont, A.; Tamarat, R.; Guicheux, J.; Weiss, P.; et al. A biomaterial-assisted mesenchymal stromal cell therapy alleviates colonic radiation-induced damage. *Biomaterials* **2017**, *115*, 40–52. [CrossRef]
89. Chapel, A.; Francois, S.; Douay, L.; Benderitter, M.; Voswinkel, J. New insights for pelvic radiation disease treatment: Multipotent stromal cell is a promise mainstay treatment for the restoration of abdominopelvic severe chronic damages induced by radiotherapy. *World J. Stem Cells* **2013**, *5*, 106–111. [CrossRef]
90. Johannesen, T.; Lien, H.; Hole, K.; Lote, K. Radiological and clinical assessment of long-term brain tumour survivors after radiotherapy. *Radiother. Oncol. J. Eur. Soc. Ther. Radiol. Oncol.* **2003**, *69*, 169–176. [CrossRef]
91. Tofilon, P.; Fike, J. The radioresponse of the central nervous system: A dynamic process. *Radiat. Res.* **2000**, *153*, 357–370. [CrossRef]
92. Ali, F.; Hussain, M.; Gutiérrez, C.; Demireva, P.; Ballester, L.; Zhu, J.; Blanco, A.; Esquenazi, Y. Cognitive disability in adult patients with brain tumors. *Cancer Treat. Rev.* **2018**, *65*, 33–40. [CrossRef] [PubMed]
93. Taphoorn, M.; Klein, M. Cognitive deficits in adult patients with brain tumours. *Lancet. Neurol.* **2004**, *3*, 159–168. [CrossRef]
94. Greene-Schloesser, D.; Robbins, M.; Peiffer, A.; Shaw, E.; Wheeler, K.; Chan, M. Radiation-induced brain injury: A review. *Front. Oncol.* **2012**, *2*, 73. [CrossRef] [PubMed]
95. Balentova, S.; Adamkov, M. Molecular, Cellular and Functional Effects of Radiation-induced brain injury: A review. *Int. J. Mol. Sci.* **2015**, *16*, 27796–27815. [CrossRef] [PubMed]
96. DeAngelis, L.; Mandell, L.; Thaler, H.; Kimmel, D.; Galicich, J.; Fuks, Z.; Posner, J. The role of postoperative radiotherapy after resection of single brain metastases. *Neurosurgery* **1989**, *24*, 798–805. [CrossRef]
97. Meyers, C.; Smith, J.; Bezjak, A.; Mehta, M.; Liebmann, J.; Illidge, T.; Kunkler, I.; Caudrelier, J.; Eisenberg, P.; Meerwaldt, J.; et al. Neurocognitive function and progression in patients with brain metastases treated with whole-brain radiation and motexafin gadolinium: Results of a randomized phase III trial. *J. Clin. Oncol.* **2004**, *22*, 157–165. [CrossRef]
98. Lonergan, P.; Martin, D.; Horrobin, D.; Lynch, M. Neuroprotective effect of eicosapentaenoic acid in hippocampus of rats exposed to gamma-irradiation. *J. Biol. Chem.* **2002**, *277*, 20804–20811. [CrossRef]
99. Ramanan, S.; Kooshki, M.; Zhao, W.; Hsu, F.; Riddle, D.; Robbins, M. The PPARalpha agonist fenofibrate preserves hippocampal neurogenesis and inhibits microglial activation after whole-brain irradiation. *Int. J. Radiat. Oncol. Biol. Phys.* **2009**, *75*, 870–877. [CrossRef]
100. Peng, X.; Huang, J.; Wang, S.; Liu, L.; Liu, Z.; Sethi, G.; Ren, B.; Tang, F. Traditional Chinese medicine in neuroprotection after brain insults with special reference to radioprotection. *Evid. Based Complement. Altern. Med.* **2018**, *2018*, 2767208. [CrossRef]

101. Acharya, M.; Rosi, S.; Jopson, T.; Limoli, C. Human neural stem cell transplantation provides long-term restoration of neuronal plasticity in the irradiated hippocampus. *Cell Transplant.* **2015**, *24*, 691–702. [CrossRef] [PubMed]
102. Wang, G.; Ren, X.; Yan, H.; Gui, Y.; Guo, Z.; Song, J.; Zhang, P. Neuroprotective effects of umbilical cord-derived mesenchymal stem cells on radiation-induced brain injury in mice. *Ann. Clin. Lab. Sci.* **2020**, *50*, 57–64. [PubMed]
103. Wang, G.; Liu, Y.; Wu, X.; Lu, Y.; Liu, J.; Qin, Y.; Li, T.; Duan, H. Neuroprotective effects of human umbilical cord-derived mesenchymal stromal cells combined with nimodipine against radiation-induced brain injury through inhibition of apoptosis. *Cytotherapy* **2016**, *18*, 53–64. [CrossRef] [PubMed]
104. Xu, Y.; Hu, W.; Liu, Y.; Xu, P.; Li, Z.; Wu, R.; Shi, X.; Tang, Y. P2Y6 Receptor-mediated microglial phagocytosis in radiation-induced brain injury. *Mol. Neurobiol.* **2016**, *53*, 3552–3564. [CrossRef]
105. Liao, H.; Wang, H.; Rong, X.; Li, E.; Xu, R.; Peng, Y. Mesenchymal stem cells attenuate radiation-induced brain injury by inhibiting Microglia Pyroptosis. *Biomed. Res. Int.* **2017**, *2017*, 1948985. [CrossRef]
106. Soria, B.; Martin-Montalvo, A.; Aguilera, Y.; Mellado-Damas, N.; López-Beas, J.; Herrera-Herrera, I.; López, E.; Barcia, J.; Alvarez-Dolado, M.; Hmadcha, A.; et al. Human mesenchymal stem cells prevent neurological complications of radiotherapy. *Front. Cell. Neurosci.* **2019**, *13*, 204. [CrossRef]
107. Thomas, J.; Parker Kerrigan, B.; Hossain, A.; Gumin, J.; Shinojima, N.; Nwajei, F.; Ezhilarasan, R.; Love, P.; Sulman, E.; Lang, F. Ionizing radiation augments glioma tropism of mesenchymal stem cells. *J. Neurosurg.* **2018**, *128*, 287–295. [CrossRef]
108. Graves, P.; Siddiqui, F.; Anscher, M.; Movsas, B. Radiation pulmonary toxicity: From mechanisms to management. *Semin. Radiat. Oncol.* **2010**, *20*, 201–207. [CrossRef]
109. Xu, T.; Zhang, Y.; Chang, P.; Gong, S.; Shao, L.; Dong, L. Mesenchymal stem cell-based therapy for radiation-induced lung injury. *Stem Cell Res. Ther.* **2018**, *9*, 18. [CrossRef]
110. Wynn, T. Integrating mechanisms of pulmonary fibrosis. *J. Exp. Med.* **2011**, *208*, 1339–1350. [CrossRef]
111. Balli, D.; Ustiyan, V.; Zhang, Y.; Wang, I.; Masino, A.; Ren, X.; Whitsett, J.; Kalinichenko, V.; Kalin, T. Foxm1 transcription factor is required for lung fibrosis and epithelial-to-mesenchymal transition. *EMBO J.* **2013**, *32*, 231–244. [CrossRef] [PubMed]
112. Xia, C.; Chang, P.; Zhang, Y.; Shi, W.; Liu, B.; Ding, L.; Liu, M.; Gao, L.; Dong, L. Therapeutic effects of bone marrow-derived mesenchymal stem cells on radiation-induced lung injury. *Oncol. Rep.* **2016**, *35*, 731–738. [CrossRef] [PubMed]
113. Hao, Y.; Ran, Y.; Lu, B.; Li, J.; Zhang, J.; Feng, C.; Fang, J.; Ma, R.; Qiao, Z.; Dai, X.; et al. Therapeutic effects of human umbilical cord-derived mesenchymal stem cells on canine radiation-induced lung injury. *Int. J. Radiat. Oncol. Biol. Phys.* **2018**, *102*, 407–416. [CrossRef] [PubMed]
114. Dong, L.; Jiang, Y.; Liu, Y.; Cui, S.; Xia, C.; Qu, C.; Jiang, X.; Qu, Y.; Chang, P.; Liu, F. The anti-fibrotic effects of mesenchymal stem cells on irradiated lungs via stimulating endogenous secretion of HGF and PGE2. *Sci. Rep.* **2015**, *5*, 8713. [CrossRef]
115. Klein, D.; Steens, J.; Wiesemann, A.; Schulz, F.; Kaschani, F.; Röck, K.; Yamaguchi, M.; Wirsdörfer, F.; Kaiser, M.; Fischer, J.; et al. Mesenchymal stem cell therapy protects lungs from radiation-induced endothelial cell loss by restoring superoxide dismutase 1 expression. *Antioxid. Redox Signal.* **2017**, *26*, 563–582. [CrossRef]
116. Xu, S.; Liu, C.; Ji, H. Concise review: Therapeutic potential of the mesenchymal stem cell derived secretome and extracellular vesicles for radiation-induced lung injury: Progress and hypotheses. *Stem Cells Transl. Med.* **2019**, *8*, 344–354. [CrossRef]
117. Zhang, C.; Zhu, Y.; Wang, J.; Hou, L.; Li, W.; An, H. CXCR4-overexpressing umbilical cord mesenchymal stem cells enhance protection against radiation-induced lung injury. *Stem Cells Int.* **2019**, *2019*, 2457082. [CrossRef]
118. Liu, D.; Kong, F.; Yuan, Y.; Seth, P.; Xu, W.; Wang, H.; Xiao, F.; Wang, L.; Zhang, Q.; Yang, Y.; et al. Decorin-modified umbilical cord mesenchymal stem cells (MSCs) attenuate radiation-induced lung injuries via regulating inflammation, fibrotic factors, and immune responses. *Int. J. Radiat. Oncol. Biol. Phys.* **2018**, *101*, 945–956. [CrossRef]
119. Chen, H.; Xiang, H.; Xu, W.; Li, M.; Yuan, J.; Liu, J.; Sun, W.; Zhang, R.; Li, J.; Ren, Z.; et al. Manganese superoxide dismutase gene-modified mesenchymal stem cells attenuate acute radiation-induced lung injury. *Hum. Gene Ther.* **2017**, *28*, 523–532. [CrossRef]
120. Kursova, L.; Konoplyannikov, A.; Pasov, V.; Ivanova, I.; Poluektova, M.; Konoplyannikova, O. Possibilities for the use of autologous mesenchymal stem cells in the therapy of radiation-induced lung injuries. *Bull. Exp. Biol. Med.* **2009**, *147*, 542–546. [CrossRef]
121. Kim, J.; Jung, Y. Radiation-induced liver disease: Current understanding and future perspectives. *Exp. Mol. Med.* **2017**, *49*, e359. [CrossRef] [PubMed]
122. Huang, Y.; Chen, S.; Fan, C.; Ting, L.; Kuo, C.; Chiou, J. Clinical parameters for predicting radiation-induced liver disease after intrahepatic reirradiation for hepatocellular carcinoma. *Radiat. Oncol.* **2016**, *11*, 89. [CrossRef] [PubMed]
123. Lawrence, T.; Robertson, J.; Anscher, M.; Jirtle, R.; Ensminger, W.; Fajardo, L. Hepatic toxicity resulting from cancer treatment. *Int. J. Radiat. Oncol. Biol. Phys.* **1995**, *31*, 1237–1248. [CrossRef]
124. Liang, S.; Huang, X.; Zhu, X.; Zhang, W.; Cai, L.; Huang, H.; Li, Y.; Chen, L.; Liu, M. Dosimetric predictor identification for radiation-induced liver disease after hypofractionated conformal radiotherapy for primary liver carcinoma patients with Child-Pugh Grade A cirrhosis. *Radiother. Oncol. J. Eur. Soc. Ther. Radiol. Oncol.* **2011**, *98*, 265–269. [CrossRef] [PubMed]
125. Fajardo, L.; Colby, T. Pathogenesis of veno-occlusive liver disease after radiation. *Arch. Pathol. Lab. Med.* **1980**, *104*, 584–588. [PubMed]
126. Munoz-Schuffenegger, P.; Ng, S.; Dawson, L. Radiation-Induced Liver Toxicity. *Semin. Radiat. Oncol.* **2017**, *27*, 350–357. [CrossRef]

127. Francois, S.; Mouiseddine, M.; Allenet-Lepage, B.; Voswinkel, J.; Douay, L.; Benderitter, M.; Chapel, A. Human mesenchymal stem cells provide protection against radiation-induced liver injury by antioxidative process, vasculature protection, hepatocyte differentiation, and trophic effects. *Biomed. Res. Int.* **2013**, *2013*, 151679. [CrossRef]
128. Radwan, R.; Mohamed, H. Nigella sativa oil modulates the therapeutic efficacy of mesenchymal stem cells against liver injury in irradiated rats. *J. Photochem. Photobiol. B Biol.* **2018**, *178*, 447–456. [CrossRef]
129. Mouiseddine, M.; François, S.; Souidi, M.; Chapel, A. Intravenous human mesenchymal stem cells transplantation in NOD/SCID mice preserve liver integrity of irradiation damage. *Methods Mol. Biol.* **2012**, *826*, 179–188. [CrossRef]
130. Chen, Y.; Zeng, Z.; Sun, J.; Zeng, H.; Huang, Y.; Zhang, Z. Mesenchymal stem cell-conditioned medium prevents radiation-induced liver injury by inhibiting inflammation and protecting sinusoidal endothelial cells. *J. Radiat. Res.* **2015**, *56*, 700–708. [CrossRef]
131. Zhang, J.; Zhou, S.; Zhou, Y.; Feng, F.; Wang, Q.; Zhu, X.; Ai, H.; Huang, X.; Zhang, X. Hepatocyte growth factor gene-modified adipose-derived mesenchymal stem cells ameliorate radiation induced liver damage in a rat model. *PLoS ONE* **2014**, *9*, e114670. [CrossRef] [PubMed]
132. Van Nimwegen, F.; Schaapveld, M.; Janus, C.; Krol, A.; Petersen, E.; Raemaekers, J.; Kok, W.; Aleman, B.; van Leeuwen, F. Cardiovascular disease after Hodgkin lymphoma treatment: 40-year disease risk. *JAMA Intern. Med.* **2015**, *175*, 1007–1017. [CrossRef] [PubMed]
133. Cheng, Y.; Nie, X.; Ji, C.; Lin, X.; Liu, L.; Chen, X.; Yao, H.; Wu, S. Long-Term Cardiovascular Risk After Radiotherapy in Women With Breast Cancer. *J. Am. Heart Assoc.* **2017**, *6*, e005633. [CrossRef] [PubMed]
134. Wang, Y.; Boerma, M.; Zhou, D. Ionizing radiation-induced endothelial cell senescence and cardiovascular diseases. *Radiat. Res.* **2016**, *186*, 153–161. [CrossRef]
135. Monceau, V.; Llach, A.; Azria, D.; Bridier, A.; Petit, B.; Mazevet, M.; Strup-Perrot, C.; To, T.; Calmels, L.; Germaini, M.; et al. Epac contributes to cardiac hypertrophy and amyloidosis induced by radiotherapy but not fibrosis. *Radiother. Oncol. J. Eur. Soc. Ther. Radiol. Oncol.* **2014**, *111*, 63–71. [CrossRef]
136. Wang, B.; Wang, H.; Zhang, M.; Ji, R.; Wei, J.; Xin, Y.; Jiang, X. Radiation-induced myocardial fibrosis: Mechanisms underlying its pathogenesis and therapeutic strategies. *J. Cell. Mol. Med.* **2020**, *24*, 7717–7729. [CrossRef]
137. Shen, Y.; Jiang, X.; Meng, L.; Xia, C.; Zhang, L.; Xin, Y. Transplantation of bone marrow mesenchymal stem cells prevents radiation-induced artery injury by suppressing oxidative stress and inflammation. *Oxidative Med. Cell. Longev.* **2018**, *2018*, 5942916. [CrossRef]
138. Tao, X.; Sun, M.; Chen, M.; Ying, R.; Su, W.; Zhang, J.; Xie, X.; Wei, W.; Meng, X. HMGB1-modified mesenchymal stem cells attenuate radiation-induced vascular injury possibly via their high motility and facilitation of endothelial differentiation. *Stem Cell Res. Ther.* **2019**, *10*, 92. [CrossRef]
139. Gao, S.; Zhao, Z.; Wu, R.; Zeng, Y.; Zhang, Z.; Miao, J.; Yuan, Z. Bone marrow mesenchymal stem cell transplantation improves radiation-induced heart injury through DNA damage repair in rat model. *Radiat. Environ. Biophys.* **2017**, *56*, 63–77. [CrossRef]
140. Chen, Z.; Hu, Y.; Hu, X.; Cheng, L. The conditioned medium of human mesenchymal stromal cells reduces irradiation-induced damage in cardiac fibroblast cells. *J. Radiat. Res.* **2018**, *59*, 555–564. [CrossRef]
141. Clara, J.; Monge, C.; Yang, Y.; Takebe, N. Targeting signalling pathways and the immune microenvironment of cancer stem cells—A clinical update. *Nat. Rev. Clin. Oncol.* **2020**, *17*, 204–232. [CrossRef] [PubMed]
142. Dvorak, H. Tumors: Wounds that do not heal-redux. *Cancer Immunol. Res.* **2015**, *3*, 1–11. [CrossRef] [PubMed]
143. Houghton, J.; Stoicov, C.; Nomura, S.; Rogers, A.; Carlson, J.; Li, H.; Cai, X.; Fox, J.; Goldenring, J.; Wang, T. Gastric cancer originating from bone marrow-derived cells. *Science.* **2004**, *306*, 1568–1571. [CrossRef] [PubMed]
144. Papaccio, F.; Paino, F.; Regad, T.; Papaccio, G.; Desiderio, V.; Tirino, V. Concise review: Cancer cells, cancer stem cells, and mesenchymal stem cells: Influence in cancer development. *Stem Cells Transl. Med.* **2017**, *6*, 2115–2125. [CrossRef] [PubMed]
145. McLean, K.; Gong, Y.; Choi, Y.; Deng, N.; Yang, K.; Bai, S.; Cabrera, L.; Keller, E.; McCauley, L.; Cho, K.; et al. Human ovarian carcinoma-associated mesenchymal stem cells regulate cancer stem cells and tumorigenesis via altered BMP production. *J. Clin. Investig.* **2011**, *121*, 3206–3219. [CrossRef] [PubMed]
146. Raghavan, S.; Snyder, C.; Wang, A.; McLean, K.; Zamarin, D.; Buckanovich, R.; Mehta, G. Carcinoma-associated mesenchymal stem cells promote chemoresistance in ovarian cancer stem cells via PDGF signaling. *Cancers* **2020**, *12*, 2063. [CrossRef] [PubMed]
147. Galland, S.; Stamenkovic, I. Mesenchymal stromal cells in cancer: A review of their immunomodulatory functions and dual effects on tumor progression. *J. Pathol.* **2020**, *250*, 555–572. [CrossRef]
148. Patel, S.; Meyer, J.; Greco, S.; Corcoran, K.; Bryan, M.; Rameshwar, P. Mesenchymal stem cells protect breast cancer cells through regulatory T cells: Role of mesenchymal stem cell-derived TGF-beta. *J. Immunol.* **2010**, *184*, 5885–5894. [CrossRef]
149. Ljujic, B.; Milovanovic, M.; Volarevic, V.; Murray, B.; Bugarski, D.; Przyborski, S.; Arsenijevic, N.; Lukic, M.; Stojkovic, M. Human mesenchymal stem cells creating an immunosuppressive environment and promote breast cancer in mice. *Sci. Rep.* **2013**, *3*, 2298. [CrossRef]
150. Gonzalez, M.; Martin, E.; Anwar, T.; Arellano-Garcia, C.; Medhora, N.; Lama, A.; Chen, Y.; Tanager, K.; Yoon, E.; Kidwell, K.; et al. Mesenchymal stem cell-induced DDR2 mediates stromal-breast cancer interactions and metastasis growth. *Cell Rep.* **2017**, *18*, 1215–1228. [CrossRef]
151. Berger, L.; Shamai, Y.; Skorecki, K.; Tzukerman, M. Tumor specific recruitment and reprogramming of mesenchymal stem cells in tumorigenesis. *Stem Cells.* **2016**, *34*, 1011–1026. [CrossRef] [PubMed]

152. Chen, Y.; Gonzalez, M.; Burman, B.; Zhao, X.; Anwar, T.; Tran, M.; Medhora, N.; Hiziroglu, A.; Lee, W.; Cheng, Y.; et al. Mesenchymal stem/stromal cell engulfment reveals metastatic advantage in breast cancer. *Cell Rep.* **2019**, *27*, 3916–3926.e5. [CrossRef] [PubMed]
153. Lazennec, G.; Jorgensen, C. Concise review: Adult multipotent stromal cells and cancer: Risk or benefit? *Stem Cells.* **2008**, *26*, 1387–1394. [CrossRef] [PubMed]
154. Ahn, J.; Coh, Y.; Lee, H.; Shin, I.; Kang, S.; Youn, H. Human adipose tissue-derived mesenchymal stem cells inhibit melanoma growth in vitro and in vivo. *Anticancer Res.* **2015**, *35*, 159–168. [PubMed]
155. Zhang, J.; Hou, L.; Zhao, D.; Pan, M.; Wang, Z.; Hu, H.; He, J. Inhibitory effect and mechanism of mesenchymal stem cells on melanoma cells. *Clin. Transl. Oncol.* **2017**, *19*, 1358–1374. [CrossRef]
156. François, S.; Usunier, B.; Forgue-Lafitte, M.; L'Homme, B.; Benderitter, M.; Douay, L.; Gorin, N.; Larsen, A.; Chapel, A. Mesenchymal stem cell administration attenuates colon cancer progression by modulating the immune component within the colorectal tumor microenvironment. *Stem Cells Transl. Med.* **2019**, *8*, 285–300. [CrossRef] [PubMed]
157. Lv, F.; Tuan, R.; Cheung, K.; Leung, V. Concise review: The surface markers and identity of human mesenchymal stem cells. *Stem Cells.* **2014**, *32*, 1408–1419. [CrossRef]
158. Phinney, D. Functional heterogeneity of mesenchymal stem cells: Implications for cell therapy. *J. Cell. Biochem.* **2012**, *113*, 2806–2812. [CrossRef]
159. Wang, J.; Liao, L.; Tan, J. Mesenchymal-stem-cell-based experimental and clinical trials: Current status and open questions. *Expert Opin. Biol. Ther.* **2011**, *11*, 893–909. [CrossRef]
160. Walczak, P.; Zhang, J.; Gilad, A.; Kedziorek, D.; Ruiz-Cabello, J.; Young, R.; Pittenger, M.; van Zijl, P.; Huang, J.; Bulte, J. Dual-modality monitoring of targeted intraarterial delivery of mesenchymal stem cells after transient ischemia. *Stroke* **2008**, *39*, 1569–1574. [CrossRef]
161. Nielsen, K.; Offersen, B.; Nielsen, H.; Vaage-Nilsen, M.; Yusuf, S. Short and long term radiation induced cardiovascular disease in patients with cancer. *Clin. Cardiol.* **2017**, *40*, 255–261. [CrossRef] [PubMed]
162. Hu, C.; Li, L. Preconditioning influences mesenchymal stem cell properties in vitro and in vivo. *J. Cell. Mol. Med.* **2018**, *22*, 1428–1442. [CrossRef] [PubMed]

Article

Radiation-Activated PI3K/AKT Pathway Promotes the Induction of Cancer Stem-Like Cells via the Upregulation of SOX2 in Colorectal Cancer

Ji-Hye Park [1,†], Young-Heon Kim [1,†], Sehwan Shim [1], Areumnuri Kim [1], Hyosun Jang [1], Su-Jae Lee [2], Sunhoo Park [1,3], Songwon Seo [4], Won Il Jang [1], Seung Bum Lee [1,*,‡] and Min-Jung Kim [1,*,‡]

1 Laboratory of Radiation Exposure & Therapeutics, National Radiation Emergency Medical Center, Korea Institute of Radiological and Medical Sciences, Seoul 01812, Korea; parkjh0420@kirams.re.kr (J.-H.P.); bumbroberto1@nate.com (Y.-H.K.); ssh3002@kirams.re.kr (S.S.); nurikim@kirams.re.kr (A.K.); hsjang@kirams.re.kr (H.J.); sunhoo@kirams.re.kr (S.P.); zzangIl@kirams.re.kr (W.I.J.)
2 Department of Life Science, Research Institute for Natural Sciences, Hanyang University, Seoul 04763, Korea; sj0420@hanyang.ac.kr
3 Department of Pathology, Korea Cancer Center Hospital, Korea Institute of Radiological and Medical Sciences, Seoul 01812, Korea
4 Laboratory of Radiation Epidermiology, National Radiation Emergency Medical Center, Korea Institute of Radiological and Medical Sciences, Seoul 01812, Korea; seo@kirams.re.kr
* Correspondence: sblee@kirams.re.kr (S.B.L.); kimmj74@kirams.re.kr (M.-J.K.); Tel.: +82-2-3399-5874 (S.B.L.); +82-2-3399-5875 (M.-J.K.)
† Authors contributed equally to this study as co-first authors.
‡ Corresponding authors contributed equally to this work.

Abstract: The current treatment strategy for patients with aggressive colorectal cancer has been hampered by resistance to radiotherapy and chemotherapy due to the existence of cancer stem-like cells (CSCs). Recent studies have shown that SOX2 expression plays an important role in the maintenance of CSC properties in colorectal cancer. In this study, we investigated the induction and regulatory role of SOX2 following the irradiation of radioresistant and radiosensitive colorectal cancer cells. We used FACS and western blotting to analyze SOX2 expression in cells. Among the markers of colorectal CSCs, the expression of CD44 increased upon irradiation in radioresistant cells. Further analysis revealed the retention of CSC properties with an upregulation of SOX2 as shown by enhanced resistance to radiation and metastatic potential in vitro. Interestingly, both the knockdown and overexpression of SOX2 led to increase in CD44+ population and induction of CSC properties in colorectal cancer following irradiation. Furthermore, selective genetic and pharmacological inhibition of the PI3K/AKT pathway, but not the MAPK pathway, attenuated SOX2-dependent CD44 expression and metastatic potential upon irradiation in vitro. Our findings suggested that SOX2 regulated by radiation-induced activation of PI3K/AKT pathway contributes to the induction of colorectal CSCs, thereby highlighting its potential as a therapeutic target.

Keywords: colorectal cancer; cancer-stem like cells; radioresistance; SOX2; PI3K/AKT

Citation: Park, J.-H.; Kim, Y.-H.; Shim, S.; Kim, A.; Jang, H.; Lee, S.-J.; Park, S.; Seo, S.; Jang, W.I.; Lee, S.B.; et al. Radiation-Activated PI3K/AKT Pathway Promotes the Induction of Cancer Stem-Like Cells via the Upregulation of SOX2 in Colorectal Cancer. *Cells* **2021**, *10*, 135. https://doi.org/10.3390/cells10010135

Received: 30 November 2020
Accepted: 8 January 2021
Published: 12 January 2021

1. Introduction

Colorectal cancer is one of the most common malignancies and the fourth leading cause of cancer death in the world. The current treatment strategy for patients with aggressive colorectal cancer has been hampered by their resistance to radiotherapy and chemotherapy [1,2]. Growing evidence indicates that the existence of a small population of cancer cells known as cancer stem-like cells (CSCs) is responsible for tumor recurrence and is the main cause of treatment resistance in many cancers, including glioma, breast, oral, and colorectal cancer. CSCs further exhibit diverse cancer-initiating properties such as self-renewal and metastatic potential [3–7]. For the identification of CSCs, several putative markers such as transmembrane glycoprotein (CD133) and the cell-surface glycoprotein

(CD44) were reported to be expressed in colorectal cancer and correlated with high-risk cases and a poorer survival rate of colorectal cancer patients [8,9]. However, the molecular subclassification of CSCs based on their cancer-promotion property in colorectal cancer needs to be understood.

The stemness program is involved in maintaining the properties of CSCs, such as self-renewal and cancer-initiation, which are the hallmarks of cancer cells. SOX2 is a member of the SRY-related HMG-box (SOX) gene family and play an essential role in the maintenance of a pluripotent state in stem cells and cell-fate determination during developmental processes [10,11]. SOX2 is one of the key molecules driving CSC properties. SOX2-expressing cancer cells, such as those in skin and bladder carcinomas, express high levels of CSC markers, depending on tissue origin, and reveal enhanced tumorigenicity [12,13]. Recent studies have shown that SOX2 is aberrantly expressed and involved in the maintenance of CSCs. The properties of CSCs, including spheroid-like growth and metastatic potential, were observed in SOX2-positive colorectal cancer with an increased expression of CSC markers such as CD44 [14,15]. Moreover, Ghisolfi et al. recently showed that environmental stresses such as radiation induced the expression of both mRNA and protein of SOX2 and regulated CSCs in hepatocellular carcinoma cells [16]. However, the induction and the signaling pathways of SOX2 upon environmental stress in colorectal CSCs remain unclear.

In this study, we investigated the induction and regulatory role of SOX2 in colorectal CSCs following radiation exposure in both radioresistant and radiosensitive colorectal cancer cell lines.

2. Materials and Methods

2.1. Antibodies and Reagents

Anti-CD44, anti-phospho-AKT, anti-AKT, anti-phospho-ERK, anti-ERK1/2, anti-phosph p38, anti-p38, anti-phospho-SAPK/JNK, and anti-JNK1/2 antibodies were purchased from Cell Signaling Technology (Cambridge, MA, USA). Anti-phospho-JNK, anti-Oct3/4 and anti-β-actin were obtained from Santa Cruz Biotechnology (Santa Cruz, CA, USA). Anti-SOX2, anti-Notch2, and anti-Snail were purchased from Abcam (Cambridge, MA, USA). Anti-β-catenin was procured from BD Bioscience (Franklin Lakes, NJ, USA). For transfection, non-targeting siRNA and commercial siRNA for SOX2 or Snail or AKT were purchased from Genolution (Genolution Pharmaceuticals, Seoul, Korea). The cells were transfected with each siRNA (50 nM) for 48 h using Lipofectamine 2000 (Invitrogen, Carlsbad, CA, USA), as described in the manufacturer's procedure.

2.2. Cell Culture and Irradiation

HT29, DLD1, HCT116, SW480, RKO, and LoVo colorectal cancer cells were purchased from the Korean Cell Line Bank (KCLB, Seoul, Korea). These cell lines from passages 4 to 10 were used for the experiments and were maintained in RPMI 1640 medium containing 1% antibiotic-antimycotic (GIBCO) supplemented with 10% FBS. The cells were cultured in a humidified 5% CO_2 atmosphere at 37 °C. Cells were exposed to radiation (0~10 Gy) using a Gammacell 3000 Elan irradiator (137Cs γ-ray source; MDS Nordion, Canada).

2.3. Flow Cytometry and Aldehyde Dehydrogenase (ALDH) Assay

Irradiated cells were gently dissociated and incubated with anti-CD44-Fluorescein (FITC) or anti-CD133-phycoerythrin (PE) antibody (eBioscience Inc., San Diego, CA, USA) for 30 min on ice. The samples were washed twice with PBS containing 0.1% BSA and EDTA, and the cells were resuspended in PBS containing 0.1% BSA and EDTA and analyzed using FACSCalibur and CellQuest programs (BD Biosciences, San Jose, CA, USA). For ALDH assay, an ALDELUOR kit (Stemcell Technologies Inc., Vancouver, BC, Canada) was used to detect the cell population with high ALDH enzymatic activity, according to the manufacturer's procedure, after which the cells were analyzed by flow cytometry. Apoptotic cells were analyzed as previously described [17].

2.4. Fluorescence-Activated Cell Sorting (FACS Sorting)

To isolate CSCs from colorectal cancer cell lines, the cells were gently dissociated and incubated with anti-CD44-FITC antibody for 30 min on ice. CD44+ positive cells were sorted by a using a FACS Vantage SE flow cytometer (BD Biosciences) equipped with FlowJo software (Tree Star, Ashland, OR, USA).

2.5. Immunocytochemistry (ICC)

The cells were fixed with 4% paraformaldehyde and permeabilized with 0.1% Triton X-100 in PBS. Following fixation, cells were incubated at 4 °C overnight with anti–CD44 or anti–SOX2 primary antibodies in PBS with 1% BSA and 0.1% Triton X-100. Stained proteins were visualized using secondary antibodies against anti-mouse immunoglobulin/FITC or anti-rabbit immunoglobulin/PE (1:400, BD Bioscience). Nuclei were counterstained with DAPI (Sigma, St. Louis, MO, USA). Following staining, cells were observed with an Olympus IX71 fluorescence microscope (Olympus, Tokyo, Japan).

2.6. Western Blot Analysis

The cells were lysed using lysis buffer (40 mM Tris-HCl [pH 8.0], 120 mM NaCl, 0.1% Nonidet-P40) supplemented with protease and phosphatase inhibitors. The protein concentration was measured using the Bradford assay (Bio-rad, Hercules, CA, USA). Equal amounts of total protein in cell lysates were separated by 10% SDS-PAGE and transferred to a nitrocellulose membrane (Amersham, UK). The membranes were blocked with 5% skim milk for 1 h at room temperature, incubated with primary antibodies (1:1000) overnight at 4 °C. Blots were developed with the appropriate horseradish peroxidase (HRP)-conjugated secondary antibody (anti-mouse IgG-HRP or anti-rabbit IgG-HRP; Cell Signaling Technology, Danvers, MA, USA), and proteins were visualized using enhanced chemiluminescence (ECL) procedures.

2.7. Apoptosis Assay

Apoptosis analysis was performed using a FITC Annexin V Apoptosis Detection Kit I (BD Biosciences, Waltham, MA, USA), according to the manufacturer's instructions. Briefly, radiation-induced apoptotic cells were collected at the indicated time points and resuspended in 1× diluted binding buffer in Kit. For staining, Annexin V-FITC and PI were added to each sample, and the mixture was incubated for 5 min at room temperature in the dark. The cells were analyzed immediately using a BD FACS CANTO II flow cytometer (BD Biosciences).

2.8. Colony Formation Assay

Colony formation assay was performed as previously described [17]. To test the effect of IR on cell viability, appropriately seeded cells were irradiated with different doses of radiation (0, 0.5, 1, 2, 3, or 4 Gy) and incubated continuously for 2 weeks. The colonies were stained with 1% crystal violet. Colonies containing >50 cells were scored as surviving cells.

2.9. Invasion and Migration Assays

Invasion assay was performed as previously reported [18]. Briefly, the cells treated with either inhibitors or siRNAs for transfection were seeded in the upper well of a Transwell chamber (8-μm pore size) that was pre–coated with 10 mg/mL growth factor-reduced Matrigel (BD Bioscience). After 72 h, non–invading cells on the upper surface of the filter were removed with a cotton swab, and the migrated cells on the lower surface of the filter were fixed and stained with a Kwik-Diff kit (Thermo Fisher Scientific, Waltham, MA, USA). Invasiveness was determined by counting cells in fields per well, and the extent of invasion was expressed as the average number of cells per microscopic field. The cells were imaged by phase contrast microscopy. For the migration assay, we used Transwell chambers with inserts that contained the same type of membrane but without the Matrigel coating.

2.10. Tumoursphere-Formation Assay

After producing a single-cell suspension, the cells were cultured in ultra-low attachment 6-well or 96-well plates in medium consisting of DMEM/F12 supplemented with 2% B27 supplements (Invitrogen), 10 ng mL^{-1} bFGF (Peprotech, Rocky Hill, NJ, USA) and 10 ng/mL EGF. The cells were cultured for 7 days, and the morphology and size of sphere were determined using an Olympus IX71 fluorescence microscope (Olympus).

2.11. Statistical Analysis

Statistical significance of the differences between mean values was calculated with pairwise comparisons using Least Significance Difference (LSD) test after a one-way Analysis of Variance (ANOVA). Due to the exploratory nature of this study, multiplicity adjustment was not made. Statistical analyses were conducted using SPSS (version 12.0; SPSS Inc., Chicago, IL, USA) or Excel (Microsoft, Redmond, WA, USA) software packages.

3. Results

3.1. Radiation Increased the Population of Radioresistant Rather Than Radiosensitive CD44+ Colorectal Cancer Cells

Experimental and clinical data show that CSCs play an important role in tumor recurrence and resistance to therapy [7–9]. To investigate the relationship between radiation resistance and population of CSCs, we first confirmed the radiation resistance in various types of colorectal cancer cells, including in previously reported radioresistant and radiosensitive cells [17]. Colorectal cancer cells such as HCT116, DLD1, and HT29 were relatively resistant to radiation by annexin V/PI staining and colony formation assay. (Figure 1A,B); however, the expression of colorectal CSC markers such as CD44, CD133, and ALDH was similar to that of radiosensitive colorectal cancer cells under untreated conditions (Figure 1C). Next, we examined the effect of radiation on the expression of colorectal CSC markers. Flow cytometry and immunocytochemistry analysis (Figure 1C,D) showed that radiation increased the expression of CD133 and ALDH in all colorectal cancer cells except LoVo cells with unchanged expression of CD133, whereas CD44 expression was selectively increased in radioresistant colorectal cancer cells such as HCT116, DLD1, and HT29 (Figure 1C). These results suggested that resistance to radiation in radioresistant colorectal cancer cells may be acquired by radiation-upregulated expression of CD44, which is one of the markers of colorectal CSCs.

3.2. Radiation-Enriched CD44+ Cells Exhibited the Properties of CSCs Including an Increase in SOX2 Expression

To delineate the role of radiation-induced CD44 expression in radioresistant colorectal cancer cells, we isolated both CD44 positive (CD44+) and negative (CD44−) cells in HCT116 and DLD1 cells following irradiation using anti-CD44-FITC antibodies by FACS, and the expression of CD44 in both CD44+ and CD44− cells is shown in Figure 2A. Since the CD44 marker correlated with the features of CSCs in colorectal cancers [19,20], we evaluated the properties of colorectal CSCs including metastatic potential and self-renewal. We observed an increase in colony formation, migration and invasion in the sorted CD44+ cells after irradiation and not in CD44− cells in both cell lines (Figure 2B–D). Interestingly, immunoblotting of stemness-related proteins revealed significant elevation in SOX2 levels among stemness-related proteins [21,22] on sorted CD44+ cells (Figure 2A). Given the evidence that SOX2 was aberrantly expressed and involved in the maintenance of CSCs in colorectal cancer [14,15], these results indicated the possibility of a functional relationship between SOX2 expression and CD44-mediated CSC property in radioresistant cells upon radiation exposure.

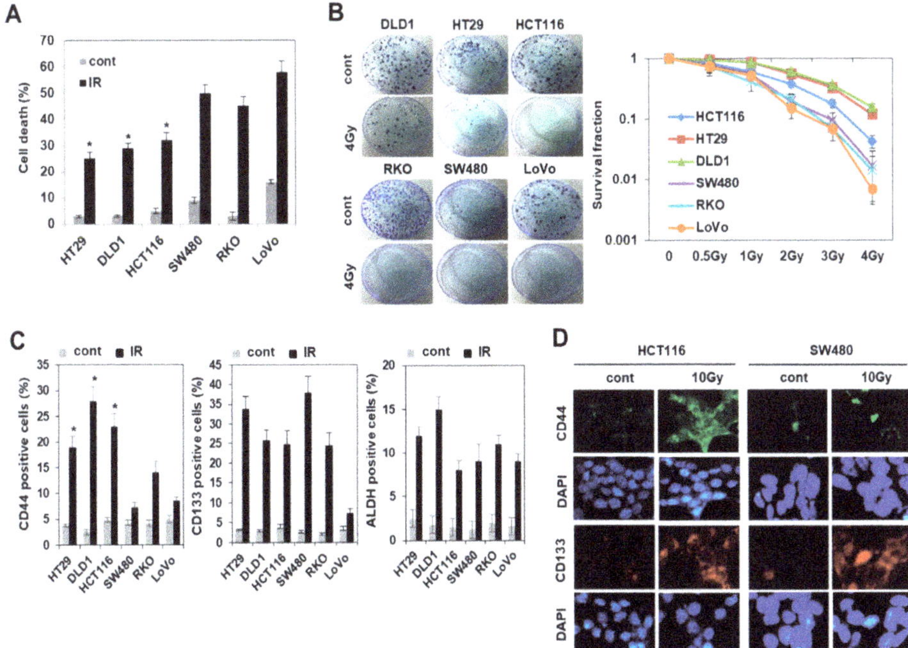

Figure 1. Enriched population of CD44 positive (CD44+) cells was found in radioresistant cells and not in radiosensitive cells of colorectal cancer in response to radiation. (**A**) The apoptotic cells on day 2 after irradiation with 10 Gy were measured by annexin V staining and flow cytometry analysis in various types of colorectal cancer cell lines including both radioresistant cells (HT29, DLD1, and HCT116) and radiosensitive cells (SW480, RKO, and LoVo). Data are shown as mean ± SD ($n = 3$) with * $p < 0.05$ for the pairwise comparisons between radioresistant cells and radiosensitive cells. (**B**) Colony formation assay was performed with indicated cells treated with 4 Gy (left panel). Graph showing quantification of relative colony numbers at different doses of IR (right panel). (**C**) Cell populations for the CD44+, CD133+, or ALDH+, which are known markers of cancer stem-like cells (CSC) in these indicated cells after radiation exposure were measured by flow cytometric analysis. The percentage of each CSC marker-expressing cell is shown as a bar graph. Data are shown as mean ± SD ($n = 3$) with * $p < 0.05$ for the pairwise comparisons between radioresistant cells and radiosensitive cells. (**D**) Cells were stained with an anti-CD44 antibody (green) and anti-CD133 antibody (red). Nuclei were counterstained with DAPI ((blue). CSCs: cancer stem-like cells.

3.3. Modulation of SOX2 Expression in Colorectal Cancer Cells Is Associated with Induction of Colorectal CSCs Following Irradiation

We further determined whether the expression of either CD44 or SOX2 in response to radiation is dependent on radioresistant colorectal cancer cells. These proteins were upregulated by irradiation in radioresistant, but not radiosensitive colorectal cancer cells (Figure 3A). To further clarify the role of radiation induced SOX2 in regulating colorectal CSCs, we examined the effect of SOX2 siRNA on the properties of CSCs in both HCT116 and DLD1 cells. Immunoblotting analysis in Figure 3B showed the efficient knockdown of SOX2 expression in both cells with SOX2 siRNA treatment. In addition, the knockdown of SOX2 attenuated the radiation-induced properties of CSCs, including the enhanced ability to migrate, invade, and form tumourspheres, and reduced CD44+ population growth (Figure 3B–D). Next, we examined the effects of SOX2 overexpression. Upon irradiation, the overexpression of SOX2 facilitated the acquisition of the properties of colorectal CSCs in radiosensitive colorectal cancer cells (SW480 and LoVo) due to an increase in CD44+ population, cell survival, migration, invasion, and tumoursphere-formation (Figure 4). Taken together, these results suggested that SOX2 regulated population growth and properties of CSCs in colorectal cancer following irradiation, and SOX2 may be a potential target for studies involving resistance to radiation.

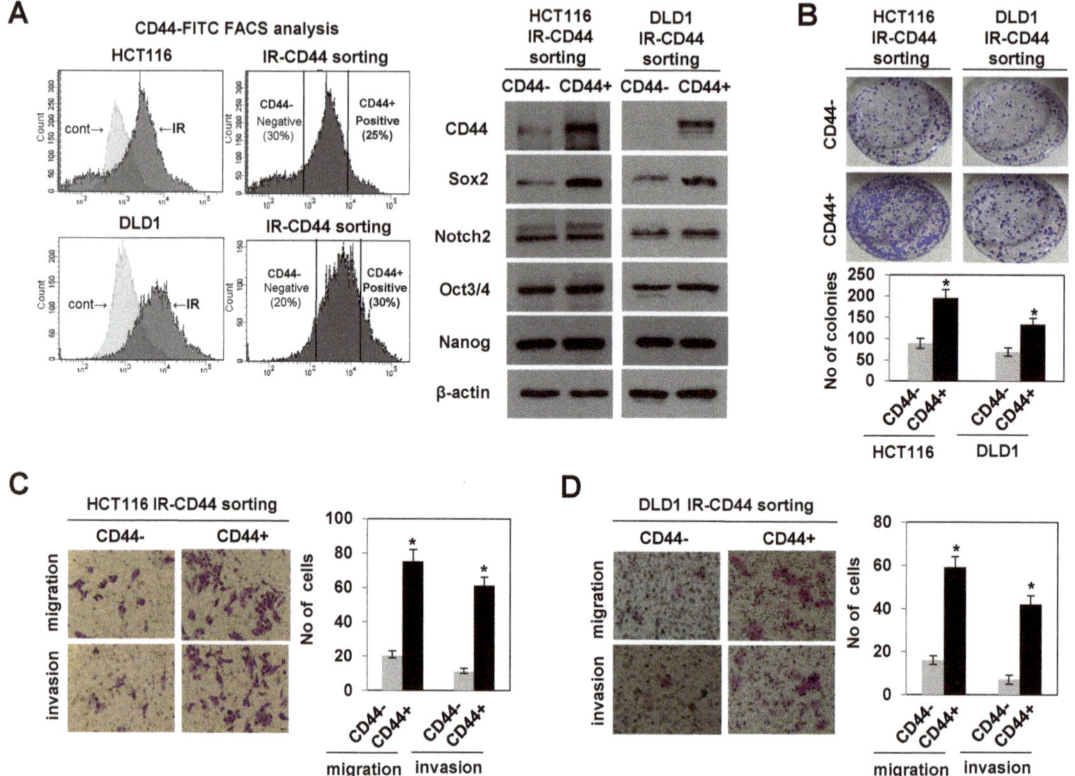

Figure 2. CD44+ cells induced by radiation exhibited the properties of cancer stem-like cells (CSCs) with an increase in SOX2 levels. (**A**) CD44+ CD44− cells on day 2 after irradiation with 10 Gy in radioresistant colorectal cancer cells (HCT116 and DLD1) were sorted (left panel). Immunoblotting for the expression of CSC-related proteins in CD44+ (positive) and CD44− (negative) in radioresistant cells (right panel). (**B**) Colony formation assay was performed with CD44+ (or CD44−) cells, and the bar graphs show the quantification of relative colony numbers in indicated cells. Data are shown as mean ± SD (n = 3) * $p < 0.05$ compared to control. (**C**,**D**) The migration and invasion analysis (left panel) and quantification of cells involved in migration and invasion (right panel) in CD44+ and CD44− cells sorted from HCT116 and DLD1 cells, respectively. All experiments were performed in triplicates. Data are shown as mean ± SD. * $p < 0.05$ compared to CD44− cell. CD44−: negative, CD44+: positive, CSCs: cancer stem-like cells.

3.4. Radiation-Induced Activation of the PI3K/AKT Pathway, but Not the MAPK Pathway Modulated SOX2-Dependent Induction of Colorectal CSCs

Next, the potential molecular mechanism involved in the SOX2-dependent induction of colorectal CSCs following irradiation was elucidated. The phenomenon of epithelial-mesenchymal transition (EMT) has emerged as a feature of CSCs in recent times [6,23]. In addition, expression levels of the master regulator of EMT such as Snail and Zeb1/2 were modulated by SOX2 protein level [24,25]. Therefore, we investigated whether EMT is associated with SOX2-dependent induction of colorectal CSCs. Immunoblotting analysis showed that among EMT-associated proteins, Snail expression was decreased in SOX2 siRNA-transfected HCT116 and DLD1 cells (Figure 5A). In addition, we found that knock-down of Snail dramatically suppressed the ability of migration and invasion, a hallmark of

EMT (Figure 5B). However, Snail did not affect the induction of properties of colorectal CSCs including CD44+ population growth, resistance to radiation, and ability of tumoursphere formation (Figure 5C,D), suggesting that the Snail-mediated EMT process might not be involved in SOX2-dependent induction of colorectal CSCs upon irradiation.

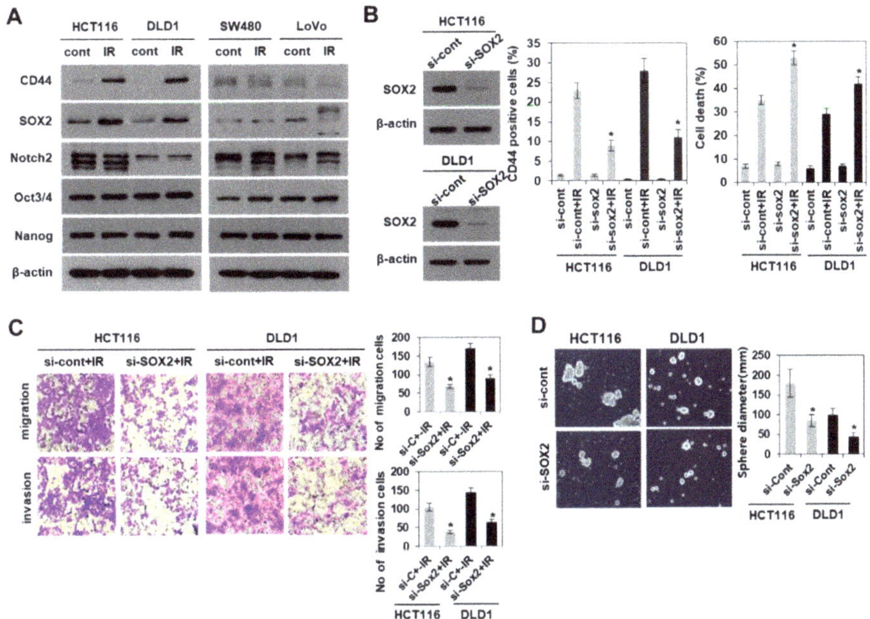

Figure 3. Knockdown of SOX2 in radioresistant colorectal cancer cells attenuated the induction of colorectal CSCs after irradiation. (**A**) Immunoblotting for the expression of CSC-related proteins on day 2 after radiation (10 Gy) in colorectal cancer cells as indicated. (**B**) siRNA-mediated SOX2 knockdown in cells was identified by western blotting (left) and CD44+ cell population (middle), or apoptotic cells were analyzed by flow cytometry (right). All experiments were performed with the SOX2 siRNA-transfected HCT116 and DLD1 cells on day 2 after radiation (10 Gy). Data are shown as mean ± SD ($n = 3$). * $p < 0.05$ compared si-Cont + IR to si-SOX2 + IR. (**C**) The images of migration and invasion on day 2 after radiation (10 Gy) of the SOX2 siRNA-transfected HCT116 and DLD1 cells were quantified. Bars indicate measurements of migration and invasion. * $p < 0.05$ compared si-Cont + IR to si-SOX2 + IR. (**D**) Tumoursphere-formation assay was performed to evaluate self-renewal ability of CSCs in SOX2 siRNA-transfected cells. Indicated cells were seeded in a non-adherent culture condition. After culturing for 7 days, the number of tumoursphere cells (>100 μm diameter) was quantified. Data are shown as mean ± SD ($n = 3$). * $p < 0.05$ versus si-Cont. IR: irradiation, Si-Cont: control siRNA, si-SOX2: SOX2 siRNA, CSCs: cancer stem-like cells.

Since SOX2 expression is important for inducing the characteristics of CSCs as observed in our study and from previous reports [14,15], we attempted to identify the upstream regulator of SOX2. To do so, we investigated the activation of mitogen-activated protein kinase (MAPK) and phosphatidylinositol 3-kinase (PI3K)/AKT pathways, known to regulate SOX2 under other conditions [26–29]. With western blotting, we found that radiation activated both MAPK and PI3K/AKT pathways except ERK activation (Figure 6A). Treatment with the pharmacological inhibitor of PI3K/AKT pathway (LY294002), but not the inhibitors of ERK (PD98059), p38 (SB203580), and SAPK/JNK (SP600125) pathways, dramatically suppressed radiation induced CD44 expression, which is a marker of colorectal CSCs and radioresistance in HCT116 and DLD1 cells (Figure 6B). The concentrations of the inhibitors used were referenced to previous studies, including our report [30]. Based on AKT silencing, we further confirmed the function of PI3K/AKT as an upstream regulator of SOX2-dependent induction of colorectal CSCs by observation of the reduced expression of SOX2 and CD44, as well as CSC properties, such as radioresistance, in vitro metastatic potential, and tumoursphere formation (Figure 6C–E). Moreover, immunocytochemistry further supported the AKT-dependent expression of both CD44 and SOX2 in irradiated HCT116 cells with or without AKT siRNA (Figure 6G). Collectively, these results suggested that radiation enhanced PI3K/AKT/SOX2 axis promoted the induction of colorectal CSCs in radioresistant colorectal cancer cells.

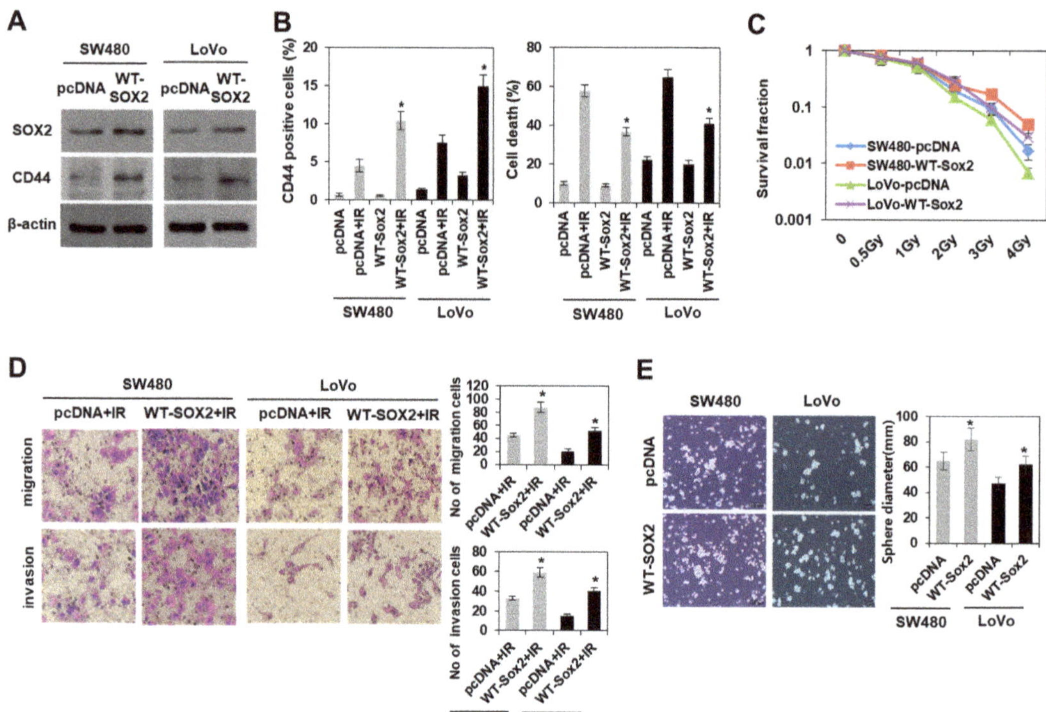

Figure 4. SOX2 overexpression in radiosensitive colorectal cancer cells facilitated the induction of colorectal CSCs following irradiation. (**A**) Immunoblotting of SOX2 and CD44 in SOX2-overexpressing radiosensitive colorectal cancer cells (SW480 and LoVo) on day 2 after irradiation with 10 Gy. (**B**) Analysis of CD44+ cell population (left panel) and apoptotic cells (right panel) by flow cytometry in SOX2-overexpressing SW480 and LoVo cells. Data are shown as mean ± SD ($n = 3$). * $p < 0.05$ compared pcDNA + IR to WT-SOX2 + IR. (**C**) Colony formation assay was performed with SOX2-overexpressing SW480 and LoVo cells, and graph showing the quantification of relative colony numbers at different doses of radiation. Data are shown as mean ± SD ($n = 3$). (**D**) The images of migration and invasion on day 2 after radiation (10 Gy) of SOX2-overexpressing SW480 and LoVo cells were quantified. Bars indicate the measurements of migration and invasion. * $p < 0.05$ compared pcDNA + IR to WT-SOX2 + IR. (**E**) Tumoursphere-formation assay was performed to evaluate self-renewal of CSCs in SOX2-overexpressing SW480 and LoVo cells. Indicated cells were seeded in a non-adherent culture condition. After culturing for 7 days, the images and size of tumoursphere cells were analyzed. Data are shown as mean ± SD ($n = 3$). * $p < 0.05$ versus Cont.

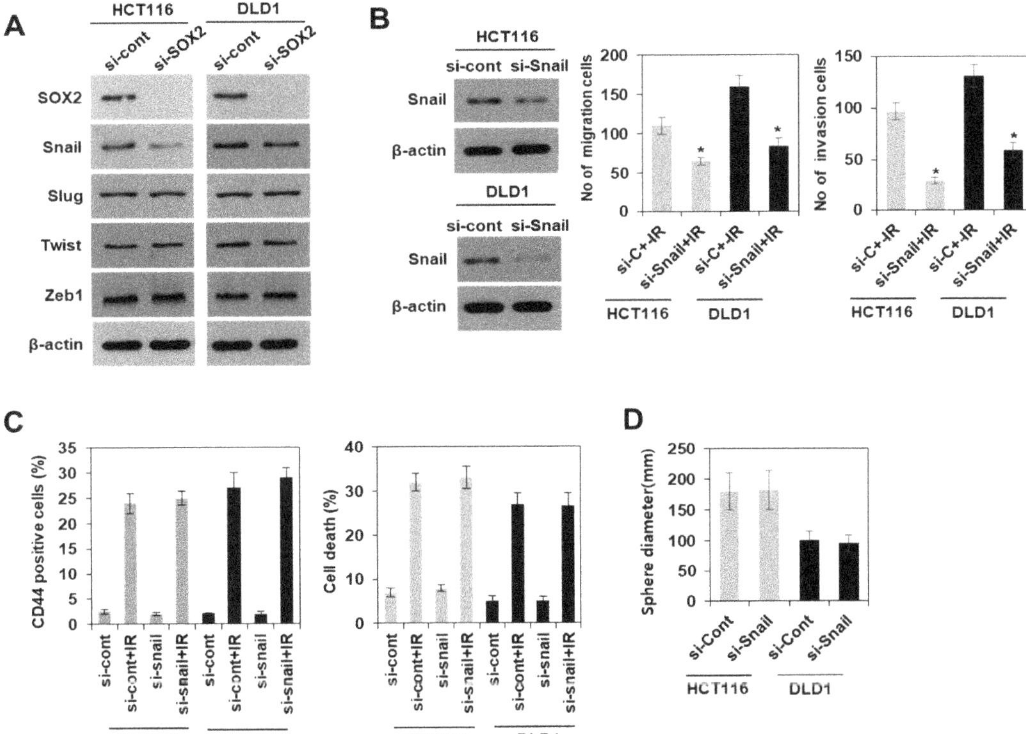

Figure 5. SOX2-dependent induction of colorectal CSCs was not associated with Snail-promoted ability of migration and invasion after irradiation. (**A**) Immunoblotting for SOX2 and EMT regulator (Snail, Slug, Twist, and Zeb1) in SOX2 siRNA-transfected radioresistant colorectal cancer cells (HCT116 and DLD1). (**B**) siRNA-mediated Snail knockdown cells were identified by western blotting (left) and the migration and invasion (right) on day 2 after radiation (10 Gy), of the Snail siRNA-transfected HCT116 and DLD1 cells were quantified. Bars indicate measurements of migration and invasion. * $p < 0.05$ compared si-Cont + IR to si-Snail + IR. (**C**) Analysis of CD44+ cell population (left panel) and apoptotic cells (right panel) by flow cytometry from irradiated cells with 10 Gy on day 2. Data are shown as mean ± SD ($n = 3$). (**D**) Tumoursphere-formation assay was performed to evaluate self-renewal of CSCs in Snail siRNA-transfected cells. Indicated cells were seeded in a non-adherent culture condition. After culturing for 7 days, the size of tumoursphere cells was measured. Data are shown as mean ± SD ($n = 3$). EMT: epithelial-mesenchymal transition, si-Cont: control siRNA, si-Snail: Snail siRNA, CSCs: cancer stem-like cells.

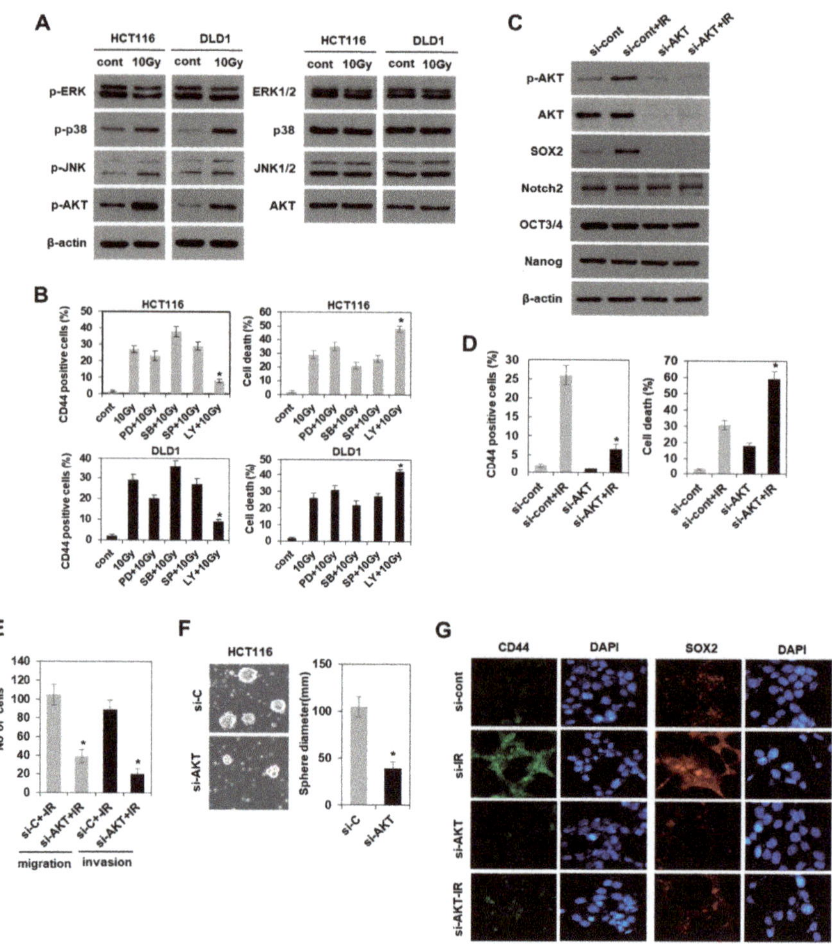

Figure 6. SOX2-dependent induction of colorectal CSCs was modulated by radiation-activated PI3K/AKT pathway, but not MAPK pathway. (**A**) Immunoblotting for mitogen-activated protein kinase (MAPK) pathway activation (p-ERK, ERK1/2, p-p38, p38, p-SAPK/JNK, JNK1/2) and phosphatidylinositol 3-kinase (PI3K)/AKT pathway activation (p-AKT, AKT) on day 2 after radiation (10 Gy) in radioresistant colorectal cancer cells (HCT116 and DLD1). (**B**) Analysis of CD44+ cell population by flow cytometry in HCT116 (left panel) and DLD1 (right panel) cells. The cells were exposed to radiation in the absence or presence of a pharmacological inhibitor of ERK pathway (PD98059, 10 μm), p38 pathway (SB203580, 10 μm), SAPK/JNK pathway (SP600125, 10 μm) and PI3K/AKT pathway (LY294002, 10 μm) for 2 days. Data are shown as mean ± SD ($n = 3$) with * $p < 0.05$ for the pairwise comparisons of CD44+ cell population between irradiated cells with inhibitors and the reference group (i.e., 10 Gy). (**C**) Immunoblotting for the expression of CSC-related proteins (SOX2, Notch2, OCT3/4, and Nanog) on day 2 after radiation (10 Gy) in AKT siRNA-transfected HCT116 cells. (**D**) Analysis of CD44+ cell population (left panel) and apoptotic cells (right panel) by flow cytometry in AKT siRNA-transfected HCT116 cells. Data are shown as mean ± SD ($n = 3$). * $p < 0.05$ compared si-Cont + IR to si-AKT + IR. (**E**) The migration and invasion on day 2 after irradiation (10 Gy) of the AKT siRNA-transfected HCT 116 cells were quantified. Bars indicate the measurements of migration and invasion. * $p < 0.05$ compared si-Cont + IR to si-AKT + IR. (**F**) Tumoursphere-formation assay was performed to evaluate self-renewal ability of CSCs in AKT siRNA-transfected cells. Data are shown as mean ± SD ($n = 3$). * $p < 0.05$ versus si-Cont. (**G**) Cells were stained with an anti-CD44 (green) and anti-SOX2 (red) antibody. Nuclei were counterstained with DAPI (blue). Si-Cont: control siRNA, si-AKT: AKT siRNA, CSCs: cancer stem-like cells.

4. Discussion

The stemness program plays an important role in maintaining the properties of CSCs due to self-renewal, which is a hallmark of cancer-initiating cells. Recent studies have shown that SOX2 is aberrantly expressed and involved in the maintenance of properties of colorectal CSCs, including spheroid-like growth and metastatic potential [14,15]. Here, we extended these studies to demonstrate that SOX2 is regulated by the PI3K/AKT pathway and contributes to the induction of colorectal CSCs in response to radiation. By comparative analysis of radiation-induced population of CSCs in both radioresistant and radiosensitive colorectal cancer cells, we found that radioresistant cells such as HCT116 and DLD1 specifically increased the CD44+ population after irradiation, which is one of the properties of CSCs. Interestingly, we also found that the radiation-induced activation of PI3K/AKT pathway functions as an upstream regulator of SOX2-dependent induction of CSCs in colorectal cancer.

In this study, we report that radiation-enriched CD44+ cells exhibited colorectal CSC properties including resistance to radiation, enhanced in vitro metastatic potential, and a spheroid growth pattern. CD44 is a receptor of hyaluronan and is a transmembrane glycoprotein that participates in many cellular processes, including growth, survival, differentiation, and motility [31–34]. CD44 is considered a more selective marker of colorectal CSCs than CD133, because the properties of colorectal CSCs are not regulated by CD133 modulation [19,20,35,36]. Consistent with this, our comparative study between radioresistant and radiosensitive colorectal cancer cells showed that CD44 expression, but not that of CD133, was selectively increased in radioresistant colorectal cancer along with acquiring the properties of colorectal CSCs after irradiation. A recent study reported that SOX2 expression primarily coincided with CD44+ and ALDH1+ population in pancreatic CSCs [37] and CD44+ and CD24+ in colorectal cancer [14]. Indeed, we also observed that FACS-sorted CD44+ cells showed an upregulation of SOX2 expression and demonstrated its important role in modulating the CD44+ population growth and the properties of CSCs in colorectal cancer using both knockdown and overexpression of SOX2, which is consistent with previous reports. Notably, in our study, this functional relationship occurred in response to radiation, indicating that radiation affects SOX2-dependent induction of CD44+ population.

Factors that are important for self-renewal in stem cells are found to be dysregulated in human malignancies. SOX2 expression has been implicated in the control of colorectal CSC properties; however, the related signaling pathways are less understood. In a previous study, SOX2-induced CSCs in cervical and pancreatic cancer have been linked to epithelial-mesenchymal transition (EMT)-related factors [37,38]. However, these studies were considered controversial. Han et al. reported a role for SOX2 in EMT and increased in vitro metastatic potential, such as in migration and invasion in colorectal cancer [24], while Lundberg et al. reported that SOX2 mediated induction of CSC characteristics in an independent manner [14]. In our system, the ability of migration and invasion was dramatically regulated by SOX2-modulated Snail expression, known as a master regulator of EMT. However, we observed that Snail did not affect the induction of colorectal CSC properties, including CD44+ population growth, resistance to radiation, and the ability of tumoursphere formation. This suggested that the Snail-mediated EMT process might not be involved in SOX2-dependent induction of colorectal CSCs, although we could not exclude the possibility of involvement of other regulators of EMT process or factors related to tumor microenvironment [39] affected by irradiation. Therefore, it is likely that SOX2 modulates either EMT process or CSC induction through alternative pathways, at least in response to radiation. Further studies are required to clarify the relationship between EMT and CSCs induction.

An elucidation of the signaling pathways that govern the SOX2-dependent induction of CSCs is also required for devising an optimal targeted therapy. Considering that MAPK and PI3K/AKT pathways, in addition to being activated by radiation [40], were associated with resistance to therapy and tumorigenicity in cancer cells [41,42], we investigated the effect of the inhibitors of MAPK and PI3K/AKT for induction potential of CSCs. It was found that radiation activated the genes of both MAPK and PI3K/AKT pathways, consistent with previous reports, except ERK. Interestingly, the induction of SOX2 and CSC characteristics, including CD44+ cells in colorectal cancer, were only affected by the inactivation and downregulation of PI3K/AKT following irradiation. This finding is contradictory to the report by Wang et al., [43] who showed that activation of both AKT and MAPK pathways was involved in the induction of properties of colorectal CSCs, such as the colony formation ability in primary colon cancer cells. These differences can be explained using target cells with differential markers of CSCs and the response to stresses. To isolate colorectal CSCs, Wang et al. used CD133, a colorectal CSC marker, and characterized cells with or without CD133 expression under non-stress conditions for CSC properties, whereas we used CD44, which was specifically induced by radiation stress. Furthermore, the involvement of genes in the PI3K/AKT pathway in SOX2 regulation in breast and nasopharyngeal carcinoma has recently been reported [28,29]. Therefore, this is an interesting finding that radiation-activated PI3K/AKT pathway genes were essential for the SOX2-dependent induction of colorectal CSCs, and it is potentially an effective therapeutic target for CSCs in colorectal cancer activated by radiation.

Author Contributions: Designed the experiments and drafted and edited the manuscript: M.-J.K. and S.B.L. Performed the experiments, interpreted the data, and drafted the manuscript: J.-H.P. and Y.-H.K. Performed the screening of experimental conditions and provided assistance for all the experiments: S.S. (Sehwan Shim), A.K. and H.J. Provided valuable suggestions and feedback: S.-J.L. and S.P. Performed the application of statistical techniques to analyze data: S.S. (Songwon Seo) Acquisition of the financial support for publication: W.I.J. All authors have read and agreed to the published version of the manuscript.

Funding: This research was supported by a grant from the Korea Institute of Radiological and Medical Sciences (KIRAMS), funded by the Ministry of Science and ICT (MSIT), Republic of Korea (grant number 50535-2020).

Institutional Review Board Statement: Not applicable.

Informed Consent Statement: Not applicable.

Data Availability Statement: Not applicable.

Conflicts of Interest: The authors declare no conflict of interest.

References

1. Torre, L.A.; Bray, F.; Siegel, R.L.; Ferlay, J.; Lortet-Tieulent, J.; Jemal, A. Global Cancer Statistics, 2012. *CA Cancer J. Clin.* **2015**, *65*, 87–108. [CrossRef] [PubMed]
2. Brenner, H.; Kloor, M.; Pox, C.P. Colorectal Cancer. *Lancet* **2014**, *383*, 1490–1502. [CrossRef]
3. Olivares-Urbano, M.A.; Griñán-Lisón, C.; Marchal, J.A.; Núñez, M.I. CSC Radioresistance: A Therapeutic Challenge to Improve Radiotherapy Effectiveness in Cancer. *Cells* **2020**, *9*, 1651. [CrossRef] [PubMed]
4. Singh, S.K.; Hawkins, C.; Clarke, I.D.; Squire, J.A.; Bayani, J.; Hide, T.; Henkelman, R.M.; Cusimano, M.D.; Dirks, P.B. Identification of Human Brain Tumour Initiating Cells. *Nature* **2004**, *432*, 396–401. [CrossRef] [PubMed]
5. Liu, Y.; Yang, M.; Luo, J.; Zhou, H. Radiotherapy Targeting Cancer Stem Cells "Awakens" Them to Induce Tumour Relapse and Metastasis in Oral Cancer. *Int. J. Oral Sci.* **2020**, *12*, 19. [CrossRef] [PubMed]
6. Mitra, A.; Mishra, L.; Li, S. EMT, CTCs and CSCs in Tumor Relapse and Drug-Resistance. *Oncotarget* **2015**, *6*, 10697–10711. [CrossRef]
7. Zeuner, A.; Todaro, M.; Stassi, G.; De Maria, R. Colorectal Cancer Stem Cells: From the Crypt to the Clinic. *Cell Stem Cell* **2014**, *15*, 692–705. [CrossRef]
8. Szaryńska, M.; Olejniczak, A.; Kobiela, J.; Spychalski, P.; Kmieć, Z. Therapeutic Strategies against Cancer Stem Cells in Human Colorectal Cancer. *Oncol. Lett.* **2017**, *14*, 7653–7668. [CrossRef]
9. Zhou, Y.; Xia, L.; Wang, H.; Oyang, L.; Su, M.; Liu, Q.; Lin, J.; Tan, S.; Tian, Y.; Liao, Q.; et al. Cancer stem cells in progression of colorectal cancer. *Oncotarget* **2018**, *9*, 33403–33415. [CrossRef]
10. Sekido, R.; Lovell-Badge, R. Sex Determination and SRY: Down to a Wink and a Nudge? *Trends Genet.* **2009**, *25*, 19–29. [CrossRef]
11. Wegner, M. All Purpose Sox: The Many Roles of Sox Proteins in Gene Expression. *Int. J. Biochem. Cell Biol.* **2010**, *42*, 381–390. [CrossRef] [PubMed]

12. Boumahdi, S.; Driessens, G.; Lapouge, G.; Rorive, S.; Nassar, D.; Le Mercier, M.; Delatte, B.; Caauwe, A.; Lenglez, S.; Nkusi, E.; et al. SOX2 Controls Tumour Initiation and Cancer Stem-Cell Functions in Squamous-Cell Carcinoma. *Nature* **2014**, *10*, 246–250. [CrossRef] [PubMed]
13. Zhu, F.; Qian, W.; Zhang, H.; Liang, Y.; Wu, M.; Zhang, Y.; Zhang, X.; Gao, Q.; Li, Y. SOX2 Is a Marker for Stem-Like Tumor Cells in Bladder Cancer. *Stem Cell Rep.* **2017**, *9*, 429–437. [CrossRef] [PubMed]
14. Lundberg, I.V.; Edin, S.; Eklöf, V.; Öberg, Å.; Palmqvist, R.; Wikberg, M.L. SOX2 Expression Is Associated with a Cancer Stem Cell State and Down-Regulation of CDX2 in Colorectal Cancer. *BMC Cancer* **2016**, *16*, 471. [CrossRef] [PubMed]
15. Takeda, K.; Mizushima, T.; Yokoyama, Y.; Hirose, H.; Wu, X.; Qian, Y.; Ikehata, K.; Miyoshi, N.; Takahashi, H.; Haraguchi, N.; et al. Sox2 Is Associated with Cancer Stem-Like Properties in Colorectal Cancer. *Sci. Rep.* **2018**, *8*, 17639. [CrossRef]
16. Ghisolfi, L.; Keates, A.C.; Hu, X.; Lee, D.K.; Li, C.J. Ionizing Radiation Induces Stemness in Cancer Cells. *PLoS ONE* **2012**, *7*, e43628. [CrossRef]
17. Park, J.H.; Kim, Y.H.; Park, E.H.; Lee, S.J.; Kim, H.; Kim, A.; Lee, S.B.; Shim, S.; Jang, H.; Myung, J.K.; et al. Effects of Metformin and Phenformin on Apoptosis and Epithelial-Mesenchymal Transition in Chemoresistant Rectal Cancer. *Cancer Sci.* **2019**, *110*, 2834–2845. [CrossRef]
18. Cho, M.H.; Park, J.H.; Choi, H.J.; Park, M.K.; Won, H.Y.; Park, Y.J.; Lee, C.H.; Oh, S.H.; Song, Y.S.; Kim, H.S.; et al. DOT1L Cooperates with the c-Myc-p300 Complex to Epigenetically Derepress *CDH1* Transcription Factors in Breast Cancer Progression. *Nat. Commun.* **2015**, *6*, 7821. [CrossRef]
19. Du, L.; Wang, H.; He, L.; Zhang, J.; Ni, B.; Wang, X.; Jin, H.; Cahuzac, N.; Mehrpour, M.; Lu, Y.; et al. CD44 Is of Functional Importance for Colorectal Cancer Stem Cells. *Clin. Cancer Res.* **2008**, *14*, 6751–6760. [CrossRef]
20. Dalerba, P.; Dylla, S.J.; Park, I.K.; Liu, R.; Wang, X.; Cho, R.W.; Hoey, T.; Gurney, A.; Huang, E.H.; Simeone, D.M.; et al. Phenotypic Characterization of Human Colorectal Cancer Stem Cells. *Proc. Natl. Acad. Sci. USA* **2007**, *104*, 10158–10163. [CrossRef]
21. Munro, M.J.; Wickremesekera, S.K.; Peng, L.; Tan, S.T.; Itinteang, T. Cancer Stem Cells in Colorectal Cancer: A Review. *J. Clin. Pathol.* **2018**, *71*, 110–116. [CrossRef] [PubMed]
22. Meisel, C.T.; Porcheri, C.; Mitsiadis, T.A. Cancer Stem Cells, *Quo Vadis*? The Notch Signaling Pathway in Tumor Initiation and Progression. *Cells* **2020**, *9*, 1879. [CrossRef] [PubMed]
23. Brabletz, T. EMT and MET in Metastasis: Where Are the Cancer Stem Cells? *Cancer Cell* **2012**, *22*, 699–701. [CrossRef]
24. Han, X.; Fang, X.; Lou, X.; Hua, D.; Ding, W.; Foltz, G.; Hood, L.; Yuan, Y.; Lin, B. Silencing SOX2 Induced Mesenchymal-epithelial Transition and its Expression Predicts Liver and Lymph Node Metastasis of CRC Patients. *PLoS ONE* **2012**, *7*, e41335. [CrossRef]
25. Shimoda, M.; Sugiura, T.; Imajyo, I.; Ishii, K.; Chigita, S.; Seki, K.; Kobayashi, Y.; Shirasuna, K. The T-Box Transcription Factor Brachyury Regulates Epithelial-Mesenchymal Transition in Association with Cancer Stem-Like Cells in Adenoid Cystic Carcinoma Cells. *BMC Cancer* **2012**, *12*, 377. [CrossRef] [PubMed]
26. Hui, K.; Gao, Y.; Huang, J.; Xu, S.; Wang, B.; Zeng, J.; Fan, J.; Wang, X.; Yue, Y.; Wu, S.; et al. RASAL2, A RAS Gtpase-activating Protein, Inhibits Stemness and Epithelial-Mesenchymal Transition Via MAPK/SOX2 Pathway in Bladder Cancer. *Cell Death Dis.* **2017**, *8*, e2600. [CrossRef]
27. Haston, S.; Pozzi, S.; Carreno, G.; Manshaei, S.; Panousopoulos, L.; Gonzalez-Meljem, J.M.; Apps, J.R.; Virasami, A.; Thavaraj, S.; Gutteridge, A.; et al. MAPK Pathway Control of Stem Cell Proliferation and Differentiation in the Embryonic Pituitary Provides Insights into the Pathogenesis of Papillary Craniopharyngioma. *Development* **2017**, *144*, 2141–2152. [CrossRef]
28. Schaefer, T.; Wang, H.; Mir, P.; Konantz, M.; Pereboom, T.C.; Paczulla, A.M.; Merz, B.; Fehm, T.; Perner, S.; Rothfuss, O.C.; et al. Molecular and Functional Interactions Between AKT and SOX2 in Breast Carcinoma. *Oncotarget* **2015**, *6*, 43540–43556. [CrossRef]
29. Qin, J.; Ji, J.; Deng, R.; Tang, J.; Yang, F.; Feng, G.K.; Chen, W.D.; Wu, X.Q.; Qian, X.J.; Ding, K.; et al. DC120, A Novel AKT Inhibitor, Preferentially Suppresses Nasopharyngeal Carcinoma Cancer Stem-Like Cells by Downregulating Sox2. *Oncotarget* **2015**, *6*, 6944–6958. [CrossRef]
30. Kim, M.J.; Byun, J.Y.; Yun, C.H.; Park, I.C.; Lee, K.H.; Lee, S.J. c-Src-p38 Mitogen-Activated Protein Kinase Signaling Is Required for Akt Activation in Response to Ionizing Radiation. *Mol. Cancer Res.* **2008**, *6*, 1872–1880. [CrossRef] [PubMed]
31. Aruffo, A.; Stamenkovic, I.; Melnick, M.; Underhill, C.B.; Seed, B. CD44 Is the Principal Cell Surface Receptor for Hyaluronate. *Cell* **1990**, *61*, 1303–1313. [CrossRef]
32. Nagano, O.; Saya, H. Mechanism and Biological Significance of CD44 Cleavage. *Cancer Sci.* **2004**, *95*, 930–935. [CrossRef] [PubMed]
33. Cheng, C.; Sharp, P.A. Regulation of CD44 Alternative Splicing by SRm160 and Its Potential Role in Tumor Cell Invasion. *Mol. Cell. Biol.* **2006**, *26*, 362–370. [CrossRef] [PubMed]
34. Vigetti, D.; Viola, M.; Karousou, E.; Rizzi, M.; Moretto, P.; Genasetti, A.; Clerici, M.; Hascali, V.C.; De Luca, G.; Passi, A. Hyaluronan-CD44-1/2 Regulate Human Aortic Smooth Muscle Cell Motility During Aging. *J. Biol. Chem.* **2008**, *283*, 4448–4458. [CrossRef]
35. Feng, H.L.; Liu, Y.Q.; Yang, L.J.; Bian, X.C.; Yang, Z.L.; Gu, B.; Zhang, H.; Wang, C.J.; Su, X.L.; Zhao, X.M. Expression of CD133 Correlates with Differentiation of Human Colon Cancer Cells. *Cancer Biol. Ther.* **2010**, *9*, 216–223. [CrossRef]
36. Shmelkov, S.V.; Butler, J.M.; Hooper, A.T.; Hormigo, A.; Kushner, J.; Milde, T.; St Clair, R.; Baljevic, M.; White, I.; Jin, D.K.; et al. CD133 Expression Is Not Restricted to Stem Cells, and Both CD133+ and CD133− Metastatic Colon Cancer Cells Initiate Tumors. *J. Clin. Investig.* **2008**, *118*, 2111–2120. [CrossRef]

37. Herreros-Villanueva, M.; Zhang, J.S.; Koenig, A.; Abel, E.V.; Smyrk, T.C.; Bamlet, W.R.; de Narvajas, A.A.; Gomez, T.S.; Simeone, D.M.; Bujanda, L.; et al. SOX2 Promotes Dedifferentiation and Imparts Stem Cell-Like Features to Pancreatic Cancer Cells. *Oncogenesis* **2013**, *2*, e61. [CrossRef]
38. Liu, X.F.; Yang, W.T.; Xu, R.; Liu, J.T.; Zheng, P.S. Cervical Cancer Cells with Positive Sox2 Expression Exhibit the Properties of Cancer Stem Cells. *PLoS ONE* **2014**, *9*, e87092. [CrossRef]
39. François, S.; Usunier, B.; Forgue-Lafitte, M.E.; L'Homme, B.; Benderitter, M.; Douay, L.; Gorin, N.C.; Larsen, A.K.; Chapel, A. Mesenchymal Stem Cell Administration Attenuates Colon Cancer Progression by Modulating the Immune Component within the Colorectal Tumor Microenvironment. *Stem Cells Transl. Med.* **2019**, *8*, 285–300. [CrossRef]
40. Lee, S.Y.; Jeong, E.K.; Ju, M.K.; Jeon, H.M.; Kim, M.Y.; Kim, C.H.; Park, H.G.; Han, S.I.; Kang, H.S. Induction of Metastasis, Cancer Stem Cell Phenotype, and Oncogenic Metabolism in Cancer Cells by Ionizing Radiation. *Mol. Cancer* **2017**, *16*, 10. [CrossRef]
41. Narayanankutty, A. PI 3K/Akt/mTOR Pathway as a Therapeutic Target for Colorectal Cancer: A Review of Preclinical and Clinical Evidence. *Curr. Drug Targets* **2019**, *20*, 1217–1226. [CrossRef] [PubMed]
42. Lee, S.; Rauch, J.; Kolch, W. Targeting MAPK Signaling in Cancer: Mechanisms of Drug Resistance and Sensitivity. *Int. J. Mol. Sci.* **2020**, *21*, 1102. [CrossRef] [PubMed]
43. Wang, Y.K.; Zhu, Y.L.; Qiu, F.M.; Zhang, T.; Chen, Z.G.; Zheng, S.; Huang, J. Activation of Akt and MAPK Pathways Enhances the Tumorigenicity of CD133+ Primary Colon Cancer Cells. *Carcinogenesis* **2010**, *31*, 1376–1380. [CrossRef] [PubMed]

 *cells*

Article

Keap1-Nrf2 Pathway Regulates ALDH and Contributes to Radioresistance in Breast Cancer Stem Cells

Dinisha Kamble, Megharani Mahajan, Rohini Dhat and Sandhya Sitasawad *

Redox Biology Lab, National Centre for Cell Science (NCCS), Pune 411007, India; dinisha.kamble19@gmail.com (D.K.); meghanandgaon@gmail.com (M.M.); rohini.dhat@nccs.res.in (R.D.)
* Correspondence: ssitaswad@nccs.res.in; Tel.: +20-2570-8109

Abstract: Tumor recurrence after radiotherapy due to the presence of breast cancer stem cells (BCSCs) is a clinical challenge, and the mechanism remains unclear. Low levels of ROS and enhanced antioxidant defenses are shown to contribute to increasing radioresistance. However, the role of Nrf2-Keap1-Bach1 signaling in the radioresistance of BCSCs remains elusive. Fractionated radiation increased the percentage of the ALDH-expressing subpopulation and their sphere formation ability, promoted mesenchymal-to-epithelial transition and enhanced radioresistance in BCSCs. Radiation activated Nrf2 via Keap1 silencing and enhanced the tumor-initiating capability of BCSCs. Furthermore, knockdown of Nrf2 suppressed $ALDH^+$ population and stem cell markers, reduced radioresistance by decreasing clonogenicity and blocked the tumorigenic ability in immunocompromised mice. An underlying mechanism of Keap1 silencing could be via miR200a, as we observed a significant increase in its expression, and the promoter methylation of Keap1 or GSK-3β did not change. Our data demonstrate that $ALDH^+$ BCSC population contributes to breast tumor radioresistance via the Nrf2-Keap1 pathway, and targeting this cell population with miR200a could be beneficial but warrants detailed studies. Our results support the notion that Nrf2-Keap1 signaling controls mesenchymal epithelial plasticity, regulates tumor-initiating ability and promotes the radioresistance of BCSCs.

Keywords: BCSC; ALDH activity; fractionated dose of γ radiation; radioresistance; ROS; Nrf2; Keap1; miR200a; epithelial–mesenchymal transition (EMT)

Citation: Kamble, D.; Mahajan, M.; Dhat, R.; Sitasawad, S. Keap1-Nrf2 Pathway Regulates ALDH and Contributes to Radioresistance in Breast Cancer Stem Cells. *Cells* **2021**, *10*, 83. https://doi.org/10.3390/cells10010083

Received: 11 November 2020
Accepted: 28 December 2020
Published: 6 January 2021

1. Introduction

Radiotherapy (RT) is a critical factor of primary, adjuvant and palliative treatment for almost all kinds of cancers, including breast cancer. It alone is capable of lowering the 10-year risk of relapse by one half and reducing the 15-year risk of breast-cancer-related death [1]. Although profound benefits are achieved with RT due to its localized treatment, especially for ductal carcinoma and early invasive cancer, local control of the disease fails by 8–15% in radiotherapy-treated patients with advanced invasive tumors due to resistance and relapse of the tumor [2]. The reason for RT failure and the locoregional recurrence of breast cancer is the presence of a subset of radioresistant tumor cells, termed breast cancer stem cells (BCSCs), which show a difference in sensitivities to radiation [3–5]. Standard fractionated doses of radiation are sublethal for BCSCs as they typically evade radiation to develop innate or acquired resistance and establish tumor recurrence and metastasis, leading to the majority of cancer-related deaths. The molecular mechanisms that govern the emergence of aggressive radioresistance in BCSCs are yet unknown.

Low levels of reactive oxygen species (ROS) and enhanced ROS defenses appear to partially contribute to the adaptive tumor radioresistance in BCSCs [5–7]. Thus, the identification of underlying mechanisms and overcoming low ROS levels within BCSCs may be a useful method for improving radiation therapy. The transcription factor nuclear factor (erythroid-derived 2)-like 2 (Nrf2), the master regulator of antioxidant defense mechanisms, is a critical regulator of the redox balance. In the cytosol, Nrf2 activity is tightly regulated by two main inhibitors, Keap1 and GSK-3. The Neh2 and Neh6 domains of Nrf2 are the degron.

While the Neh2 domain binds the E3 ligase adapter Keap1 that presents Nrf2 for ubiquitination to a CUL3/RBX1 complex, the Neh6 domain requires previous phosphorylation by GSK-3 to bind the E3 ligase adapter b-TrCP and subsequent ubiquitination by a CUL1/RBX1 complex [8]. Inactivation of either of these regulators due to oxidative or electrophilic stress stabilizes Nrf2, which then translocates to the nucleus and binds to antioxidant response elements (ARE) in the promoter region of target genes by the formation of a heterodimer with small Maf proteins. In the nucleus, Bach1 negatively regulates nuclear Nrf2 activity by competitive-binding with small Maf proteins [9] and thereby inactivates HO1 [10,11]. Previous studies have shown elevated levels of Nrf2 as a critical regulator of chemoresistance in CSC-enriched breast tumors [12,13] and the activation of Nrf2-associated antioxidant genes, such as HO1, NQO1, Prx1, etc., that contribute to radioresistance in other cancer cells [14]. Since BCSCs contain low levels of ROS and enhanced antioxidant defense [5], the role of the Nrf2 pathway in the radioresistance of BCSCs deserves further investigation.

In this study, we observed an increase in ALDH activity, indicative of BCSCs with increased radioresistance, tumorigenesis, reduced apoptosis and the activation of signaling pathways, which promote mesenchymal–epithelial transition (MET) and migration. Additionally, enhanced tumorigenicity was observed after fractionated irradiation. Further investigation of the role of Nrf2 in radioresistance showed that Nrf2 and its associated genes HO1 and NQO1 were significantly increased after irradiation. The shRNA-mediated knockdown of Nrf2 expression led to a decrease in all of the above processes of radioresistance in BCSCs. The mechanism of Nrf2 activation was found to be regulated via Keap1 silencing, as we did not see any change in GSK-3β, as well as in Bach1, the negative regulator of Nrf2. We also did not find any change in the methylation status of the Keap1 promoter; however, a significant increase in the expression of miR200a was observed. This indicates that miR200a could be a possible mechanism of Keap1 silencing. This study provides evidence for the role of Nrf2 and its downstream genes and suggests mechanisms by which the Nrf2/Keap1 pathway induces radioresistance in BCSCs. Overall, the data indicate the contribution of $ALDH^+$ cell population to radioresistance via the Nrf2-Keap1 axis, suggesting that targeting $ALDH^+$ BCSC cell population with miR200a could be beneficial but warrants detailed studies.

2. Materials and Methods

2.1. Reagents

Antibodies recognizing NANOG (D73G4), SOX2 (D6D9), KLF4 (D1F2), HO1 (D60G11), Vimentin (D21H3) and Keap1 (D6B12) were obtained from Cell Signaling Technology (Danvers, MA, USA). E-cadherin (610404) was obtained from BD Biosciences (San Jose, CA, USA). Nrf2 (ab-89443), SLUG (ab-27568), SNAIL (ab-53519) and Bach1 (ab-115210) were purchased from Abcam, (Burlingame, CA, USA) and NQO1 (sc-32793), BAX (sc-7480), BCL2 (sc-7382) were obtained from Santa Cruz Biotechnology (Dallas, TX, USA). Nrf2-targeting shRNA lentiviral particles (sc-37030-v) were obtained from Santa Cruz Biotechnology (CA, USA). Bovine serum albumin (BSA), fibroblast growth factor (FGF) and epidermal growth factor (EGF) were purchased from Sigma-Aldrich (St. Louis, MO, USA). ALDEFLUOR Kit (01700) was obtained from STEM CELL technologies (Vancouver, Canada). Nrf2 Transcription Factor Assay Kit (Colorimetric) (ab207223) was obtained from Abcam. Annexin-V-FITC (556419) was obtained from BD Biosciences (San Jose, CA, USA).

2.2. Cell Culture and Irradiation

MCF-7 and MDA-MB-231 cells were obtained from the American Type Culture Collection (ATCC) and maintained in DMEM supplemented with 10% FBS (Gibco, MD, USA). Cells were irradiated at room temperature with a ^{60}Co-γ rays laboratory irradiator (Gamma Chamber-5000, BRIT, Mumbai) at a dose rate of 2.163 Gy/min for the time required to obtain the prescribed dose. For fractionated doses of radiation, cells were irradiated with 2Gy for three consecutive days, and for an acute dose of 6Gy, cells were irradiated once on the last day of fractionated irradiation. Corresponding controls were mock irradiated [13]. Control and irradiated cells were further incubated for 24 h postirradiation.

2.3. *Mammosphere Formation*

After irradiation, mammospheres were formed using single-cell suspensions in an ultralow attachment 6-well plate at a density of 2×10^4 cells/well in specific media for mammosphere culture containing DMEM and Nutrient Mixture F-12 medium supplemented with 20 ng/mL EGF, B27 (1:50, Life Technologies, MA, USA), 20 ng/mL FGF (Sigma-Aldrich, St. Louis, MO, USA) and penicillin/streptomycin (Invitrogen) for 4–5 days. The floating aggregates with a >50 μm diameter were selected as mammospheres, manually counted and dissociated by incubation with 1:5 of 0.25% trypsin/EDTA. The mammosphere-forming efficiency (MFE) was calculated using the following equation: MFE = No. of spheres formed/No. of cells seeded × platting efficiency.

2.4. *Colony Formation*

For the colony formation assay, MCF-7 cells (1000 cells/well) were grown on 6-well plates and maintained in a humidified chamber comprising 95% air and 5% CO_2 at 37 °C for 14 days. Cells were then fixed with 3.7% paraformaldehyde at room temperature for 10 min and stained using crystal violet solution (0.2% crystal violet and 1X PBS) at room temperature for 30 min. Stained cells were washed with 1× PBS and air-dried at room temperature. The numbers of colonies were quantified using the Image J program (version-v1.53e). The survival fraction was calculated as the number of colonies counted/the number of cells inoculated × plating efficiency at 0 Gy. Colonies consisting of 50 or more cells were counted as clonogenic survivors.

2.5. *FACS Analysis for CD44/24 and ALDEFLUOR Assay*

Cells were irradiated with the specific doses of radiation and then stained with anti-CD44-APC and anti-CD24-PE with their respective isotype controls, incubated for 40 min and analyzed on a *FACSCanto*™ flow cytometer (BD). To measure ALDH activity, cells were analyzed by an ALDEFLUOR assay kit (STEMCELL Technologies), following the manufacturer's protocol. Flow cytometry was performed on a *FACSCanto*™ flow cytometer (BD), USA and analyzed by DIVA software (BD Biosciences, USA).

2.6. *Immunoblotting*

Protein extracts were prepared, and immunoblotting was performed, as described previously [15], using the following antibodies: Nrf2, Bach1, SNAIL and SLUG (Abcam, Burlingame, CA, USA); NQO1, HO1 (Santa Cruz, TX, USA), Keap1, SOX2 and KLF4, NANOG (Cell Signaling Technology, ((Danvers, MA, USA); BAX and BCL2 (Santa Cruz, Dallas, TX, USA) at 1:1000 and GAPDH (1:10,000, Sigma). Primary antibodies were detected using secondary antibodies (1:10,000, Bio-Rad), conjugated with HRP, and protein–antibody complexes were detected by the Substrate Detection Kit (Thermo Fisher, CA, USA). Densitometry was performed using the Image Lab and Image J software. GAPDH was used as loading control for whole-cell lysates.

2.7. *qRT-PCR*

Total RNA was extracted using TRIzol (Invitrogen). cDNA was synthesized using reverse transcription, followed by quantitative real-time PCR with SYBR Green Supermix (Life Technologies), using primers for Nrf2, Keap1, HO1, NQO1, SOX2, NANOG, KLF4 and GAPDH, which were used as the normalizing control. miR200a detection was carried out using the stem-loop method, as described previously [16]. The gene-specific primers used to perform real-time qRT-PCR analysis are listed in Table S1.

2.8. Nrf2 Activity

Nuclear protein fractions of irradiated MCF-7 cells were isolated using NE-PER Nuclear and Cytoplasmic Extraction Reagents (Thermo Fisher, MA, USA), according to the manufacturer's instructions. Protein concentrations were assessed using the Bradford reagent (Bio-Rad). Nrf2 Transcription Factor Assay (Colorimetric) was performed using 20 mg of nuclear proteins (ab207223, Abcam) to detect nuclear Nrf2 and antioxidant responsive element (ARE) sequence binding at OD 450 nm, following the manufacturer's instructions.

2.9. Scratch Wound Assay

MCF-7 cells were irradiated with specific doses of radiation and incubated for 24 h. The cells were scratched with a pipette tip to create wounds. Images were taken at different planes at 0 h and 24 h at 10× magnification. Percent cell migration was calculated as described in our previous paper [17].

2.10. ROS Detection

Detection of ROS was performed, as described previously [18]. Cells were treated with 1 μmol/L 2′,7′dichlorodihydrofluorescein diacetate (DCF-DA; Invitrogen, MA, USA) for 30 min, followed by a 1× PBS wash for 2 times. The reduced DCF-DA was oxidized by intracellular ROS and converted into fluorescent 2′, 7′-dichlorofluorescein (DCF). Fluorescent signals were detected by the *FACSCanto*™ flow cytometer (BD). A total of 10,000 cells were analyzed per sample.

2.11. Apoptosis and Cell Proliferation Assays

To perform Annexin V and propidium iodide (PI) staining, irradiated MCF-7 cells and mammospheres were trypsinized, washed with 1× PBS, centrifuged and stained with the Annexin V-FITC antibody (20 min, room temperature) and PI (0.02 mg/mL; Sigma, P4170). The percentage of apoptotic cells was evaluated using the *FACSCanto*™ flow cytometer (BD). A total of 1×10^4 cells were recorded per condition in three independent experiments. In the cell proliferation assay, irradiated MCF-7 cells and mammospheres were trypsinized, washed with 1× PBS, centrifuged and stained with Ki67. The stained cells were analyzed using the *FACSCanto*™ flow cytometer (BD).

2.12. shRNA-Mediated Knockdown

To generate knockdown cell lines, MCF-7 cells were stably transduced in the presence of polybrene (5 μg/mL) with Nrf2 (sc-37030-V), shRNA lentiviruses and control scrambled shRNA particles-A (sc-108080) obtained from Santa Cruz Biotechnology, USA. Transduced cells were then selected with puromycin (2 μg/mL) for up to 4 weeks.

2.13. In Vivo *Tumorigenicity Assay*

Female SCID mice (6 to 8 weeks old; n = 5 per group) were maintained, according to the procedures and guidelines of the Institutional Animal Ethics Committee (NCCS). A total of 2×10^6 MCF-7 cells (wild type/shNrf2) were injected subcutaneously into the mammary fat pads of female SCID mice along with a 1:2 ratio of growth-factor-reduced Matrigel (BD Biosciences). These mice were also injected with β-estradiol (Sigma-Aldrich, St. Louis, MO, USA) and observed for 2 months for the development of breast tumors. All the mice were euthanized, according to the institute's ethical procedures, and tumors were collected for further analysis. The length and width of the tumors were measured using a vernier caliper and volumes were calculated using the following formula: Tumor volume = $1/2\ (\text{length} \times \text{width}^2)$.

2.14. Bisulfite Sequencing and CpG Methylation Status

Genomic DNA was extracted using the DNeasy Tissue kit (QIAGEN, MD, USA). The EZ DNA Methylation Kit (Zymo Research, CA, USA) was used for sodium bisulfite conversion, according to the manufacturer's protocol. Primers spanning two promoters of the Keap1 gene were designed using Methyl Primer Express (Thermo Scientific). Promoter 1 Forward: 5′- GAGTTTTGGYGGGGAATT-3′; Reverse: 5′-CCCTACCRCCTAAAACCAA-3′. Bisulfite-modified DNA (100 ng) was amplified in a PCR mix containing 0.4 μM of forward and reverse primer, HotStarTaq Master Mix Kit (Qiagen, Germany: 203445). Methylation status analysis was performed by Quantification Tool for Methylation Analysis (QUMA) software.

2.15. Statistical Analysis

One-way analysis of variance (ANOVA), followed by Tukey's post hoc multiple comparisons tests by Prism software (GraphPad, San Diego, CA, USA), was used to analyze statistical significance. All the data values are presented as mean ± SE, reflecting the minimum of three independent determinations. Statistical significance was determined by comparing the treatments with untreated controls, and the significant differences are indicated as * $p < 0.05$, ** $p < 0.01$ and *** $p < 0.001$.

3. Results

3.1. Fractionated Doses of Radiation Selectively Increase E-BCSC Population While Decreasing M-BCSC Population

Recent studies indicate that BCSCs exist in two phenotypes, i.e., epithelial (E-BCSC) and mesenchymal (M-BCSC), and BCSC plasticity plays a crucial role in future strategies for therapeutic resistance [19]. E-BCSCs characterized as $ALDH^+$ population are proliferative, locate in the tumor's hypoxic region and show the MET phenotype. On the other hand, M-BCSCs that express the $CD44^+/24^-$ phenotype are primarily quiescent, located on the invasive front and have the EMT phenotype. Previous studies have shown an increase in $CD44^+/24^-$ cells and high $ALDH^+$ characteristics of tumor-initiating or cancer stem cells in breast tumors and established cell lines after irradiation [20–23]. In our study, fractionated irradiation with 2 Gy x 3 days of γ-rays increased the population of $ALDH^+$ cells (Figure 1A) but decreased $CD44^+/24^-$ cells (Figure 1B) in MCF-7 and MDA-MB-231 cells and their corresponding mammospheres. Since mammospheres render an enriched BCSC population [24], we characterized these mammospheres by quantifying embryonic stem cell markers, SOX2 and NANOG. Compared to the MCF-7 cells, MCF-7-derived mammospheres express significantly high levels of SOX2 and NANOG, indicating the enriched BCSC population (Figure S1). Similar to ALDH activity, the expression of embryonic stem cell markers, i.e., SOX2 and NANOG in MCF-7 cells and mammospheres, and mammosphere formation efficiency (MFE) in MCF-7 cells was also increased upon exposure to fractionated doses of radiation (Figure 1C,D). Collectively, these results suggest that exposure of fractionated doses to radiation induces the E-BCSC phenotype in mammospheres, which may contribute to radioresistance and promote tumor recurrence.

3.2. Fractionated Doses of Radiation Induce Cellular Plasticity by Regulating EMT

An increase in E-BCSC signature in our study prompted us to further analyze the EMT markers. Fractionated irradiation caused the induction of MET, as levels of the epithelial marker E-cadherin were observed to be increased and the levels of mesenchymal markers Vimentin, SLUG and SNAIL were found to be decreased significantly only in BCSC-enriched mammospheres but not in MCF-7 cells (Figure 2A,B), thus inducing plasticity toward epithelial phenotype.

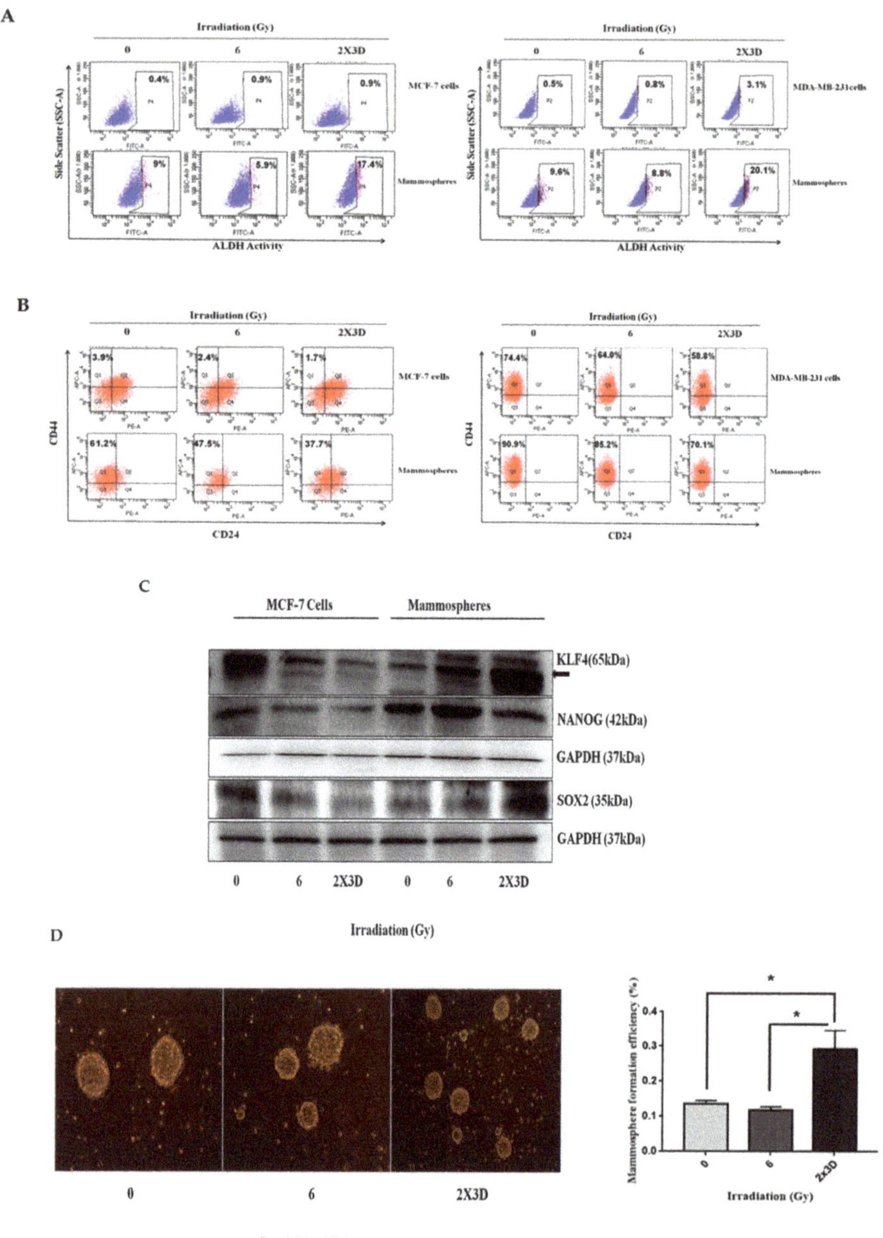

Figure 1. Effect of fractionated doses of radiation on breast cancer stem cell (BCSC) population induction and epithelial–mesenchymal transition (EMT). (**A**) BCSC population was identified in MCF-7 (left) and MDA-MB-231 cells (right) irradiated with a fractionated and acute dose of radiation by assessing ALDH activity and (**B**) CD44/CD24 markers using flow cytometry in MCF-7 (left) and MDA-MB-231 cells (right) after irradiation. (**C**) Expression of stem cell markers, i.e., NANOG and SOX2, was analyzed by Western blotting. (**D**) Phase-contrast images depict the effect of a fractionated and acute dose of radiation on sphere formation. All values are given as the mean ± SE, * $p < 0.05$, ** $p < 0.01$, *** $p < 0.001$ vs. control. All images are representative of three independent experiments.

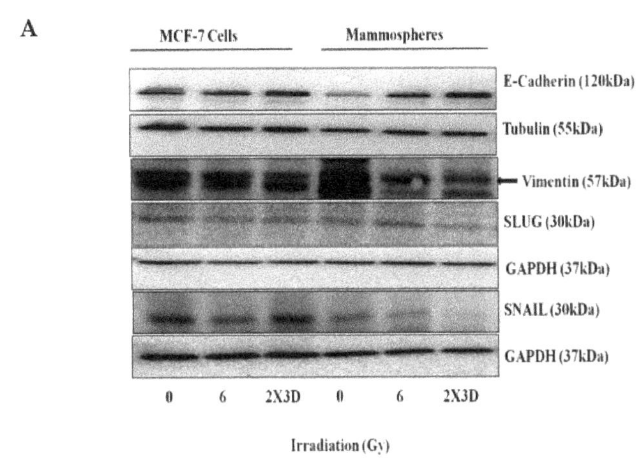

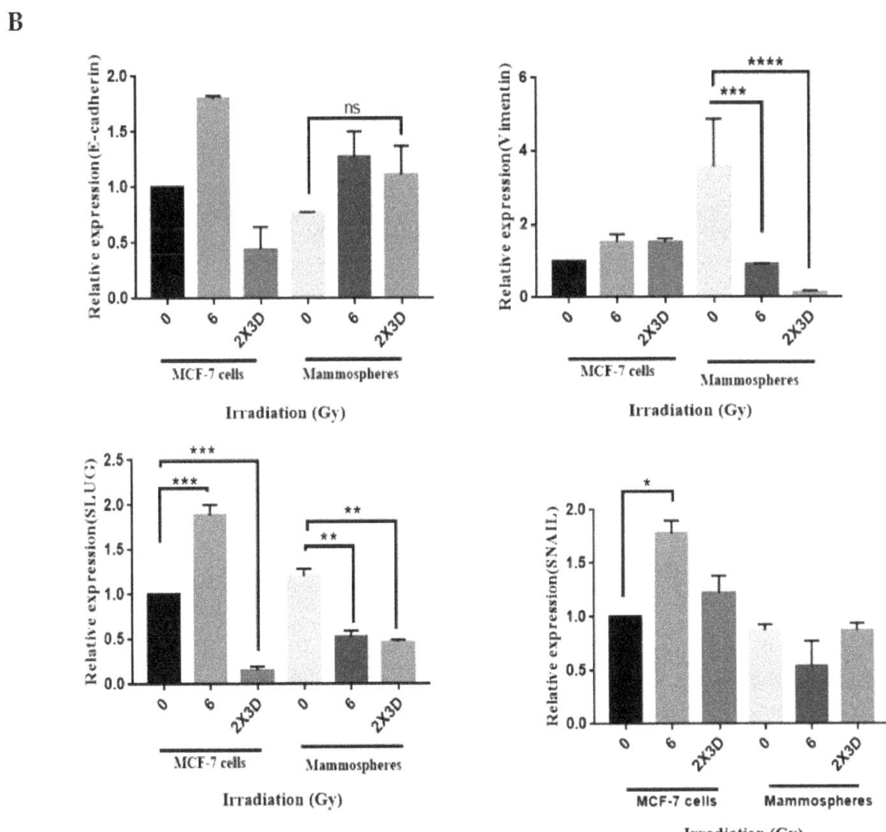

Figure 2. Effect of fractionated doses of radiation on EMT. (**A**) Expression of EMT markers, i.e., E-cadherin, Vimentin, SLUG and SNAIL, were analyzed by Western blotting and (**B**) qRT-PCR of E-cadherin, Vimentin, SLUG and SNAIL. GAPDH is used as loading control. All values are given as the mean ± SE, * $p < 0.05$, ** $p < 0.01$, *** $p < 0.001$, ****$p < 0.0001$ vs. control. All images are representative of three independent experiments.

3.3. BCSCs with High ALDH+ Activity Display Radioresistance upon Exposure to Fractionated Irradiation

Although controversial, previous findings suggest that BCSCs might be less sensitive to irradiation than cancer cells in in vitro assays [4,5]. We used a clonogenic cell survival assay to analyze the relative radioresistance of BCSCs. A single-cell suspension of MCF-7 cells was plated and irradiated with an acute dose (6 Gy) and fractionated doses (2 Gy $\times$ 3 days) of γ-rays. Our clonogenic survival assay demonstrated significantly higher radioresistance in MCF-7 and MDA-MB-231 cells and their corresponding mammospheres upon exposure to fractionated doses of radiation compared to controls (Figure 3A,B). Not only did the number of the colonies formed increase significantly after fractionated irradiation but also proliferative capacity, as indicated by Ki67 staining, was higher in these cells (Figure 3C). Ionizing radiation significantly increased the proportion of these CSCs and also showed enhanced proliferation shortly after treatment, further resulting in rapid tumor repopulation [25]. As there was an increase in the proliferation in cancer cells and mammospheres after fractionated irradiation, we further assessed apoptosis and the expression of anti- and proapoptotic genes, BCL2 and BAX. Although there was no significant change in the Annexin V^+ apoptotic population in MCF-7 cells and mammospheres after fractionated irradiation compared to their respective controls (Figure 3D), a significant increase in the BCL2/BAX ratio was observed at the protein levels, further supporting radioresistance in these cells (Figure 3E).

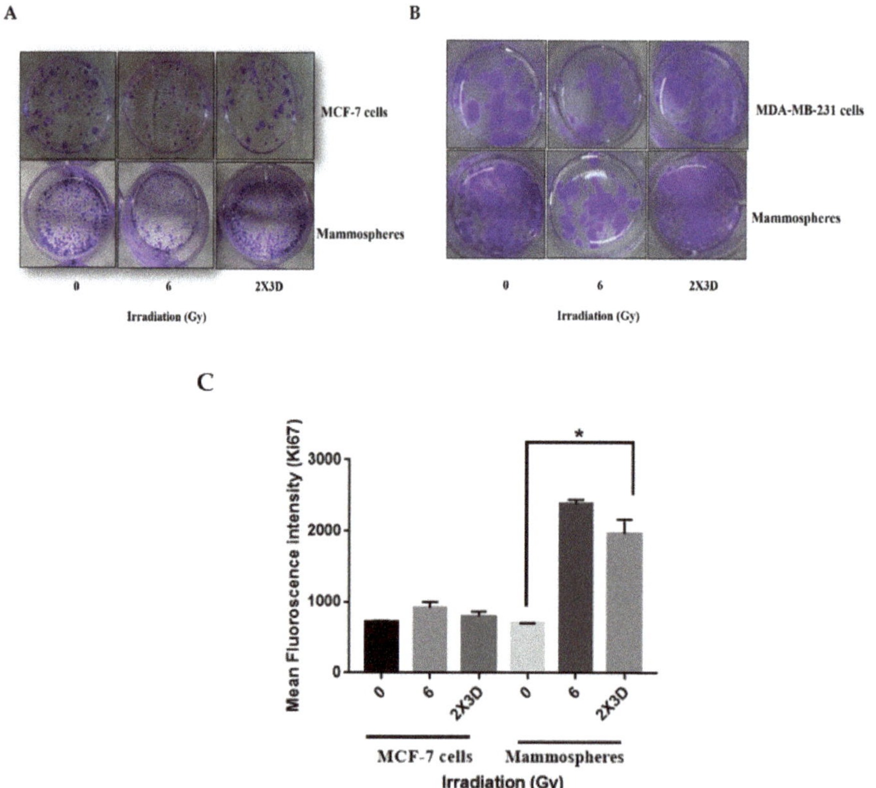

Figure 3. *Cont.*

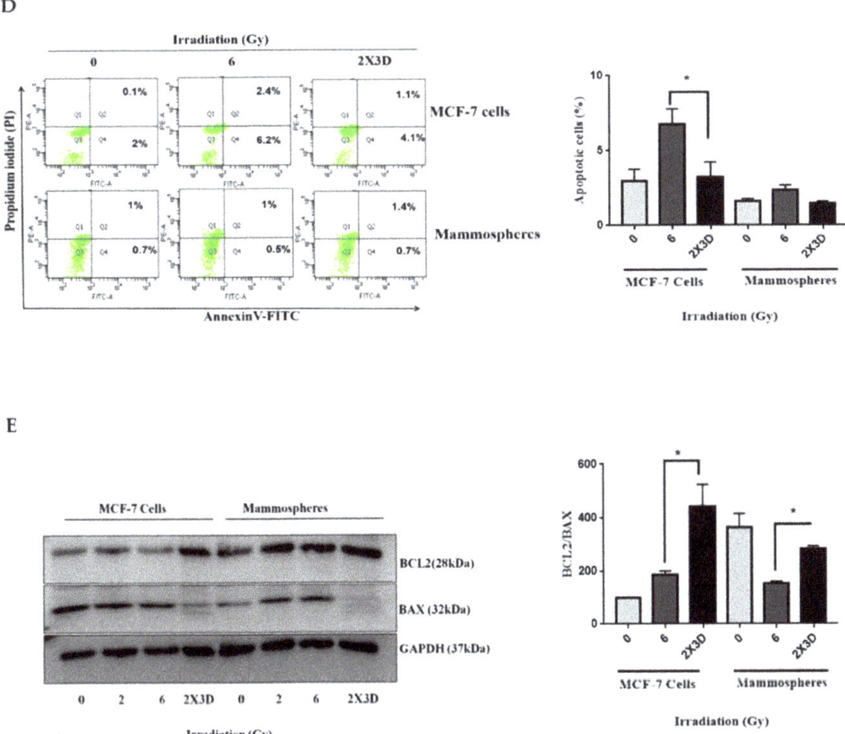

Figure 3. Fractionated doses of radiation enhance radiation resistance and reduce apoptosis in BCSCs. (**A**) Clonogenic assay was carried out for up to 14 days. The representative images show an increase in the colony formation of MCF-7 cells. (**B**) MDA-MB-231 and their corresponding mammospheres after irradiation with fractionated doses. (**C**) Cell proliferation was measured by analyzing the expression of Ki67 using flow cytometry. (**D**) The dot plots depict Annexin V-FITC and PI staining by flow cytometry. The horizontal (x) axis represents Annexin V-FITC and the vertical (y) axis represents PI staining. The bar graph represents the percentage of apoptotic cells as Annexin-V-FITC-positive cells (early apoptotic cells) and the percentage of Annexin-V-FITC- and PI-positive cells (late apoptotic cells). (**E**) BCL2 and BAX levels were analyzed by Western blotting. GAPDH was used as loading control. The representative bar graph shows the ratio of BCL2 and BAX. All values are given mean ± SE; * $p < 0.05$,; fractionated dose irradiation vs. acute irradiation.

3.4. The Emergence of Radioresistance Is Associated with High Migratory Potential and Tumorigenicity in Cancer Cells

To analyze whether breast cancer cells irradiated with fractionated doses of radiation have functional characteristics of BCSCs, we examined their cell migration potential in vitro and tumorigenic properties in vivo. Compared to the controls and an acute dose, a significant increase in migration efficiency was observed in cells irradiated with the fractionated doses of radiation in the scratch wound assay (Figure 4A). Further, tumors in mice derived from MCF-7 cells irradiated with fractionated doses of radiation weighed significantly more than tumors derived from nonirradiated or acute-dose-irradiated MCF-7 cells (Figure 4B,C). Consistent with the in vitro results, analysis of the xenograft tumors derived from tumor cells irradiated with fractionated doses also showed enhanced ALDH activity (Figure 4D). Overall, these data demonstrate that fractionated dose exposure enhances migration potential in vitro and increases tumorigenicity by elevating the $ALDH^+$ population in vivo.

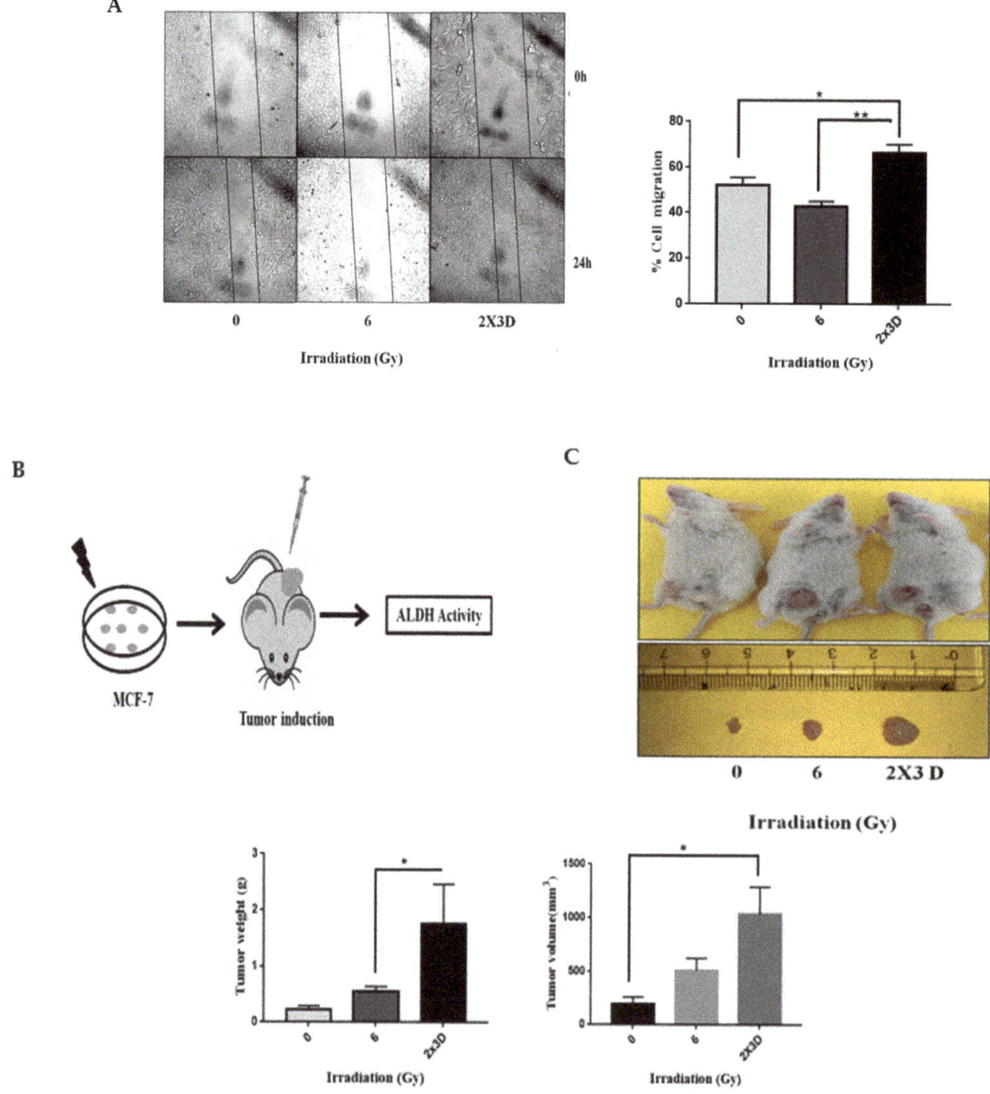

Figure 4. *Cont.*

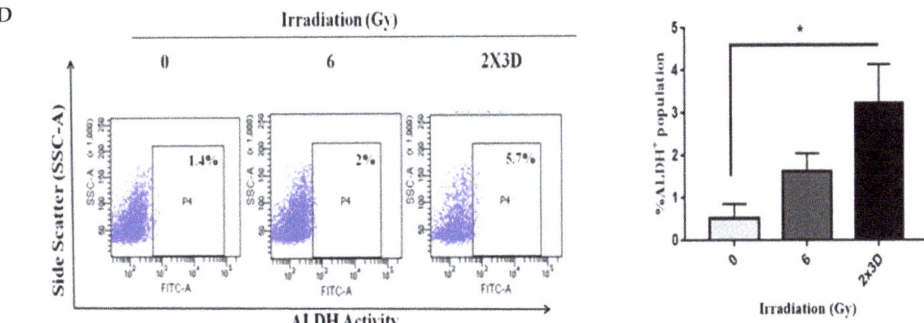

Figure 4. Fractionated doses of radiation enhance cell migration in vitro and tumor xenograft volume in vivo by increasing BCSC population. (**A**) Migration capacity was analyzed by scratch wound assay in confluent monolayers of irradiated MCF-7 cells and was expressed as % of gap closure of irradiated wells. (**B**) The flow diagram illustrates irradiated MCF-7 cells subcutaneously injected in SCID mice (n = 5). Tumors were dissected and dissociated in single cells, and ALDH activity was analyzed. (**C**) The image demonstrates isolated tumors. The bar graph represents tumor weight and volume of the xenograft tumors derived from MCF-7 control and irradiated cells. (**D**) ALDH activity was determined in isolated tumors using flow cytometry. All values are represented as mean ± SE. * $p < 0.05$; ** $p < 0.01$; vs. fractionated dose irradiation.

3.5. *Keap1-Nrf2 and not Bach1-Nrf2 Signaling Plays a Role in the Maintenance of Radioresistant ALDH+ BCSCs*

Diehn et al. [5] showed that CSCs in breast tumors contain low ROS levels and enhanced ROS defenses compared to their nontumorigenic progeny, and these differences appear to be critical for maintaining stem cell function, which could contribute to tumor radioresistance. Previous studies have shown the involvement of Nrf2 in chemoresistance in BCSCs [12,13], hence we hypothesized that Nrf2 could also play a significant role in the radioresistance of BCSCs. We first determined the levels of ROS in MCF-7 cells and mammospheres irradiated with fractionated doses of radiation. We did not see any change in the ROS levels in these cells compared to their respective controls. However, an acute dose of radiation increased the levels of ROS in MCF-7 cells as well as in mammospheres (Figure 5A). Western blot and qRT-PCR analysis revealed that Nrf2 expression (Figure 5B,C), activity (Figure 5D), as well as its targets HO1 and NQO1 (Figure 5E,F), increased significantly when treated with fractionated doses of radiation. We observed a significant decrease in the expression of Keap1, and there was no change in the expression of Bach1, MCF-7 and MDA-MB-231 cells and their corresponding mammospheres irradiated with fractionated doses (Figure 5G,H), indicating that Keap1-mediated Nrf2 degradation is impaired, leading to the stabilization of Nrf2 and its nuclear accumulation [10,12]. The reduced level of ROS in our study could therefore be attributed to the activation of the antioxidant defense mechanism.

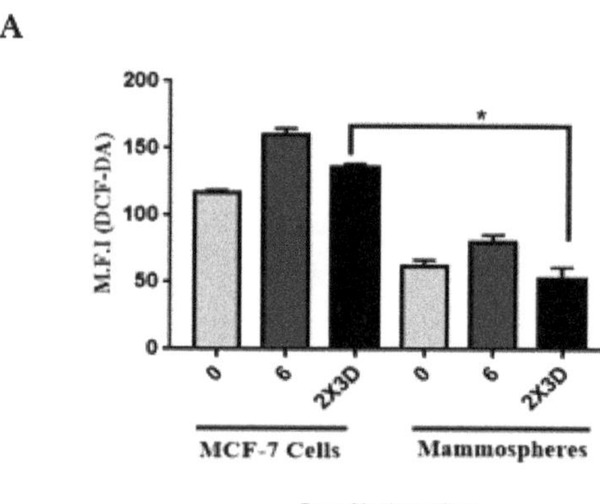

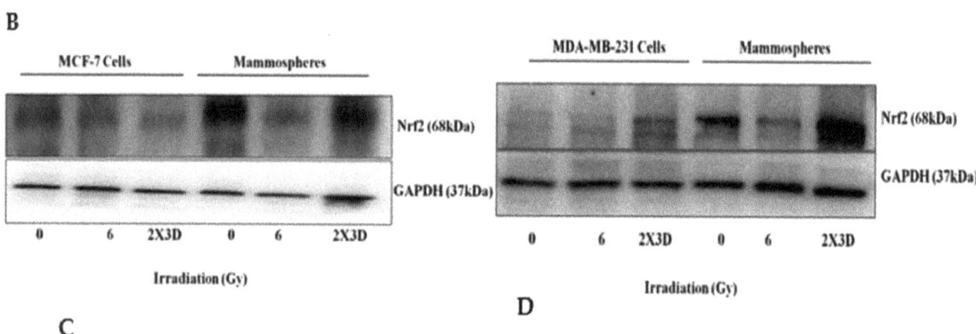

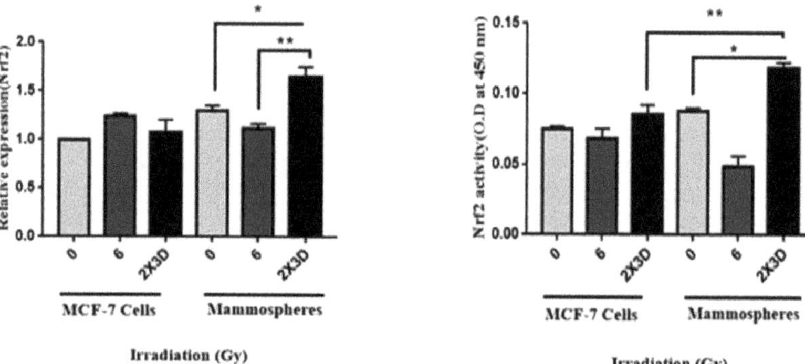

Figure 5. *Cont.*

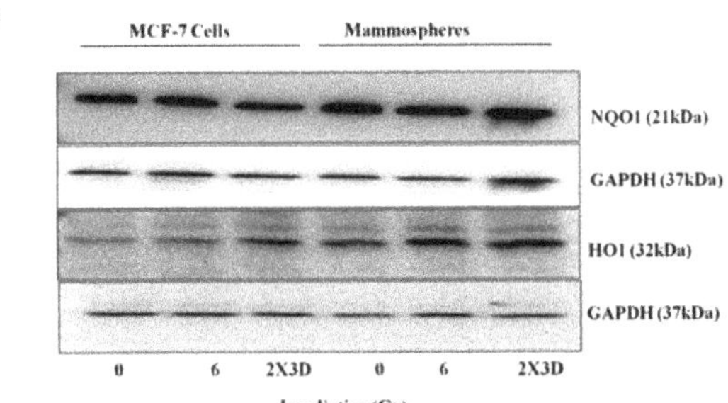

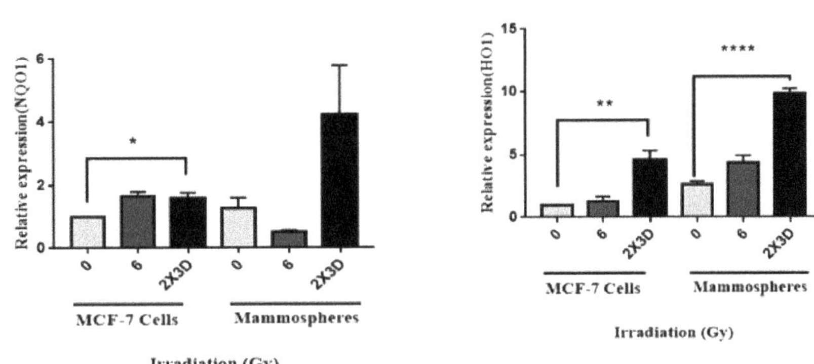

Figure 5. *Cont.*

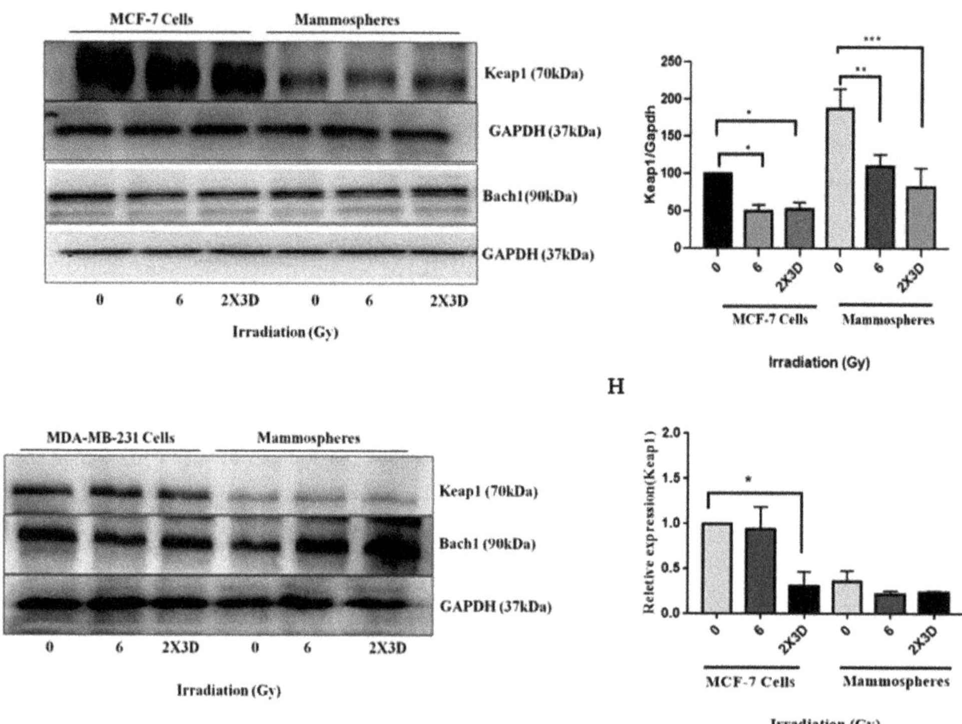

Figure 5. Fractionated doses of radiation generate low ROS and upregulate Nrf2 in BCSCs. (**A**) The bar graph represents ROS generation, assessed by DCF-DA staining using flow cytometry in MCF-7 cells and the corresponding CSC-enriched spheroids in control and irradiated cells (mean ± SE. * $p < 0.05$, fractionated-dose-irradiated MCF-7 cells vs. mammospheres). (**B**) Western blot analysis and (**C**) qRT-PCR illustrating the expression of Nrf2 in MCF-7 cell and MDA-MB-231 and their mammospheres. GAPDH served as loading control. (**D**) The bar graph represents the quantification of Nrf2 activity in irradiated MCF-7 cells and mammospheres. (**E**) The blots depict the Nrf2 targets HO1 and NQO1 by Western blotting. (**F**) The bar graph depicts the transcript levels of HO1 and NQO1 in irradiated MCF-7 cells and mammospheres by qRT-PCR. (**G**) Keap1 and Bach1 expression in irradiated MCF-7 cells (upper) and MDA-MB-231 (lower) and their mammospheres using Western blot analysis. (**H**) Transcript levels of Keap1 by qRT-PCR. GAPDH served as loading control. Mean from three independent experiments. All values are given mean ± SE. * $p < 0.05$, ** $p < 0.01$, *** $p < 0.001$; vs. fractionated dose irradiation.

3.6. *Inhibition of Nrf2 Concealed Radioresistance, Tumorigenesis and Induced Apoptosis via Reducing BCSC Population*

To further investigate the role of Nrf2 in radioresistance, Nrf2 was knocked down in MCF-7 cells (shNrf2). These cells showed a 55% reduction in Nrf2 transcripts levels (Figure S2). A 50% reduction in the population of ALDH$^+$ cells was observed in Nrf2-knockdown mammospheres and MCF-7 cells after fractionated irradiation (Figure 6A). As a phenotypic effect, stable silencing of Nrf2 also resulted in the inhibition of mammosphere formation efficiency by two-fold in MCF-7 cells (Figure 6B). A reduction in the levels of SOX2, KLF4 and NANOG in these knockdown cells after irradiation indicated the role of Nrf2 in the suppression of BCSC population (Figure 6C). Tumorigenicity in SCID mice was decreased after injection of the irradiated Nrf2 knockdown cells. A significant decrease in tumor size (Figure 6D) as well as the percentage of ALDH$^+$ population was observed in these tumors compared to the corresponding control (Figure 6E). Further, a reduction in clonogenicity (Figure 6F) and a significantly higher number of Annexin-V-/PI-positive cells were observed compared to their respective controls in shNrf2 mammospheres and MCF-7 cells irradiated with fractionated doses (Figure 6G). Thus, these results suggest that Nrf2 plays a crucial role in the acquisition of radiation resistance in BCSCs.

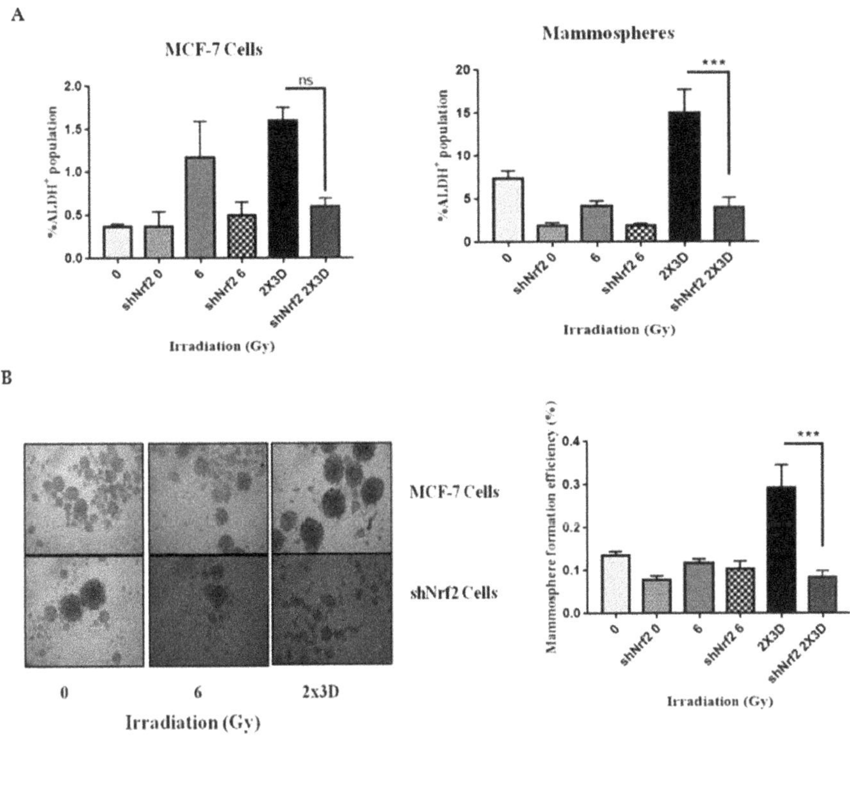

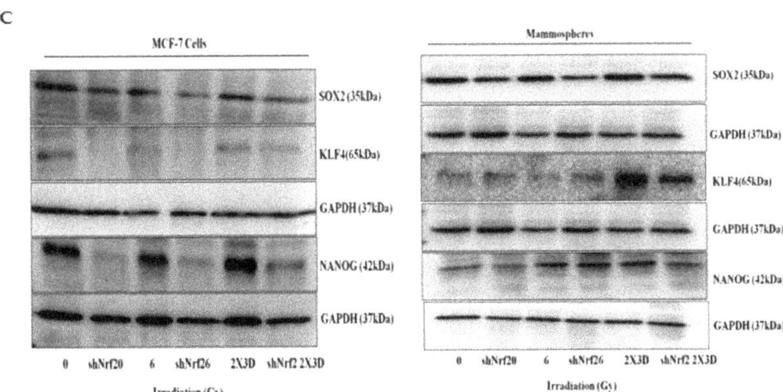

Figure 6. *Cont.*

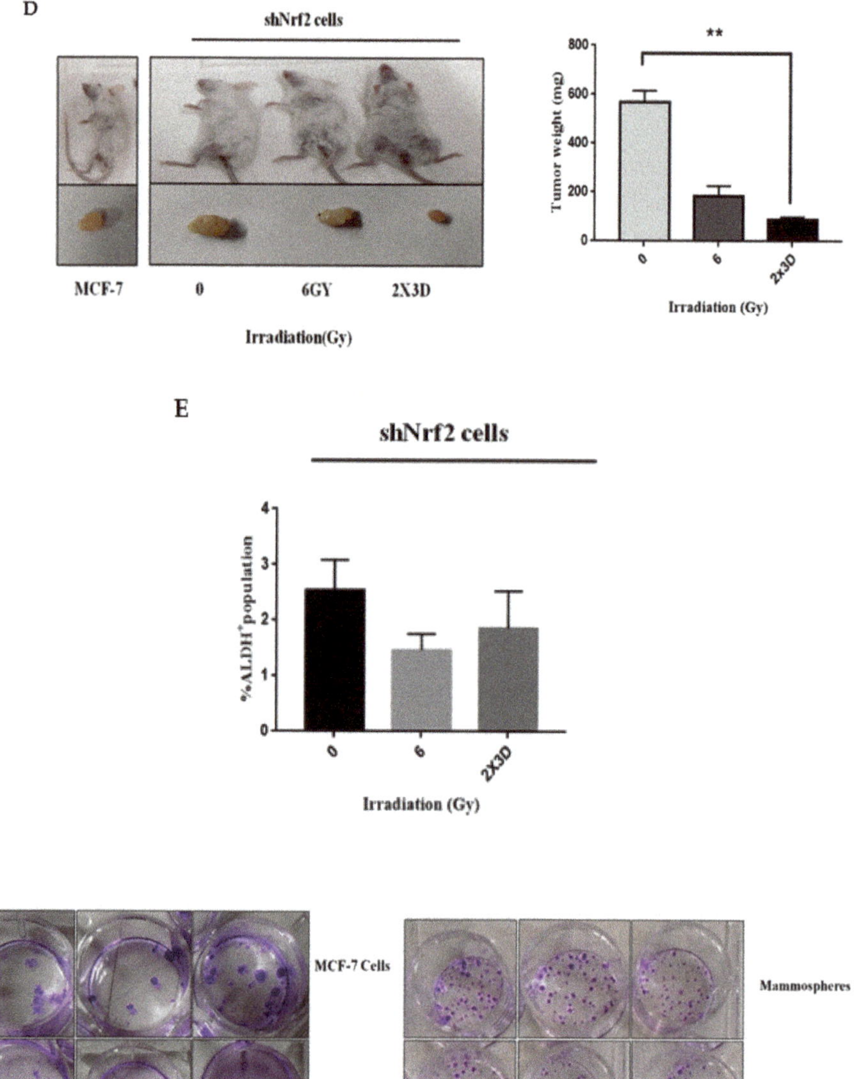

Figure 6. *Cont.*

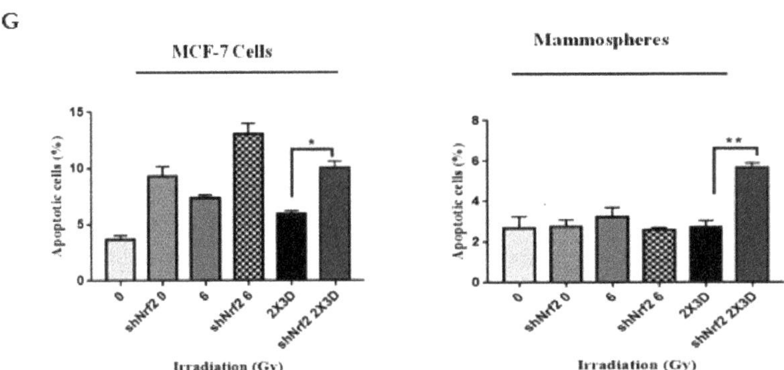

Figure 6. Inhibition of Nrf2 radiosensitizes breast cancer cells by inducing apoptosis and suppressing BCSC population after radiation treatment. (**A**) BCSC population measured by ALDH activity using flow cytometry in shNrf2 MCF-7 cells and mammospheres. (**B**) Phase-contrast images depict the effect of fractionated and acute doses of radiation on sphere formation in shNrf2 MCF-7 cells. The bar graph represents mammosphere formation efficiency for the same. (**C**) Expression of stem cell markers, i.e., SOX2, KLF4 and NANOG, was analyzed in shNrf2 MCF-7 cells and mammospheres by Western blotting, GAPDH is used as loading control. All values are given as the mean ± SE, *** $p < 0.001$ vs. fractionated-dose-irradiated shNrf2 cells. (**D**) The image demonstrates isolated tumors of the xenograft derived from shNrf2 MCF-7 control and irradiated cells. (**E**) ALDH activity was measured in shNrf2-derived tumors. (**F**) The representative images show a decrease in the colony formation of shNrf2 MCF-7 cells and mammospheres upon fractionated dose radiation treatment. (**G**) The bar graphs depict the percentage of apoptotic cells in shNrf2 MCF-7 cells and mammospheres. All values are given as the mean ± SE, * $p < 0.05$, ** $p < 0.01$; vs. fractionated dose irradiation shNrf2 cells. All images are representative of three independent experiments.

3.7. miR200a and not Promoter Methylation of Keap1 is Involved in Radioresistance of BCSC

Since we observed a significant decrease in the expression of Keap1 at mRNA and protein levels, we further investigated its regulation at the epigenetic level, especially the methylation status of the Keap1 promoter by bisulfite sequencing [26]. We did not observe any change in the methylation status of the CpGs region in the Keap1 promoter, indicating that Keap1 promoter methylation may not be the key event in Nrf2 stabilization (Figure 7A,B). We next examined the role of the miR-200 family as it targets a conserved region in the Keap1 3′-UTR [27]. We observed no change in the expression of miR-141 but a significant increase in the expression of miR-200a, 1.4-fold in mammospheres and 1.85-fold in MCF-7 cells irradiated with the fractionated dose of radiation by RT-PCR (Figure 7C,D). Collectively, these results indicate that Keap1 downregulation could be due to increased miR200a; however, more studies are required to confirm the role of miR200a in this context.

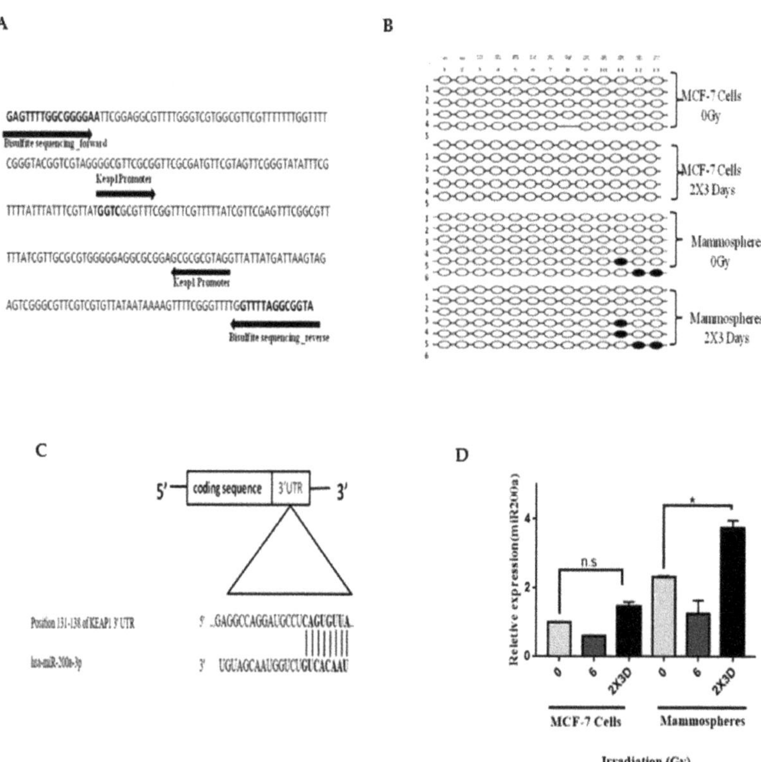

Figure 7. Promoter methylation and the role of miRNA200 in Keap1 regulation. (**A**) Primers' design for bisulfite sequencing. The original genomic sequence of the Keap1 promoter region is shown. The Keap1 promoter contains 13 CpGs sites. (**B**) Keap1 promoter methylation by Quantification Tool for Methylation Analysis (QUMA) analysis: ○, unmethylated CpGs; •, methylated CpGs. (**C**) Predicted binding sites between miR200a and Keap1 at 3′ UTR. (**D**) miR200a expression level in control and fractionated-dose-irradiated MCF-7 cells and mammospheres. All images are representative of three independent experiments. All values are given mean ± SE; * $p < 0.05$; vs. fractionated dose irradiation.

4. Discussion

Radiation can induce cancer cell death by generating ROS and DNA damage; however, it is inefficient in targeting CSCs, which are largely responsible for therapy resistance, tumorigenesis and tumor recurrence [28–30]. Our study demonstrates that fractionated doses of radiation enhanced the E-BCSC marker ALDH$^+$ and transcription factors of embryonic stem cells in BCSC-enriched mammospheres, indicating the E-BCSC phenotype, which is proliferative in nature. BCSC plasticity plays a crucial role in therapy resistance. BCSCs exhibit plasticity, which transitions between quiescent mesenchymal- (M-BCSCs) and proliferative epithelial-like (E-BCSCs) states [31]. An increase in E-BCSCs such as ALDH+ population and E-cadherin, indicative of MET, and a decrease in M-BCSCs such as CD44$^+$/24$^-$ population, the mesenchymal markers Vimentin, SNAIL and SLUG, demonstrated that fractionated doses of radiation increase the epithelial type of BCSCs [24,31,32]. Thus, these results support the notion that BCSC markers are not restricted to a particular population but change according to their plasticity based on the therapy. Hence, plasticity from M- BCSCs to E-BCSCs contributes to radioresistance. Since NANOG, SOX2 and KLF4 are essential for converting tumor cells into aggressive stem-like cells, an increase in the expression of these markers in our study after irradiation further supports the increased cancer stem cell population.

Emerging evidence indicates that Nrf2 plays a crucial role in CSC survival and resistance [33]. It is shown to be involved in chemotherapeutic drug resistance due to enhanced antioxidant capacity and detoxification of anticancer agents [14,34,35]. However, the involvement of the Nrf2-Keap1 axis in radioresistance of BCSCs is poorly understood. A strong association between low levels of ROS and enhanced antioxidant defense in BCSC radioresistance reported by Diehn et al. [5] prompted us to further investigate the role of Nrf2. Enhanced expression of Nrf2 and its downstream genes HO1 and NQO1 after irradiation in breast cancer cells and their corresponding mammospheres ascertains the involvement of Nrf2 in radioresistance. A recent report has shown that Nrf2 enhances $ALDH^+$ E-BCSCs [24]. This supports our results, as we have observed a decrease in the $ALDH^+$ E-BCSCs after Nrf2 inhibition. A decrease in embryonic stem cell markers, colony and sphere formation ability and reduced tumorigenicity after Nrf2 knockdown further indicate that Nrf2 is involved in the reprogramming process, and Nrf2 signaling is an important target for radiation resistance of BCSCs.

In the current study, Nrf2 appears to be regulated by Keap1 as we observed a decrease in the Keap1 levels with no change in the expression of either GSK-3β (Figure S3) [36] or Bach1. Additionally, as Bach1 binds to HO1 [10,11], an increase in the levels of HO1 in our study further confirms that Bach1 does not play a role in the regulation of Nrf2. Loss of Keap1 function is shown to mediate Nrf2 stabilization and is often associated with reduced drug sensitivity in several cancers [37–39]. A reduction in Keap1 expression with a concomitant increase in the expression of Nrf2 and its downstream targets HO1 and NQO1 clearly demonstrates the role played by Keap1 in Nrf2 regulation in the facilitation of acquired radioresistance. Hence, we tried to understand the mechanism of Keap1 regulation in this study.

Besides mutations through cysteine residues, epigenetic mechanisms, particularly the promoter hypermethylation [26], and miRNAs are the main regulators of Keap1. We did not see any change in the promoter methylation status of Keap1 after irradiation, which suggested that irradiation may regulate Keap1 post-transcriptionally rather than epigenetically. Hence, we further studied the role of the miR200 family as it is known to be involved in the regulation of Keap1. A significant increase in the transcript levels of miR200a indicates its role in the regulation of Keap1 in the radioresistance of BCSCs. Furthermore, reports from other studies have shown that miR200a suppresses the expression of transcriptional factors ZEB1/2 and inhibits the transition from the epithelial-to-mesenchymal phenotype [40]. This further strengthens and supports our studies where miR200a could be responsible for the inhibition of Keap1 as well as EMT in BCSC-enriched mammospheres.

In conclusion, the current study provides interesting insights into the mechanism by which fractionated doses of radiation increases radioresistance in the BCSC population. Our results indicate the enrichment of the E-BCSC phenotype. The regulation of Nrf2 in irradiated conditions occurs via the downregulation of Keap1 and not by GSK3β or Bach1. We provide mechanistic insight into the regulation of Keap1, possibly via post-transcriptional modification through miR200a and not via promoter methylation. Although the current study is limited to only the higher expression of miR200a, and given its potential for therapeutic purposes, additional mechanistic studies regarding its role in Keap1 inhibition and thus radioresistance is highly warranted. Nevertheless, alteration in the Nrf2-Keap1 pathway establishes relationships between radioresistance and BCSCs.

Supplementary Materials: The following are available online at https://www.mdpi.com/2073-4409/10/1/83/s1, Figure S1: Characterization of mammospheres. Figure S2: Nrf2 expression in Nrf2 knockdown cells. Figure S3: p-GSK3β levels in fractionated-dose-irradiated MCF-7 cells and mammospheres. Table S1: List of primers.

Author Contributions: D.K.: conceptualization, methodology, resources, formal analysis, validation and original draft preparation; M.M.: contribution in shNrf2 cell line generation; R.D.: contribution in promoter methylation studies; S.S.: investigation, conceptualization, formal analysis, visualization, validation, writing—original draft preparation and writing, review and editing, supervision, funding acquisition. All authors have read and agreed to the published version of the manuscript.

Funding: This work was supported by grants from: BRNS, D.A.E. (Department of Atomic Energy). M.M. and R.D. are SRF of Council of Scientific and Industrial Research (CSIR), Govt. of India.

Institutional Review Board Statement: All animal experiments were carried out in accordance with the procedures and guidelines of the Institutional Animal Ethics Committee (NCCS) for animal experiments and approved by the Institute Animal Ethical Committee (IAEC) for Animal Experiments at the National Centre for Cell Science, S.P Pune University, Pune, India (IAEC/2016/B-264).

Informed Consent Statement: Not applicable.

Data Availability Statement: The data presented in this study are available on request from the corresponding author.

Acknowledgments: We sincerely acknowledge B.S Patro, Bio-Organic Division, BARC, Mumbai, India for his technical support. We wish to thank Sarojini Singh for proof reading the manuscript. We also acknowledged Director (NCCS) for providing all institutional facilities.

Conflicts of Interest: The authors declare no conflict of interest.

References

1. Darby, S.C.; McGale, P.; Correa, C.R.; Taylor, C.A.; Arriagada, R.; Clarke, M.; Cutter, D.; Davies, C.; Ewertz, M.; Godwin, J.; et al. Effect of radiotherapy after breast-conserving surgery on 10-year recurrence and 15-year breast cancer death: Meta-analysis of individual patient data for 10 801 women in 17 randomised trials. *Lancet* **2011**, *378*, 1707–1716. [CrossRef]
2. Speers, C.; Zhao, S.; Liu, M.; Bartelink, H.; Pierce, L.J.; Feng, F.Y. Development and Validation of a Novel Radiosensitivity Signature in Human Breast Cancer. *Clin. Cancer Res.* **2015**, *21*, 3667–3677. [CrossRef]
3. Duru, N.; Fan, M.; Candas, D.; Menaa, C.; Liu, H.-C.; Nantajit, D.; Wen, Y.; Xiao, K.; Eldridge, A.; Chromy, B.A.; et al. HER2-Associated Radioresistance of Breast Cancer Stem Cells Isolated from HER2-Negative Breast Cancer Cells. *Clin. Cancer Res.* **2012**, *18*, 6634–6647. [CrossRef] [PubMed]
4. Phillips, T.M.; McBride, W.H.; Pajonk, F. The Response of CD24−/low/CD44+ Breast Cancer–Initiating Cells to Radiation. *J. Natl. Cancer Inst.* **2006**, *98*, 1777–1785. [CrossRef] [PubMed]
5. Diehn, M.; Cho, R.W.; Lobo, N.A.; Kalisky, T.; Dorie, M.J.; Kulp, A.N.; Qian, D.; Lam, J.S.; Ailles, L.E.; Wong, M.; et al. Association of reactive oxygen species levels and radioresistance in cancer stem cells. *Nature* **2009**, *458*, 780–783. [CrossRef] [PubMed]
6. Mizuno, T.; Suzuki, N.; Makino, H.; Furui, T.; Morii, E.; Aoki, H.; Kunisada, T.; Yano, M.; Kuji, S.; Hirashima, Y.; et al. Cancer stem-like cells of ovarian clear cell carcinoma are enriched in the ALDH-high population associated with an accelerated scavenging system in reactive oxygen species. *Gynecol. Oncol.* **2015**, *137*, 299–305. [CrossRef] [PubMed]
7. Wu, T.; Harder, B.G.; Wong, P.K.; Lang, J.E.; Zhang, D.D. Oxidative stress, mammospheres and Nrf2-new implication for breast cancer therapy? *Mol. Carcinog.* **2014**, *54*, 1494–1502. [CrossRef] [PubMed]
8. Cuadrado, A.; Manda, G.; Hassan, A.; Alcaraz, M.J.; Barbas, C.; Daiber, A.; Ghezzi, P.; León, R.; López, M.G.; Oliva, B.; et al. Transcription Factor NRF2 as a Therapeutic Target for Chronic Diseases: A Systems Medicine Approach. *Pharmacol. Rev.* **2018**, *70*, 348–383. [CrossRef]
9. Dhakshinamoorthy, S.; Jain, A.K.; Bloom, D.A.; Jaiswal, A.K. Bach1 Competes with Nrf2 Leading to Negative Regulation of the Antioxidant Response Element (ARE)-mediated NAD(P)H: Quinone Oxidoreductase 1 Gene Expression and Induction in Response to Antioxidants. *J. Biol. Chem.* **2005**, *280*, 16891–16900. [CrossRef]
10. Reichard, J.F.; Motz, G.T.; Puga, A. Heme oxygenase-1 induction by NRF2 requires inactivation of the transcriptional repressor BACH1. *Nucleic Acids Res.* **2007**, *35*, 7074–7086. [CrossRef]
11. Hintze, K.J.; Katoh, Y.; Igarashi, K.; Theil, E.C. Bach1 Repression of Ferritin and Thioredoxin Reductase1 Is Heme-sensitive in Cells and in Vitro and Coordinates Expression with Heme Oxygenase1, beta-Globin, and NADP(H) Quinone (Oxido) Reductase1. *J. Biol. Chem.* **2007**, *282*, 34365–34371. [CrossRef] [PubMed]
12. Ryoo, I.; Choi, B.; Kwak, M. Activation of NRF2 by p62 and proteasome reduction in sphere-forming breast carcinoma cells. *Oncotarget* **2015**, *6*. [CrossRef] [PubMed]
13. Achuthan, S.; Santhoshkumar, T.R.; Prabhakar, J.; Nair, S.A.; Pillai, M.R. Drug-induced Senescence Generates Chemoresistant Stemlike Cells with Low Reactive Oxygen Species. *J. Biol. Chem.* **2011**, *286*, 37813–37829. [CrossRef] [PubMed]
14. Zhou, S.; Ye, W.; Shao, Q.; Zhang, M.; Liang, J. Nrf2 is a potential therapeutic target in radioresistance in human cancer. *Crit. Rev. Oncol.* **2013**, *88*, 706–715. [CrossRef] [PubMed]
15. Arkat, S.; Umbarkar, P.; Singh, S.; Sitasawad, S.L. Mitochondrial Peroxiredoxin-3 protects against hyperglycemia induced myocardial damage in Diabetic cardiomyopathy. *Free Radic. Biol. Med.* **2016**, *97*, 489–500. [CrossRef]
16. Varkonyi-Gasic, E.; Wu, R.; Wood, M.; Walton, E.F.; Hellens, R.P. Protocol: A highly sensitive RT-PCR method for detection and quantification of microRNAs. *Plant Methods* **2007**, *3*, 12. [CrossRef]
17. Dasgupta, A.; Sawant, M.A.; Kavishwar, G.; Lavhale, M.; Sitasawad, S. AECHL-1 targets breast cancer progression via inhibition of metastasis, prevention of EMT and suppression of Cancer Stem Cell characteristics. *Sci. Rep.* **2016**, *6*, 38045. [CrossRef]
18. Kumar, S.; Kain, V.; Sitasawad, S.L. Cardiotoxicity of calmidazolium chloride is attributed to calcium aggravation, oxidative and nitrosative stress, and apoptosis. *Free Radic. Biol. Med.* **2009**, *47*, 699–709. [CrossRef]

19. Luo, M.; Shang, L.; Brooks, M.D.; Jiagge, E.; Zhu, Y.; Buschhaus, J.M.; Conley, S.; Fath, M.A.; Davis, A.; Gheordunescu, E.; et al. Targeting Breast Cancer Stem Cell State Equilibrium through Modulation of Redox Signaling. *Cell Metab.* **2018**, *28*, 69–86. [CrossRef]
20. Shao, J.; Fan, W.; Ma, B.; Wu, Y. Breast cancer stem cells expressing different stem cell markers exhibit distinct biological characteristics. *Mol. Med. Rep.* **2016**, *14*, 4991–4998. [CrossRef]
21. Yang, Y.; Ma, H.; Zhang, J.; Zhu, L.; Wang, C.; Yang, Y. Unraveling the roles of CD44/CD24 and ALDH1 as cancer stem cell markers in tumorigenesis and metastasis. *Sci. Rep.* **2017**, *7*, 1–15. [CrossRef]
22. Chekhun, V.F.; Lukianova, N.Y.; Chekhun, S.V.; Bezdieniezhnykh, N.O.; Zadvorniy, T.V.; Borikun, T.V.; Polishchuk, L.Z.; Klyusov, O.M. Association of CD44+CD24−/Low with Markers of Aggressiveness and Plasticity of Cell Lines and Tumors of Patients with Breast Cancer. *Exp. Oncol.* **2017**, *39*, 203–211. [CrossRef]
23. Paula, A.D.C.; Lopes, C. Implications of Different Cancer Stem Cell Phenotypes in Breast Cancer. *Anticancer Res.* **2017**, *37*, 2173–2183. [CrossRef] [PubMed]
24. Xie, G.; Zhan, J.; Tian, Y.-H.; Liu, Y.; Chen, Z.; Ren, C.; Sun, Q.; Lian, J.; Chen, L.; Ruan, J.; et al. Mammosphere cells from high-passage MCF7 cell line show variable loss of tumorigenicity and radioresistance. *Cancer Lett.* **2012**, *316*, 53–61. [CrossRef] [PubMed]
25. Zielske, S.P.; Spalding, A.C.; Wicha, M.S.; Lawrence, T.S. Ablation of Breast Cancer Stem Cells with Radiation. *Transl. Oncol.* **2011**, *4*, 227–233. [CrossRef] [PubMed]
26. Muscarella, L.A.; Barbano, R.; D'Angelo, V.; Copetti, M.; Coco, M.; Balsamo, T.; La Torre, A.; Notarangelo, A.; Troiano, M.; Parisi, S.; et al. Regulation ofKEAP1expression by promoter methylation in malignant gliomas and association with patient's outcome. *Epigenetics* **2011**, *6*, 317–325. [CrossRef]
27. Sun, X.; Zuo, H.; Liu, C.; Yang, Y. Overexpression of miR-200a protects cardiomyocytes against hypoxia-induced apoptosis by modulating the kelch-like ECH-associated protein 1-nuclear factor erythroid 2-related factor 2 signaling axis. *Int. J. Mol. Med.* **2016**, *38*, 1303–1311. [CrossRef]
28. Sia, J.; Szmyd, R.; Hau, E.; Gee, H.E. Molecular Mechanisms of Radiation-Induced Cancer Cell Death: A Primer. *Front. Cell Dev. Biol.* **2020**, *8*, 41. [CrossRef]
29. Sato, K.; Shimokawa, T.; Imai, T. Difference in Acquired Radioresistance Induction Between Repeated Photon and Particle Irradiation. *Front. Oncol.* **2019**, *9*, 1–12. [CrossRef]
30. Pajonk, F.; Vlashi, E.; McBride, W.H. Radiation Resistance of Cancer Stem Cells: The 4 R's of Radiobiology Revisited. *Stem Cells* **2010**, *28*, 639–648. [CrossRef]
31. Scheel, C.; Weinberg, R.A. Cancer stem cells and epithelial–mesenchymal transition: Concepts and molecular links. *Semin. Cancer Biol.* **2012**, *22*, 396–403. [CrossRef] [PubMed]
32. Beerling, E.; Seinstra, D.; De Wit, E.; Kester, L.; Van Der Velden, D.; Maynard, C.; Schäfer, R.; Van Diest, P.; Voest, E.; Van Oudenaarden, A.; et al. Plasticity between Epithelial and Mesenchymal States Unlinks EMT from Metastasis-Enhancing Stem Cell Capacity. *Cell Rep.* **2016**, *14*, 2281–2288. [CrossRef] [PubMed]
33. Ryoo, I.-G.; Choi, B.-H.; Ku, S.-K.; Kwak, M.-K. High CD44 expression mediates p62-associated NFE2L2/NRF2 activation in breast cancer stem cell-like cells: Implications for cancer stem cell resistance. *Redox Biol.* **2018**, *17*, 246–258. [CrossRef] [PubMed]
34. Harder, B.; Jiang, T.; Wu, T.; Tao, S.; De La Vega, M.R.; Tian, W.; Chapman, E.; Zhang, D.D. Molecular mechanisms of Nrf2 regulation and how these influence chemical modulation for disease intervention. *Biochem. Soc. Trans.* **2015**, *43*, 680–686. [CrossRef]
35. Ueda, S.; Takanashi, M.; Sudo, K.; Kanekura, K.; Kuroda, M. miR-27a ameliorates chemoresistance of breast cancer cells by disruption of reactive oxygen species homeostasis and impairment of autophagy. *Lab. Investig.* **2020**, *100*, 863–873. [CrossRef]
36. Culbreth, M.; Aschner, M. GSK-3β, a double-edged sword in Nrf2 regulation: Implications for neurological dysfunction and disease. *F1000Research* **2018**, *7*, 1043. [CrossRef]
37. Huang, H.; Wu, Y.; Fu, W.; Wang, X.; Zhou, L.; Xu, X.; Huang, F.; Wu, Y. Downregulation of Keap1 contributes to poor prognosis and Axitinib resistance of renal cell carcinoma via upregulation of Nrf2 expression. *Int. J. Mol. Med.* **2019**, *43*, 2044–2054. [CrossRef]
38. Zhang, P.; Singh, A.; Yegnasubramanian, S.; Esopi, D.; Kombairaju, P.; Bodas, M.; Wu, H.; Bova, S.G.; Biswal, S. Loss of Kelch-Like ECH-Associated Protein 1 Function in Prostate Cancer Cells Causes Chemoresistance and Radioresistance and Promotes Tumor Growth. *Mol. Cancer Ther.* **2010**, *9*, 336–346. [CrossRef]
39. Tian, Y.; Wu, K.; Liu, Q.; Han, N.; Zhang, L.; Chu, Q.; Chen, Y. Modification of platinum sensitivity by KEAP1/NRF2 signals in non-small cell lung cancer. *J. Hematol. Oncol.* **2016**, *9*, 1–14. [CrossRef]
40. Wu, H.-T.; Zhong, H.-T.; Li, G.-W.; Shen, J.-X.; Ye, Q.-Q.; Zhang, M.-L.; Liu, J. Oncogenic functions of the EMT-related transcription factor ZEB1 in breast cancer. *J. Transl. Med.* **2020**, *18*, 1–10. [CrossRef]

MDPI

Review

Mesenchymal Stem Cell-Derived Exosomes: Biological Function and Their Therapeutic Potential in Radiation Damage

Xiaoyu Pu [1,†], Siyang Ma [1,†], Yan Gao [1], Tiankai Xu [1], Pengyu Chang [1,2,*] and Lihua Dong [1,3,*]

1 Jilin Provincial Key Laboratory of Radiation Oncology & Therapy, Department of Radiation Oncology & Therapy, The First Bethune Hospital of Jilin University, Changchun 130021, China; puxy19@mails.jlu.edu.cn (X.P.); masy19@mails.jlu.edu.cn (S.M.); gao_yan@jlu.edu.cn (Y.G.); xutk16@mails.jlu.edu.cn (T.X.)
2 Key Laboratory of Organ Regeneration & Transplantation of the Ministry of Education, Department of Radiation Oncology & Therapy, The First Bethune Hospital of Jilin University, Changchun 130021, China
3 National Health Commission Key Laboratory of Radiobiology, School of Public Health, Jilin University, Changchun 130021, China
* Correspondence: changpengyu@jlu.edu.cn (P.C.); dlh@jlu.edu.cn (L.D.); Tel.: +86-431-8878-3840 (P.C. & L.D.)
† These authors contributed equally to this work.

Abstract: Radiation-induced damage is a common occurrence in cancer patients who undergo radiotherapy. In this setting, radiation-induced damage can be refractory because the regeneration responses of injured tissues or organs are not well stimulated. Mesenchymal stem cells have become ideal candidates for managing radiation-induced damage. Moreover, accumulating evidence suggests that exosomes derived from mesenchymal stem cells have a similar effect on repairing tissue damage mainly because these exosomes carry various bioactive substances, such as miRNAs, proteins and lipids, which can affect immunomodulation, angiogenesis, and cell survival and proliferation. Although the mechanisms by which mesenchymal stem cell-derived exosomes repair radiation damage have not been fully elucidated, we intend to translate their biological features into a radiation damage model and aim to provide new insight into the management of radiation damage.

Keywords: mesenchymal stem cell; exosome; radiation damage

Citation: Pu, X.; Ma, S.; Gao, Y.; Xu, T.; Chang, P.; Dong, L. Mesenchymal Stem Cell-Derived Exosomes: Biological Function and Their Therapeutic Potential in Radiation Damage. *Cells* **2021**, *10*, 42. https://doi.org/10.3390/cells10010042

Received: 31 October 2020
Accepted: 24 December 2020
Published: 30 December 2020

Publisher's Note: MDPI stays neutral with regard to jurisdictional claims in published maps and institutional affiliations.

1. Introduction

Mesenchymal stem cells (MSCs) are multipotent stem cells that can be isolated from human tissues or organs, such as the bone marrow, adipose tissue, umbilical cord, lung, spleen, liver or kidney [1]. Despite being derived from multiple sources, MSCs display similar biological phenotypes and functions [2,3]. Because of their autocrine and paracrine actions, MSCs have been shown to possess potency in repairing tissue damage [4]. Critically, delivery of only a small population of MSCs can result in accelerated damage repair in the host [5–7]. In addition, exosomes are crucial components that account for the paracrine action of MSCs [8–10]. For example, they exchange genetic material across cells by transferring bioactive molecules [11]. Similar to other cellular exosomes, MSC-exosomes are extracellular vesicles with a lipid bilayer structure and an average diameter of 100 nm [1,12]. They carry bioactive molecules, including miRNAs, lncRNAs, lipids and cytokines [1], thus providing a context for researching the biological functions of MSC-exosomes.

Treating diseases with MSC-exosomes has shown promise in the field of regenerative medicine, and numerous studies exploring the therapeutic effects of MSC-exosomes on neurological, immunological and cardiovascular diseases have been published [13]. In summary, the benefits of delivering MSC-exosomes in disease models mainly include the attenuation of inflammation, promotion of angiogenesis and improvement in the survival and proliferation of stem or progenitor cells within injured tissues or organs [14]. In fact, such benefits can be achieved with MSCs as well. Although it has also been shown that MSCs can exert therapeutic effects on radiation damage, the therapeutic potential of MSC-exosomes has not been widely explored in this field. Nevertheless, in a previous study, irradiated cells exhibited enhanced uptake of exosomes because of an increase in the formation of the integrin and tetraspanin complex CD29/CD81 on the cell surface [15], thus indicating the specific role of exosomes in mediating biological processes in injured cells. Moreover, MSC-exosomes were found to protect against acute or chronic radiation damage via their miRNA cargo, suggesting that irradiated cells might utilize MSC-exosomes to increase their resistance to ionizing irradiation [16–18]. For example, a study showed that exosomal miRNA-210 could elicit efficient DNA damage repair by controlling the transcriptional activity of HIF-1, thus enhancing cellular radio-resistance [17,19]. In this review, we explore the pro-regenerative properties of MSC-exosomes in the field of radiation damage and aim to provide new insight into the management of radiation damage by using MSC-exosomes.

2. Biological Features of MSC-Exosomes

MSCs are crucial sources of exosomes in humans. Consistent with other cell-derived exosomes, MSC-exosomes are generated through a sequential process including the invagination of lysosomal microparticles and fusion and excretion from parental cells [20]. Lysosomal microparticles first invaginate their membranes to generate endosomes, which then fuse with each other to form multivesicular bodies that contain intraluminal vesicles. Next, the outer membrane of the mature multivesicular body fuses with the plasma membrane of a cell and is ultimately transported out, constituting an exosome [20].

Exosomes consist of lipid bilayer membrane structures with diameters ranging from 40 nm to 160 nm (an average of 100 nm) [12]. They express various markers, including CD9, CD81, CD63, TSG101, flotillin, ceramide, and Alix [12], and have a density of 1.15–1.19 g/mL in sucrose gradients [21]. MSC-exosomes contain at least 170 different miRNAs [22] and 304 proteins [23], along with an indefinite number of DNAs, mRNAs and metabolites [12]. Because they contain a large number of bioactive molecules, MSC-exosomes have attracted great interest in the field of regenerative medicine. Accordingly, numerous studies have attempted to assess whether the infusion of MSC-exosomes can serve as an alternative strategy to repair tissue damage, and emerging results have mostly revealed that MSC-exosomes have therapeutic effects similar to those of their parental MSCs [24]. Moreover, MSC-exosomes have several advantages over MSCs. (i) MSC-exosomes are long-lasting and can be stored at −80 °C without affecting their biological functions [17], whereas cryopreserved MSCs exhibit impaired immunoregulatory and pro-regenerative properties compared with fresh MSCs [25]. (ii) The membranes of MSC-exosomes are enriched in sphingomyelin, cholesterol, ceramide and lipid raft proteins, enabling MSC-exosomes to spread in vivo regardless of biological barriers, such as the blood-brain barrier [26], for example, even when they are delivered via an intravenous injection, MSC-exosomes can be detected in injured neurons in the brain [27]. (iii) Infusion of MSC-exosomes elicits minimal immune rejection due to their complete lack of expression of major histocompatibility complex (MHC) molecules [28,29], which prevents their rapid clearance by host immune cells. For instance, MSC-exosomes were found to remain in a recipient for a significantly longer time than MSCs after infusion [28,30], indicating that they can perform their biological functions in vivo for a relatively long time. (iv) Infusion of MSC-exosomes can avoid several stem cell-associated challenges, such as the risk of spontaneous tumorigenesis induced by MSCs [31,32]. (v) The potential secretion of exosomes by MSCs can be impacted by various factors. For example, maintaining MSCs in a physiological state in an in vitro culture system can impact their production of exosomes with a specific phenotype in terms of biological activity [33]. Notably, although incubating MSCs with an IFN-γ plus TNF-α mixture in vitro reduced their proliferation, the production of exosomes was not adversely affected [28,34]. Moreover, this process improved the immunosuppressive function of the MSC-exosomes. This prompts speculation that exosomes with high bioactivity can be purposefully obtained by preconditioning

MSCs in vitro prior to injection to treat inflammatory diseases. Therefore, determining the components of MSC-exosomes that are able to produce high therapeutic efficacy is particularly critical.

The miRNA and protein cargo contained in MSC-exosomes are effective in promoting damage repair. Moreover, they jointly regulate the regenerative process in damaged tissue. In a colitis model, MSC-exosomes were revealed to reduce macrophage-induced inflammation by transporting metallothionein-2, an upstream protein that blocks activation of the NF-κB pathway [28]. However, this anti-inflammatory effect of MSC-exosomes was not completely lost even when blocking metallothionein-2 in vivo and in vitro [28], demonstrating that other components in MSC-exosomes also exert bioactive effects in this process. Therefore, exosomal miRNA-146a in MSCs might alleviate experimental colitis by targeting the *TRAF6* and *IRAK1* genes [35], preventing NF-κB activation along with the subsequent production of TNF-α and IL-6 [35]. Consistently, several other MSC-exosomal miRNAs such as miRNA-30b-3p [36], miRNA-223-3p [37], and miRNA-126 [38,39] were found to be responsible for suppressing pro-inflammatory responses. They also exhibit potent effects in promoting tissue regeneration and angiogenesis. Overall, we need to understand the mechanisms by which MSC-exosomes repair tissue damage.

3. Therapeutic Functions of MSC-Exosomes

3.1. Immunomodulation

To our knowledge, commoditized MSCs have been approved for treating some autoimmune diseases in a clinical setting; however, the incidence of infection secondary to infusion of allogenic MSCs has been reported to be 29.5% [40]. This has prompted us to find an alternative approach. Exosomes are thought to be superior over MSCs with regard to treatment-related safety [41]. A previous work suggested that MSC-exosomes improved the in vitro survival and function of neutrophils from patients with severe congenital neutropenia, thus increasing the potential efficacy of MSC-exosomes against acute infection [42]. Moreover, studies have suggested that MSC-exosomes exhibit effects in managing autoimmune or inflammatory diseases [37,43] (Table 1).

Table 1. Mesenchymal Stem Cell (MSC) Exosomes Perform Immunoregulation Effects.

Models In Vivo/Vitro	Exosome Source	Immunoregulation Effects	Ref.
		In Vivo Models	
Inflammatory bowel disease mice	Human bone marrow MSCs	Increased levels of anti-inflammatory cytokines IL-10; Decreased levels of pro-inflammatory IL-1β, IL-6, IFN-γ and TNF-α; Promotion of the M2b macrophage polarization.	[28]
Inflammatory bowel disease mice	Human Umbilical Cord MSCs	Increased levels of IL-10; Decreased levels of IL-1β, IL-6, TNF-α, iNOS, and IL-7; Inhibition of macrophages infiltration into the colon tissues.	[43]
Inflammatory bowel disease mice	Human Umbilical Cord MSCs	Increased levels of IL-10 and IP-10; Decreased levels of TNF-α, IL-1β, IL-6.	[44]
Myocardial ischemia/reperfusion mice	Human Umbilical Cord MSCs	Increased levels of anti-inflammatory cytokine IL-10 and decreased levels of proinflammatory cytokines TNF-α and IL-6 via exosomal miRNA-181a; Promotion of Treg cell development.	[45]
Autoimmune encephalomyelitis rats	Rat bone marrow MSCs	Increased M2-related anti-inflammatory cytokines of IL-10 and TGF-β; Decreased M1-related proinflammatory TNF-α and IL-12 levels; Promotion of the M2 phenotype microglia polarization.	[46]
Autoimmune encephalomyelitis mice	Mice adipose MSCs	Decreased Tbx21 and Gata3 expression, as the crucial regulator for Th1 and Th2 cell responses; Decreased Rorc and Elf4 expression, as the activator and inhibitor for Th17 differentiation.	[47]
Acute lung injury mice	Rat bone marrow MSCs	Decreased expression of pulmonary TNF-α, IL-1β and IL-6 via inhibiting TLR4-NF-κB signaling pathway.	[48]
Neuroinflammation rats	Rat bone marrow MSCs	Reduced oxidative stress responses; Decreased levels of IL-1β, IL-6 and TNF-α; Inhibition of neuronal degeneration and apoptosis.	[49]
Autoimmune Hepatitis mice	Mice bone marrow MSCs	Attenuation of liver inflammation via carrying miRNA-223-3p; Increased Treg/Th17 ratio; Decreased levels of IL-1β and IL-6.	[37]
Inflammatory arthritis mice	Mice bone marrow MSCs	Decreased $CD8^{+}$T cell frequency and $CD8^{+}/CD4^{+}$ T cell ratio in vivo; No reduction in $CD8^{+}$T cell and $CD8^{+}/CD4^{+}$ T cell ratio in vitro; Promotion of Treg populations.	[50]

Table 1. *Cont.*

Models In Vivo/Vitro	Exosome Source	Immunoregulation Effects	Ref.
		In Vitro Co-Culture Models of Immunocytes with Exosomes	
Bone marrow-derived dendritic cells (BMDCs)	Mice adipose MSCs	Inhibition of the BMDCs proliferation; Transformation of immature and mature DCs into tolerogenic DCs	[51]
Macrophages	Human adipose MSCs	Shift of macrophages from M1 to M2 phenotype polarization via shuttling functional miRNA-146.	[34]
Peripheral blood mononuclear cells (PBMCs)	Human bone marrow MSCs	Increased secretion of IL-10 and TGF-β1 from PBMCs; Promotion of the Treg cells differentiation.	[52]
PBMCs	Human bone marrow MSCs	Inhibition of the B lymphocytes proliferation and differentiation; Decreased levels of immunoglobulin secretion.	[47]
Neutrophils	Human adipose MSCs	Augment ROS production of neutrophils; Promotion of lifespan and function of neutrophils.	[42]

In vitro, MSC-exosomes exert an immunomodulatory function, mainly by regulating the commitment of immune cells or altering their inflammatory cytokine secretion profiles [28]. For example, in the presence of IFN-γ and TNF-α, MSCs generate exosomes that induce macrophages to switch from an M1- to an M2-like phenotype, and exosomal miRNAs, including miRNA-146 and miRNA-34, greatly contribute to this process [34]. Mechanistically, miRNA-146 upregulate expression of M2-associated genes such as *TRAF6* and *IRAK1* by targeting NF-κB signaling [53], and miRNA-34 targets *Notch1* to suppress transcription of genes encoding M1-related pro-inflammatory cytokines, such as IL-6 and TNF-α [54]. The MSC-exosomal miRNA-181a has been consistently found to enhance the production of M2-related cytokines, including IL-10 and TGF-β while reducing production of the M1-related cytokines TNF-α, IL-6 and IL-12 by macrophages [46,55]. In addition to altering the secretion profile of macrophages, miRNA-181a induces Treg cell generation by suppressing expression of the *c-Fos* gene, which functionally counteracts the Foxp3-dominant transcriptional program associated with Treg cell development [45]. Nevertheless, a coculture experiment revealed that Treg cell induction by MSC-exosomes was less efficient than that by MSCs, indicating that some other factors contribute to this process. In fact, soluble factors from MSCs, including IDO, PGE2 and IL-10 strongly induce Treg cell generation [56]; however, except for IL-10, they are not present in MSC-exosomes [28,50].

In vivo, MSC-exosomes control immunomodulatory processes in an antigen-presenting cell (APC)-mediated manner [52]. For example, dendritic cells (DCs) serve as critical mediators of the effects of MSC-exosomes on Treg induction. Mechanistically, MSC-exosomes induce mature DCs to acquire immune tolerogenic phenotypes [51]. A critical function of tolerogenic DCs is inducing Treg cell generation in vivo [57]. Tolerogenic DCs secrete high levels of anti-inflammatory cytokines, such as IL-10 and TGF-β, and express low levels of costimulatory molecules, thus inducing naïve $CD4^+$ T cells to commit to differentiation into Tregs [51]. In addition to DCs, MSC-exosomes are able to restrict B cell maturation, which decreases the production of immunoglobulin-G (IgG) [50,58]. To a certain extent, the above effects of MSC-exosomes will assist in attenuating the immune responses driven by other T subsets, such as Th1, Th2, Th17 cells or $CD8^+$ T cells [33,59–61] (Figure 1). Indeed, studies have shown that incubating mouse adipose tissue-derived MSC-exosomes with mouse splenic immunocytes in vitro significantly downregulates expression of genes encoding Tbx21, Gata3 and Rorc, which centrally control the commitment of Th1, Th2 and Th17 cells, respectively [47]. $CD8^+$ T cells that delivered human umbilical cord-derived MSC-exosomes to GVDH mice significantly decreased the number of $CD8^+$ T cells along with the ratio of $CD8^+$ T cells to $CD4^+$ T cells in the peripheral blood [59]. However, intriguingly, reduced numbers of $CD4^+$ or $CD8^+$ T cells did not occur when conditioned by MSC-exosomes in vitro, which suggests, at least, that MSC-exosomes modulate host immune responses independently of their direct effect on impairing the survival of $CD4^+$ or $CD8^+$ T cells [50] (Figure 1). In other words, the mechanisms by which MSC-exosomes induce immunomodulation in vivo are more complicated than those observed in vitro, and the details need to be further elucidated.

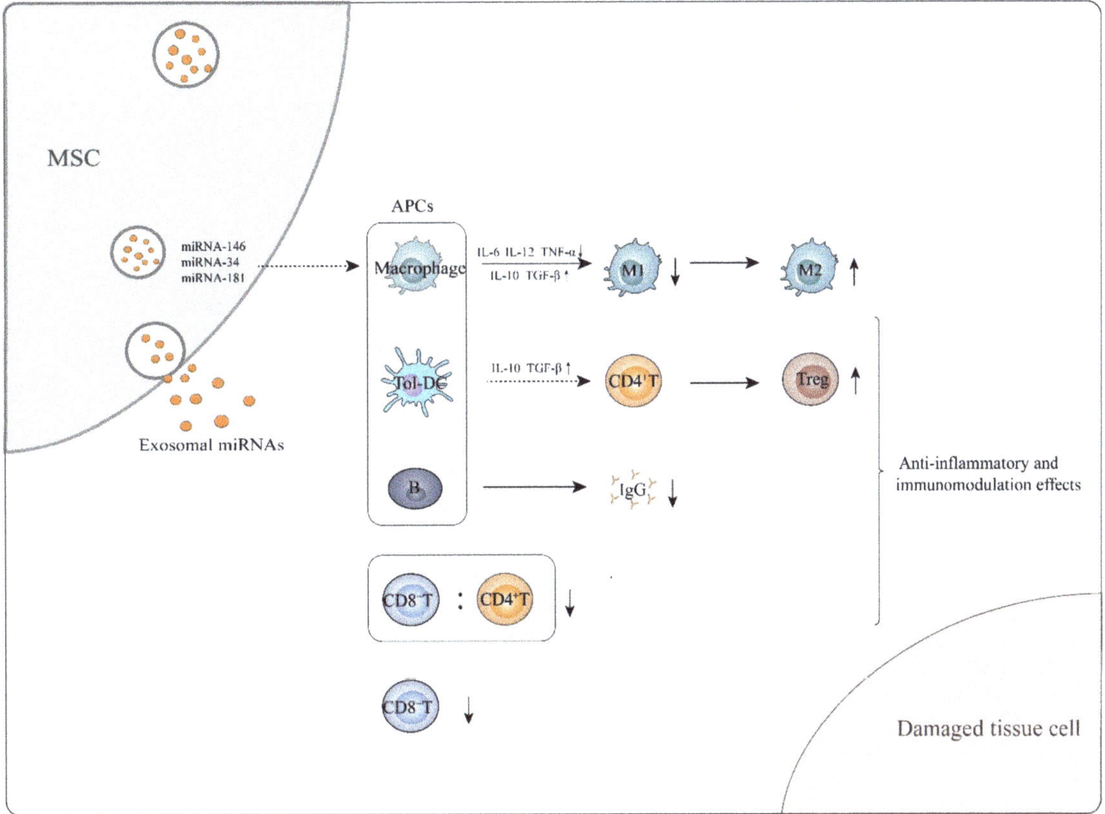

Figure 1. MSC-exosomes exert anti-inflammatory and immunomodulatory effects, which aid in attenuating tissue cell damage. MSC-exosomes perform these functions mainly through interaction of exosomal miRNAs and APCs. They can induce immature and mature DCs to differentiate into tolerogenic DCs, mediating naïve CD4$^+$ T cell differentiation into Tregs. In addition, MSC-exosomes are able to induce macrophages to transform from the M1 to the M2 phenotype while enhancing secretion of M2-related cytokines such as IL-10 and TGF-β and decreasing M1-related cytokine TNF-α, IL-6 and IL-12 levels. With regard to B cells, MSC-exosomes inhibit the maturation and function of B lymphocytes and cause a decrease in IgG secretion. MSC-exosomes also decrease the CD8$^+$ T cell number and the CD8$^+$/CD4$^+$ T cell ratio in the peripheral blood of in vivo mouse models.

3.2. Angiogenesis

Angiogenesis has a crucial role in tissue regeneration after damage. In this process, endothelial cells invade injured tissues to form buds and ultimately establish a capillary network. It is well accepted that MSCs exert therapeutic effects on ischemic diseases by directly producing or stimulating endogenous factors, such as vascular endothelial growth factor (VEGF), hepatocyte growth factor (HGF) and stromal-derived factor-1 (SDF-1) [62,63]. These factors facilitate angiogenesis in damaged tissues [64]. Recent evidence has revealed that MSC-exosomes also have pro-angiogenic properties [60,65] (Table 2).

Table 2. MSC Exosomes Facilitate Angiogenesis in Various Disorders.

Models	Exosome Source	MSCs Disposure	Carry	Angiogenic Mechanisms	Ref.
In Vivo Models					
Severe combined immunodeficiency mice	Human adipose MSCs	PDGF-stimulated	c-kit with its ligand SCF	Increasing matrix metalloproteinases content and enhancing the angiogenic potential via c-kit/SCF.	[55]
Hind limb ischemia mice	Human placental MSCs	NO-stimulated	miR-126	Promoting angiogenesis by increasing VEGF and miR-126 levels.	[39]
Cutaneous burn mice	Human umbilical cord MSCs	Blue light-stimulated	miR-135b-5p; miR-499a-3p	Promoting angiogenic activity via the upregulation of functional miR-135b-5p; miR-499a-3p.	[66]
Athymic-nude mice	Human dental pulp MSCs	HIF-1-overexpressed	Jagged 1	Enhancing Jagged 1-mediated angiogenesis through Notch signaling pathway	[67]
Bone fracture mice	Human umbilical cord MSCs	Hypoxia- treated	miR-126	Promoting angiogenesis and bone fracture healing through HIF-1α/miR-126 and SPRED1/Ras/Erk signaling pathways.	[38]
Nude mice	Human adipose MSCs	Hypoxia- treated	-	Promoting angiogenesis by increasing the expression of VEGF and activating the PKA signaling pathway.	[68]
Nude mice	Human adipose MSCs	Hypoxia- treated	-	Promoting angiogenesis at least partially through upregulating VEGF/VEGF-R signaling pathway.	[69]
Calvarial defect rats	Human bone marrow MSCs	DMOG-stimulated	-	Promoting angiogenesis by activation of the AKT/mTOR signaling pathway	[70]
Acute myocardial infarction rats	Human umbilical cord MSCs	Akt-transfected	-	Accelerating angiogenesis via upregulating PDGF-D expression.	[71]
Immunodeficient mice	Human adipose MSCs	-	miRNA-125a	Promoting angiogenesis by transferring miR-125a to endothelial cells and repressing angiogenic inhibitor delta-like 4.	[65]
Femoral head osteonecrosis mice	Human bone marrow MSCs	-	miRNA-224-3p	Promoting angiogenesis by downregulating exosomal microRNA-224-3p.	[72]
Femora fracture rats	Human bone marrow MSCs	-	-	Promoting angiogenesis and osteogenesis via activation of the HIF-1α/VEGF and the BMP-2/Smad1/RUNX2 signaling pathways.	[73]
Femoral fracture rats	Human umbilical cord MSCs	-	-	Promoting angiogenesis and fracture healing through increasing the expression of VEGF and HIF-1α.	[74]
Cutaneous wound rats	Human umbilical cord MSCs	-	-	Promoting angiogenesis via activating the Wnt4/β-catenin signaling pathway in a dose-dependent manner.	[75]
Auricle ischemic injury mice	Human placental MSCs	-	-	Stimulating angiogenic activity in endothelial cells via upregulating their responsiveness to proangiogenic growth factors.	[60]
Acute myocardial infarction rats	Rat bone marrow MSCs	-	-	Enhancing the density of new functional capillary and promoting blood flow recovery; Inhibiting proliferation and function of T cells; Reducing infarct size and preserving cardiac systolic and diastolic performance	[64]

Table 2. *Cont.*

Models	Exosome Source	MSCs Disposure	Carry	Angiogenic Mechanisms	Ref.
		In Vitro Co-Culture of Human Cells with Exosomes			
Human umbilical vein endothelial cells and fibroblasts	Human bone marrow MSCs	-	Transcriptionfactor STAT3	Inducing proliferation and migration of fibroblasts; Promoting the expression of growth factors, like HGF, IGF1, NGF and SDF1; Activating Akt, Erk, and STAT3 signaling pathways, which are both involved in angiogenesis.	[63]
Human umbilical vein endothelial cells and fibroblasts	Human bone marrow MSCs	-	Wnt3a protein	Enhancing fibroblast proliferation, migration; Promoting angiogenesis via carrying Wnt3a protein in vitro.	[76]

Based on proteomic analysis, MSC-exosomes contain various factors that are involved in angiogenesis, such as platelet-derived growth factor (PDGF), fibroblast growth factor (FGF), epidermal growth factor (EGF) and proteins associated with NF-κB activation [61]. PDGF, FGF and EGF function as common factors in mediating angiogenesis [77], and the role of proteins associated with NF-κB activation in mediating angiogenesis should be addressed. To our knowledge, NF-κB activation is conventionally associated with inducing pro-inflammatory responses. Nonetheless, intriguingly, blocking NF-κB activation abrogated tube formation by endothelial cells in vitro [61]. Consistent with this finding, exosomes from bone marrow MSCs were found to activate STAT3 signaling cascades in target cells, thus upregulating expression of genes encoding HGF, insulin-like growth factor-1 (IGF1), nerve growth factor (NGF) and SDF-1 [63]. Similarly, several other studies have revealed the mechanisms by which MSC-exosomes induce angiogenesis. For example, exosomes from umbilical cord MSCs reportedly activate Wnt/β-catenin to increase angiogenesis [75], but those from bone marrow MSCs promote angiogenesis by activating the HIF-1α/VEGF axis in target cells [73]. Furthermore, bone marrow MSCs increase the survival of pulmonary endothelial cells via exosomal miRNA-21-5p, which targets the antioncogenes *PDCD4* and *PTEN* in a mouse model of ischemia/reperfusion [78]. In addition, placental MSC-exosomes are capable of upregulating expression of genes encoding Ang2 and Tie2 by endothelial cells [60]. The details of the angiogenic features of MSC-exosomes are provided in Table 2 and Figure 2.

Similar to their immunomodulatory features, the pro-angiogenic potency of MSC-exosomes can be impacted by foreign stimuli [55,69]. For example, it was found that MSCs conditioned with PDGF showed increased production of exosomes containing angiogenic molecules, such as c-kit and stem cell factor [55]. However, preconditioning MSCs with pro-inflammatory cytokines, such as TNF-α and IL-6, increased the exosomal cargo content of miRNA-196a-5p and miRNA-17-5p, inactivating the PI3K-AKT, MAPK and VEGF-related pathways and impairing angiogenesis [79]. In addition to such bioactive substances, environmental factors impact the proangiogenic properties of MSC-exosomes. In fact, exposure of MSCs to blue light resulted in the increased content of miRNA-135b-5p and miRNA-499a-3p as exosomal cargo, which promoted angiogenesis in vitro by repressing myocyte enhancer factor 2C (MEF2C) [66]. Consistent with this finding, hypoxia enhanced the cargo content of miRNA-126 in MSC-exosomes. In this context, miRNA-126 was able to stimulate SPRED1/Ras/Erk/HIF-1α, thus increasing angiogenesis in injured tissues [38]. More intriguingly, HIF-1α was found to upregulate expression of the gene encoding RAB22A, which participates in vesicle formation in cells [80]. This event partially illustrates why HIF-1α-overexpressing MSCs can increase their production of exosomes [38]. Functionally, exosomes from cells such MSCs promote angiogenesis by activating the Jagged-1/Notch signaling pathway [55]. As the pro-angiogenic potency of MSC-exosomes can be improved by using the above methods, we can purposely generate them and utilize them to treat ischemic diseases (Table 2).

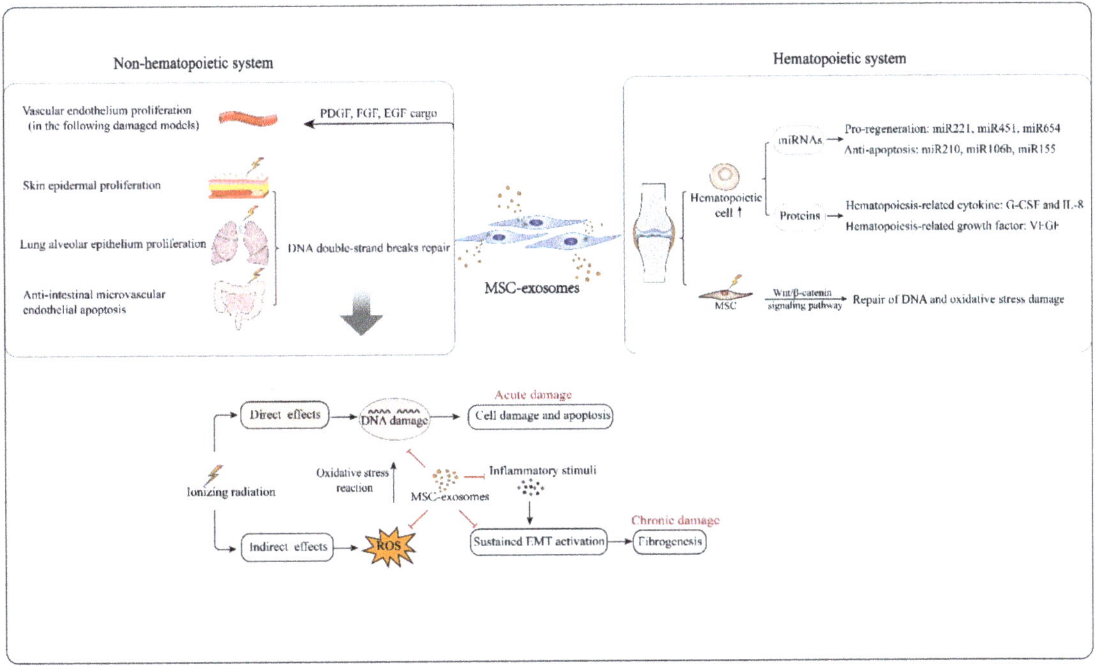

Figure 2. MSC-exosomes are capable of protecting against radiation-induced damage to hematopoietic and nonhematopoietic systems. In hematopoietic reconstruction post irradiation, MSC-exosomes enhance hematopoietic cell survival and proliferation by carrying functional molecules, such as the pro-regeneration miRNAs miRNA221, miRNA451 and miRNA654, the anti-apoptosis-related miRNAs miRNA210, miRNA106b and miRNA155, the hematopoiesis-related cytokines G-CSF and IL-8, and the hematopoiesis-related growth factor VEGF. In addition, MSC-exosomes can protect irradiated bone marrow MSCs from radiation-induced DNA and oxidative stress damage by activating the Wnt/β-catenin signaling pathway. With regard to the non-hematopoietic system, MSC-exosomes reduce apoptosis of skin epidermal, lung alveolar epithelium and intestinal epithelium cells, as MSC-exosomal miRNAs likely mediate repair of DNA double-strand breaks in damaged cells. Oxidative stress reaction and DNA damage are the major processes in radiation damage. MSC-exosomes can overcome these crucial events effectively, and have potential to suppress the development of acute and chronic radiation damage from several aspects. MSC-exosomes also facilitate vascular endothelium proliferation owing to their bioactive cargo molecules, such as PDGF, FGF and EGF.

3.3. Epithelial Recovery

Apart from their effects on immunomodulation and angiogenesis, a growing body of evidence has revealed the therapeutic effects of MSC-exosomes on epithelial injuries. In summary, MSC-exosomes increase the proliferation and survival of epithelial cells.

MSC-exosomes accelerate epithelial recovery in wounded tissues via their miRNA cargo. By using different disease models, recent studies have reported some specific roles of MSC-exosomal miRNAs in mediating epithelial recovery, such as that of miRNA-135a in increasing epithelial cell migration by suppressing expression of the gene encoding LATS2 during cutaneous wound healing [81] and that of miRNA-126 in activating the PI3K-AKT and MAPK pathways during cutaneous healing in a rat model of diabetes [82]. Moreover, exosomes may carry specific cargo such as foreign miRNA products. For example, in a study of MSCs genetically modified to overexpress a variety of miRNAs, including miRNA-100, miRNA-146a, miRNA-21, miRNA221 and miRNA-143, it was found that these exosomes enhance DNA synthesis, thus promoting the proliferation of vaginal epithelial cells [83]. In an acute lung injury model, exosomes from miRNA-30b-3p-overexpressing

MSCs protected type II alveolar epithelial cells against apoptosis by downregulating serum amyloid A3 (SAA3) [36].

In fact, increasing cell proliferation and survival by activating PI3K/Akt and MAPK are typical effects of both MSCs and their exosomes. Other effects of MSC-exosomes during epithelial recovery should be mentioned, including antioxidation. To our knowledge, oxidation is a harmful occurrence that impairs cell survival. In a renal injury model, MSC-exosomes inhibited apoptosis in tubular epithelial cells by reducing the level of reactive oxygen species (ROS) [84]. As documented, mitochondrial dysfunction is an important biological event that is closely associated with lung disease pathogenesis and/or progression [85]. Mechanistically, MSC-exosomes improve the mitochondrial function of lung epithelial cells by targeting division/fusion-related genes such as *rhot1*, *mfn1* and *opa1* [86]. Simultaneously, exosomes have been shown to carry functional mitochondria and promote mitochondrial transfer events [87], further demonstrating that MSC-exosomes have the potential to alleviate mitochondrial damage and control the progression of tissue damage.

MSC-exosomes also inhibit the epithelial-mesenchymal transition (EMT) [88–90], which is critical in inducing tissue fibrosis, resulting in pathological rather than functional restoration of damaged tissue. Although the underlying mechanisms by which MSC-exosomes restrict fibrotic development are not clear, MSC-exosomes inhibit activation of the TGF-β1/Smad pathway [91] while enhancing expression of zona occludens protein-1 in epithelial cells, which is related to cellular tight junctions [92]. Therefore, MSC-exosomes at least reduce epithelial depletion due to transformation, thereby maintaining the integrity of the epithelium and suppressing tissue fibrosis. Collectively, MSC-exosomes promote epithelial recovery by facilitating regeneration, inhibiting apoptosis and reducing EMT depletion.

4. Role of MSC-Exosomes in Repairing Radiation Damage

Despite the use of advanced treatment techniques, radiation damage is common and often unavoidable in cancer patients during or after receiving radiotherapy. The actions of ionizing radiation on biological molecules can be segmented into direct and indirect effects. DNA damage in cells can be induced by the direct effects of ionizing radiation, and it can also be caused by the oxidative stress reaction mediated by reactive oxygen species (ROS) generated by indirect effects. Ionizing irradiation-induced DNA double-strand breaks, oxidative stress, vascular damage, and subsequent inflammation are typical events in the acute phase of the pathogenesis of radiation damage, and if these events are not well managed, fibrosis occurs as a pathogenic feature in the chronic phase [55].The potential use of MSCs in repairing radiation-induced acute damage in the hematopoietic system, liver, lung, gastrointestinal tract, or skin has been explored [88,93–95], and the results indicate that MSCs have several therapeutic features including increased proliferation and survival of tissue/organ-specific stem/progenitor cells, the promotion of angiogenesis, anti-inflammation and oxidation, and the reduction of fibrotic pathogenesis [96]. The above findings indicate that MSC-exosomes have similar potencies to those of MSCs in repairing tissue or organ damage due to disease. Moreover, recent advances have demonstrated the repair of radiation damage by MSC-exosomes. In the following sections, we elaborate on the therapeutic effects of MSC-exosomes on radiation damage in the hematopoietic system and nonhematopoietic system (Figure 2).

Hematopoietic cells are sensitive to radiation exposure, which can lead to bone marrow failure. Several studies have shown that MSC-exosomes are capable of repairing radiation-induced hematopoietic system injury, but the exact mechanism is unclear. A few studies have suggested that the following processes may contribute to the relevant mechanism. (i) MSC-exosomes can transfer miRNAs with pro-regenerative or anti-apoptotic effects to irradiated hematopoietic cells. For example, intravenous delivery of human bone marrow MSC-derived extracellular vesicles (MSC-EVs, mainly comprising exosomes and microvesicles) swiftly normalized the counts of peripheral blood cells in mice that received whole-body irradiation because their cargo content, including miRNA-221, miRNA-451 and miRNA-654-3p, promoted the proliferation of irradiated marrow cells and miRNA210-5p, miRNA106b-3p and miRNA155-5p prevented radiation-induced hematopoietic cell apoptosis [17]. (ii) MSC-exosomes can restore hematopoiesis by stimulating secretion of hematopoiesis-related cytokines. A previous study has suggested that human placental MSCs rescue radiation-induced hematopoiesis in mice by secreting human hematopoiesis-related cytokines, including G-CSF, MCP-1, IL-6 and IL-8 [16], and this effect can be observed with MSC-exosomes as well [18]. In fact, the data from a recent study show that MSC-exosomes are capable of inducing production of high levels of hematopoiesis-related cytokines such as G-CSF, IL-6, IL-8 and VEGF by macrophages in vitro [18]. (iii) MSC-exosomes have several other features that cause the remodeling of hematopoietic cells. For example, incubation with MSC-exosomes enhances the activity of macrophages, which are regarded as the key regulators of demand-adapted hematopoiesis [89]. MSC-exosomes are also able to directly restore irradiated bone marrow MSCs, which are considered to be potent contributors to hematopoiesis. One critical mechanisms involves the alleviation of DNA damage and oxidative stress via Wnt/β-catenin signaling pathway activation [90].

In addition to remodeling the hematopoietic system, MSC-exosomes are capable of protecting the skin, gastrointestinal system, respiratory system and other systems against radiation damage. At the micro level, radiation damage is essentially attributed to the large number of oxygen free radicals generated by ionizing radiation, which subsequently result in DNA double-strand breaks. Previous studies have reported that MSCs play a key role in alleviating DNA damage and oxidative stress damage [97]. MSC-exosomes, the functional role of which depend on their cargo derived from cells of origin, exert similar remodeling effects [98–100]. For example, in an oxidative stress-induced skin injury model, MSC-exosome treatment decreased ROS generation and subsequent DNA damage and improved the antioxidant capacities of damaged cells through NRF2 signaling [100]. Other studies have found that after MSC-exosome treatment in an ischemic renal disease model, damaged renal cells showed reduced oxidative stress marker (MDA) levels, increased anti-oxidant marker (SOD and CAT) levels, and significantly reduced DNA damage parameters [98]. However, the underlying molecular mechanism is poorly understood. Notably, studies have shown that some miRNAs that are contained in MSC-exosomes such as miRNA210 are able to repair DNA double-strand breaks [17,19], which suggests that exosomes may exert remodeling functions in a noncoding RNA-mediated epigenetic manner. This may explain the decrease in the apoptosis of skin epidermal cells, lung alveolar epithelium, intestinal epithelium and various parenchymal cells after MSC-exosome exposure [60,101–103]. Such noncoding RNA cargo may affect nonhomologous end-joining (NHEJ), which is common and essential in mammalian cell DSB repair [104]. Current evidence suggests that MSC-exosomes treatment is beneficial for the repair of oxidative stress-induced damage [22,105,106], although the exact functional components remain to be revealed. In addition, it has been found that intravenously injected MSC-EVs (including exosomes) are highly distributed in parenchymal organs such as the liver and spleen in a whole-body irradiation mouse model [105]. This may provide the context for the development of cures for radiation-induced parenchymal organ injury. Microvascular endothelial apoptosis has been recognized as the primary process that initiates radiation-induced injury [106]. Studies have found that local MSC-exosome treatment can facilitate the proliferation of vascular endothelial cells by activating the Wnt4/β-catenin pathway [75,103]. MSC-exosomes also have the potential to cure radiation-induced injury partly due to their potent pro-angiogenic factor cargo such as PDGF, FGF and EGF [61], which induce endothelial proliferation and differentiation in vitro and neovascularization in vivo [107,108].

On the other hand, exosomal targeting of cells is mediated by members of the integrin and tetraspanin families or other associated molecules based on their expression [109,110]. An experimental study demonstrated that radiation contributes to increased formation of the integrin and tetraspanin complex CD29/CD81 on the cell surface, thus enhancing uptake of exosomes by irradiated cells [15]. This further illustrates the potential use of

MSC-exosomes in the treatment of radiation damage. In general, the repair effect of MSC-exosomes on radiation damage in multiple systems is partly attributed to their bioactive cargo, which predominately consists of noncoding RNAs and functional proteins. These molecules influence the expression of target genes associated with radiation-induced damage or tissue regeneration due to epigenetic regulation. Overall, more experimental studies are required to further explore the molecular mechanisms involved.

5. MSC-Exosomes in Repairing Radiation Damage: Perspective and Challenges

With regard to radiation-induced damage, it has been revealed that MSCs play a crucial role in tissue damage treatment and prevention. Moreover, the superior properties and improved safety of MSC-exosomes make them novel candidates for curing radiation-induced damage. They exert therapeutic effects mainly by facilitating angiogenesis, promoting cellular regeneration, and probably by enhancing the repair function through immunomodulatory effects. More importantly, there are several methods that can be used to enhance the efficacy of remodeling damaged tissue. On the one hand, exosomes secreted by MSCs with genetic modifications are a promising alternative treatment, such as exosomes derived from SDF1-overexpressing MSCs for microvascular regeneration [111]. On the other hand, MSCs can be pretreated in vitro before exosomes are collected, such as with hypoxia-treated MSC-exosomes in ischemia-related disease [55]. Last, but equally important, the tropism of exosomes can be improved by increasing expression of specific receptors on the surface of the original MSCs. Current studies on the treatment of radiation-induced damage by MSC-exosomes are mostly based on the acute phase, whereas little work has been performed on the treatment of chronic radiation-induced damage by MSC-exosomes. Notably, evidence suggests that MSC-exosomes reverse EMT of endometrial epithelial cells via the TGF-β1/Smad pathway [91] and of tubular epithelial cells via enhanced tight junctions [92]. In general, sustained EMT is a critical mechanism that underlies the fibrotic pathology of tissue [112]. Thus, it can be reasonably inferred that MSC-exosome treatment has potential for preventing tissue fibrosis in the chronic phase of tissue damage. Therefore, despite limited evidence of the repair role of MSC-exosomes in chronic radiation-induced damage, it is important that researchers make further efforts to explore their therapeutic and underlying potential in chronic radiation-induced damage. This will provide a new context for the future application of MSC-exosomes to treating chronic radiation damage (Figure 2).

In fact, there are several deficiencies with regard to managing diseases by using MSC-exosomes. (i) One concern is the challenges due to the instability of contents of exosomes. For example, studies have shown that the amount of exosomal miRNA cargo is influenced by the irradiation dose and pH value of the culture medium [113,114]. The precise experimental conditions for exosomes are more difficult to control compared to MSCs. (ii) Another concern is the lack of a uniform standard for the purification and quantification of exosomes from conditioned media. Overall, it is difficult to determine the equivalent dose of exosomes in dose-dependent experimental studies, which may lead to different conclusions as results can be affected by exosome content and impurities. Therefore, it is appropriate to find an ideal method for constructing a precise equivalent dose of exosomes for experimental purpose. Although the effects of MSC-exosomes in various disease models have been clearly shown, the exact components and mechanisms of therapy are not entirely clear. miRNAs and functional proteins may play major roles, yet the role of MSC-exosomes in tumor growth and metastasis remains controversial. Previous studies have shown that MSC-exosomes can promote tumor growth in vivo [115], but a recent study revealed that MSC-exosomes enhance radiotherapy-induced tumor cell death in primary and metastatic tumor foci through synergistic and bystander effects [116]. Urgent issues for cancer patients receiving radiotherapy include the adjuvant antitumor effect and resistance to radiation damage. There is a great need for researchers to elaborate on the role of MSC-exosomes in regenerative medicine for the treatment of radiation damage.

6. Conclusions

MSC-exosomes show potential for repairing radiation damage. Current data reveal that MSC-exosomes have therapeutic potential due to their anti-inflammatory effects and promotion of angiogenesis and epithelial survival, which are crucial biological processes in the remodeling of radiation damage. In addition, the immunomodulatory effects of MSC-exosomes probably enhance their tissue repair function. Overall, MSC-exosomes have good prospects for the treatment of radiation injury and this may inspire future research in this field.

Author Contributions: X.P. and S.M. wrote the manuscript. X.P., Y.G. and T.X. proposed and drew the figures. P.C. and L.D. reread and corrected the manuscript. All authors have read and agreed to the published version of the manuscript.

Funding: This study is supported by the National Natural Science Foundation of China (No. 81773353 and No. 81874254); the Jilin Scientific and Technological Development Program (No. 20190201204JC); and the Foundation of Scientific Research Planning Project of the 13th Five-year Plan of Jilin Provincial Department of Education (Grant No. JJKH20201043KJ).

Conflicts of Interest: The authors declare no conflict of interest.

References

1. Lai, R.C.; Yeo, R.W.Y.; Lim, S.K. Mesenchymal stem cell exosomes. *Semin. Cell Dev. Biol.* **2015**, *40*, 82–88. [CrossRef] [PubMed]
2. Barberini, D.J.; Freitas, N.P.P.; Magnoni, M.S.; Maia, L.; Listoni, A.J.; Heckler, M.C.; Sudano, M.J.; Golim, M.A.; Landim-Alvarenga, F.D.C.; Amorim, R.M. Equine mesenchymal stem cells from bone marrow, adipose tissue and umbilical cord: Immunophenotypic characterization and differentiation potential. *Stem Cell Res. Ther.* **2014**, *5*, 25. [CrossRef] [PubMed]
3. Strioga, M.; Viswanathan, S.; Darinskas, A.; Slaby, O.; Michalek, J. Same or Not the Same? Comparison of Adipose Tissue-Derived Versus Bone Marrow-Derived Mesenchymal Stem and Stromal Cells. *Stem Cells Dev.* **2012**, *21*, 2724–2752. [CrossRef] [PubMed]
4. Samsonraj, R.M.; Raghunath, M.; Nurcombe, V.; Hui, J.H.; Van Wijnen, A.J.; Cool, S. Concise Review: Multifaceted Characterization of Human Mesenchymal Stem Cells for Use in Regenerative Medicine. *Stem Cells Transl. Med.* **2017**, *6*, 2173–2185. [CrossRef]
5. Creane, M.; Howard, L.; O'Brien, T.; Coleman, C.M. Biodistribution and retention of locally administered human mesenchymal stromal cells: Quantitative polymerase chain reaction–based detection of human DNA in murine organs. *Cytotherapy* **2017**, *19*, 384–394. [CrossRef]
6. Ferrand, J.; Noël, D.; Lehours, P.; Prochazkova-Carlotti, M.; Chambonnier, L.; Ménard, A.; Mégraud, F.; Varon, C. Human bone marrow-derived stem cells acquire epithelial characteristics through fusion with gastrointestinal epithelial cells. *PLoS ONE* **2011**, *6*, e19569. [CrossRef]
7. Geng, W.; Tang, H.; Luo, S.; Lv, Y.; Liang, D.; Kang, X.; Hong, W. Exosomes from miRNA-126-modified ADSCs promotes func-tional recovery after stroke in rats by improving neurogenesis and suppressing microglia activation. *Am. J. Transl. Res.* **2019**, *11*, 780–792.
8. Morrison, T.J.; Jackson, M.V.; Cunningham, E.K.; Kissenpfennig, A.; McAuley, D.F.; O'Kane, C.M.; Krasnodembskaya, A. Mesenchymal Stromal Cells Modulate Macrophages in Clinically Relevant Lung Injury Models by Extracellular Vesicle Mitochondrial Transfer. *Am. J. Respir. Crit. Care Med.* **2017**, *196*, 1275–1286. [CrossRef]
9. Moon, G.J.; Sung, J.H.; Kim, D.H.; Kim, E.H.; Cho, Y.H.; Son, J.P.; Cha, J.M.; Bang, O.Y. Application of Mesenchymal Stem Cell-Derived Extracellular Vesicles for Stroke: Biodistribution and MicroRNA Study. *Transl. Stroke Res.* **2018**, *10*, 509–521. [CrossRef]
10. Mathieu, M.; Martin-Jaular, L.; Lavieu, G.; Théry, C. Specificities of secretion and uptake of exosomes and other extracellular vesicles for cell-to-cell communication. *Nat. Cell Biol.* **2019**, *21*, 9–17. [CrossRef]
11. Valadi, H.; Ekström, K.; Bossios, A.; Sjöstrand, M.; Lee, J.J.; Tvall, J.O.L.O. Exosome-mediated transfer of mRNAs and microRNAs is a novel mechanism of genetic exchange between cells. *Nat. Cell Biol.* **2007**, *9*, 654–659. [CrossRef]
12. Kalluri, R.; LeBleu, V.S. The biology, function, and biomedical applications of exosomes. *Science* **2020**, *367*, eaau6977. [CrossRef]
13. Elahi, F.M.; Farwell, D.G.; Nolta, J.A.; Anderson, J.D. Preclinical translation of exosomes derived from mesenchymal stem/stromal cells. *Stem Cells* **2020**, *38*, 15–21. [CrossRef] [PubMed]
14. Riazifar, M.; Mohammadi, M.R.; Pone, E.J.; Yeri, A.; Lässer, C.; Ségaliny, A.I.; McIntyre, L.L.; Shelke, G.V.; Hutchins, E.; Hamamoto, A.; et al. Stem Cell-Derived Exosomes as Nanotherapeutics for Autoimmune and Neurodegenerative Disorders. *ACS Nano* **2019**, *13*, 6670–6688. [CrossRef] [PubMed]
15. Hazawa, M.; Tomiyama, K.; Saotome-Nakamura, A.; Obara, C.; Yasuda, T.; Gotoh, T.; Tanaka, I.; Yakumaru, H.; Ishihara, H.; Tajima, K. Radiation increases the cellular uptake of exosomes through CD29/CD81 complex formation. *Biochem. Biophys. Res. Commun.* **2014**, *446*, 1165–1171. [CrossRef] [PubMed]

16. Pinzur, L.; Akyuez, L.; Levdansky, L.; Blumenfeld, M.; Volinsky, E.; Aberman, Z.; Reinke, P.; Ofir, R.; Volk, H.-D.; Gorodetsky, R. Rescue from lethal acute radiation syndrome (ARS) with severe weight loss by secretome of intramuscularly injected human placental stromal cells. *J. Cachexia Sarcopenia Muscle* **2018**, *9*, 1079–1092. [CrossRef] [PubMed]
17. Wen, S.; Dooner, M.; Cheng, Y.; Papa, E.; Del Tatto, M.; Pereira, M.; Deng, Y.; Goldberg, L.; Aliotta, J.; Chatterjee, D.; et al. Mesenchymal stromal cell-derived extracellular vesicles rescue radiation damage to murine marrow hematopoietic cells. *Leukemia* **2016**, *30*, 2221–2231. [CrossRef]
18. Kink, J.A.; Forsberg, M.H.; Reshetylo, S.; Besharat, S.; Childs, C.J.; Pederson, J.D.; Gendron-Fitzpatrick, A.; Graham, M.; Bates, P.D.; Schmuck, E.G.; et al. Macrophages Educated with Exosomes from Primed Mesenchymal Stem Cells Treat Acute Radiation Syndrome by Promoting Hematopoietic Recovery. *Biol. Blood Marrow Transplant.* **2019**, *25*, 2124–2133. [CrossRef]
19. Grosso, S.; Doyen, J.; Parks, S.K.; Bertero, T.; Paye, A.; Cardinaud, B.; Gounon, P.; Lacas-Gervais, S.; Noël, A.; Pouysségur, J.; et al. MiR-210 promotes a hypoxic phenotype and increases radioresistance in human lung cancer cell lines. *Cell Death Dis.* **2013**, *4*, e544. [CrossRef]
20. He, C.; Zheng, S.; Luo, Y.; Wang, B. Exosome Theranostics: Biology and Translational Medicine. *Theranostics* **2018**, *8*, 237–255. [CrossRef]
21. Théry, C.; Amigorena, S.; Raposo, G.; Clayton, A. Isolation and Characterization of Exosomes from Cell Culture Supernatants and Biological Fluids. *Curr. Protoc. Cell Biol.* **2006**, *30*, 3.22.1–3.22.29. [CrossRef]
22. Ferguson, S.W.; Wang, J.; Lee, C.J.; Liu, M.; Neelamegham, S.; Canty, J.M.; Nguyen, J. The microRNA regulatory landscape of MSC-derived exosomes: A systems view. *Sci. Rep.* **2018**, *8*, 1419. [CrossRef]
23. Xun, C.; Ge, L.; Tang, F.; Wang, L.; Zhuo, Y.; Long, L.; Qi, J.; Hu, L.; Duan, D.; Chen, P.; et al. Insight into the proteomic profiling of exosomes secreted by human OM-MSCs reveals a new potential therapy. *Biomed. Pharmacother.* **2020**, *131*, 110584. [CrossRef] [PubMed]
24. Shao, L.; Zhang, Y.; Lan, B.; Wang, J.; Zhang, Z.; Zhang, L.; Xiao, P.; Meng, Q.; Geng, Y.-J.; Yu, X.-Y.; et al. MiRNA-Sequence Indicates That Mesenchymal Stem Cells and Exosomes Have Similar Mechanism to Enhance Cardiac Repair. *BioMed Res. Int.* **2017**, *2017*, 1–9. [CrossRef] [PubMed]
25. Moll, G.; Alm, J.J.; Davies, L.C.; Von Bahr, L.; Heldring, N.; Stenbeck-Funke, L.; Hamad, O.A.; Hinsch, R.; Ignatowicz, L.; Locke, M.; et al. Do Cryopreserved Mesenchymal Stromal Cells Display Impaired Immunomodulatory and Therapeutic Properties? *Stem Cells* **2014**, *32*, 2430–2442. [CrossRef] [PubMed]
26. Tan, S.S.; Yin, Y.; Lee, T.; Lai, R.C.; Yeo, R.W.Y.; Zhang, B.; Choo, A.; Lim, S.K. Therapeutic MSC exosomes are derived from lipid raft microdomains in the plasma membrane. *J. Extracell. Vesicles* **2013**, *2*. [CrossRef]
27. Alvarez-Erviti, L.; Seow, Y.; Yin, H.; Betts, C.; Lakhal, S.; Wood, M.J.A. Delivery of siRNA to the mouse brain by systemic injection of targeted exosomes. *Nat. Biotechnol.* **2011**, *29*, 341–345. [CrossRef]
28. Liu, H.; Liang, Z.; Wang, F.; Zhou, C.; Zheng, X.; Hu, T.; He, X.; Wu, X.; Lan, P. Exosomes from mesenchymal stromal cells reduce murine colonic inflammation via a macrophage-dependent mechanism. *JCI Insight* **2019**, *4*, 4. [CrossRef] [PubMed]
29. Marino, J.; Paster, J.; Benichou, G. Allorecognition by T Lymphocytes and Allograft Rejection. *Front. Immunol.* **2016**, *7*, 582. [CrossRef]
30. Gonzalez-Rey, E.; Anderson, P.; Gonzalez, M.A.; Rico, L.; Buscher, D.; Delgado, M. Human adult stem cells derived from adipose tissue protect against experimental colitis and sepsis. *Gut* **2009**, *58*, 929–939. [CrossRef]
31. Volarevic, V.; Markovic, B.S.; Gazdic, M.; Volarevic, A.; Jovicic, N.; Arsenijevic, N.; Armstrong, L.; Djonov, V.; Lako, M.; Stojkovic, M. Ethical and Safety Issues of Stem Cell-Based Therapy. *Int. J. Med. Sci.* **2018**, *15*, 36–45. [CrossRef] [PubMed]
32. Røsland, G.V.; Svendsen, A.; Torsvik, A.; Sobala, E.; McCormack, E.; Immervoll, H.; Mysliwietz, J.; Tonn, J.C.; Goldbrunner, R.; Lønning, P.E.; et al. Long-term Cultures of Bone Marrow–Derived Human Mesenchymal Stem Cells Frequently Undergo Spontaneous Malignant Transformation. *Cancer Res.* **2009**, *69*, 5331–5339. [CrossRef] [PubMed]
33. Hladik, D.; Höfig, I.; Oestreicher, U.; Beckers, J.; Matjanovski, M.; Bao, X.; Scherthan, H.; Atkinson, M.J.; Rosemann, M. Long-term culture of mesenchymal stem cells impairs ATM-dependent recognition of DNA breaks and increases genetic instability. *Stem Cell Res. Ther.* **2019**, *10*, 1–12. [CrossRef] [PubMed]
34. Domenis, R.; Cifù, A.; Quaglia, S.; Pistis, C.; Moretti, M.; Vicario, A.; Parodi, P.C.; Fabris, M.; Niazi, K.R.; Soon-Shiong, P.; et al. Pro inflammatory stimuli enhance the immunosuppressive functions of adipose mesenchymal stem cells-derived exosomes. *Sci. Rep.* **2018**, *8*, 13325. [CrossRef] [PubMed]
35. Wu, H.; Fan, H.; Shou, Z.; Xu, M.; Chen, Q.; Ai, C.; Dong, Y.; Liu, Y.; Nan, Z.; Wang, Y.; et al. Extracellular vesicles containing miR-146a attenuate experimental colitis by targeting TRAF6 and IRAK. *Int. Immunopharmacol.* **2019**, *68*, 204–212. [CrossRef]
36. Yi, X.; Wei, X.; Lv, H.; An, Y.; Li, L.; Lu, P.; Yang, Y.; Zhang, Q.; Yi, H.; Chen, G. Exosomes derived from microRNA-30b-3p-overexpressing mesenchymal stem cells protect against lipopolysaccharide-induced acute lung injury by inhibiting SAA. *Exp. Cell Res.* **2019**, *383*, 111454. [CrossRef]
37. Lu, F.B.; Chen, D.Z.; Chen, L.; Hu, E.D.; Wu, J.L.; Li, H.; Gong, Y.W.; Lin, Z.; Wang, X.D.; Li, J.; et al. Attenuation of Experimental Auto-immune Hepatitis in Mice with Bone Mesenchymal Stem Cell-Derived Exosomes Carrying MicroRNA-223-3p. *Mol. Cells* **2019**, *42*, 906–918. [CrossRef]
38. Liu, W.; Li, L.; Rong, Y.; Qian, D.; Chen, J.; Zhou, Z.; Luo, Y.; Jiang, D.; Cheng, L.; Zhao, S.; et al. Hypoxic mesenchymal stem cell-derived exosomes promote bone fracture healing by the transfer of miR-126. *Acta Biomater.* **2020**, *103*, 196–212. [CrossRef]

39. Du, W.; Zhang, K.; Zhang, S.; Wang, R.; Nie, Y.; Tao, H.; Han, Z.; Liang, L.; Wang, D.; Liu, J.; et al. Enhanced proangiogenic potential of mesenchymal stem cell-derived exosomes stimulated by a nitric oxide releasing polymer. *Biomaterials* **2017**, *133*, 70–81. [CrossRef]
40. Liang, J.; Zhang, H.; Kong, W.; Deng, W.; Wang, D.; Feng, X.; Zhao, C.; Hua, B.; Wang, H.; Sun, L. Safety analysis in patients with autoimmune disease receiving allogeneic mesenchymal stem cells infusion: A long-term retrospective study. *Stem Cell Res. Ther.* **2018**, *9*, 1–10. [CrossRef]
41. Rani, S.; Ryan, A.E.; Griffin, M.D.; Ritter, T. Mesenchymal Stem Cell-derived Extracellular Vesicles: Toward Cell-free Therapeutic Applications. *Mol. Ther.* **2015**, *23*, 812–823. [CrossRef] [PubMed]
42. Mahmoudi, M.; Taghavi-Farahabadi, M.; Namaki, S.; Baghaei, K.; Rayzan, E.; Rezaei, N.; Hashemi, S.M. Exosomes derived from mesenchymal stem cells improved function and survival of neutrophils from severe congenital neutropenia patients in vitro. *Hum. Immunol.* **2019**, *80*, 990–998. [CrossRef] [PubMed]
43. Mao, F.; Wu, Y.; Tang, X.; Kang, J.; Zhang, B.; Yan, Y.; Qian, H.; Zhang, X.; Xu, W. Exosomes Derived from Human Umbilical Cord Mesenchymal Stem Cells Relieve Inflammatory Bowel Disease in Mice. *BioMed Res. Int.* **2017**, *2017*, 5356760. [CrossRef] [PubMed]
44. Wu, Y.; Qiu, W.; Xu, X.; Kang, J.; Wang, J.; Wen, Y.; Tang, X.; Yan, Y.; Qian, H.; Zhang, X.; et al. Exosomes derived from human um-bilical cord mesenchymal stem cells alleviate inflammatory bowel disease in mice through ubiquitination. *Am. J. Transl. Res.* **2018**, *10*, 2026–2036.
45. Wei, Z.; Qiao, S.; Zhao, J.; Liu, Y.; Li, Q.; Wei, Z.; Dai, Q.; Kang, L.; Xu, B. miRNA-181a over-expression in mesenchymal stem cell-derived exosomes influenced inflammatory response after myocardial ischemia-reperfusion injury. *Life Sci.* **2019**, *232*, 116632. [CrossRef]
46. Li, Z.; Liu, F.; He, X.; Yang, X.; Shan, F.; Feng, J. Exosomes derived from mesenchymal stem cells attenuate inflammation and demyelination of the central nervous system in EAE rats by regulating the polarization of microglia. *Int. Immunopharmacol.* **2019**, *67*, 268–280. [CrossRef]
47. Fathollahi, A.; Hashemi, S.M.; Hoseini, M.H.M.; Yeganeh, F. In vitro analysis of immunomodulatory effects of mesenchymal stem cell- and tumor cell -derived exosomes on recall antigen-specific responses. *Int. Immunopharmacol.* **2019**, *67*, 302–310. [CrossRef]
48. Liu, J.; Chen, T.; Lei, P.; Tang, X.; Huang, P. Exosomes Released by Bone Marrow Mesenchymal Stem Cells Attenuate Lung Injury Induced by Intestinal Ischemia Reperfusion via the TLR4/NF-κB Pathway. *Int. J. Med. Sci.* **2019**, *16*, 1238–1244. [CrossRef]
49. El-Mahalaway, A.M.; El-Azab, N.E. The potential neuroprotective role of mesenchymal stem cell-derived exosomes in cer ebellar cortex lipopolysaccharide-induced neuroinflammation in rats: A histological and immunohistochemical study. *Ultrastruct. Pathol.* **2020**, 1–15. [CrossRef]
50. Cosenza, S.; Toupet, K.; Maumus, M.; Luz-Crawford, P.; Blanc-Brude, O.; Jorgensen, C.; Noël, D. Mesenchymal stem cells-derived exosomes are more immunosuppressive than microparticles in inflammatory arthritis. *Theranostics* **2018**, *8*, 1399–1410. [CrossRef]
51. Shahir, M.; Hashemi, S.M.; Asadirad, A.; Varhram, M.; Kazempour-Dizaji, M.; Folkerts, G.; Garssen, J.; Adcock, I.M.; Mortaz, E. Effect of mesenchymal stem cell-derived exosomes on the induction of mouse tolerogenic dendritic cells. *J. Cell. Physiol.* **2020**, *235*, 7043–7055. [CrossRef] [PubMed]
52. Du, Y.-M.; Zhuansun, Y.-X.; Chen, R.; Lin, L.; Lin, Y.; Li, J.-G.; Yu-Mo, D.; Yong-Xun, Z.; Rui, C.; Ying, L.; et al. Mesenchymal stem cell exosomes promote immunosuppression of regulatory T cells in asthma. *Exp. Cell Res.* **2018**, *363*, 114–120. [CrossRef] [PubMed]
53. Vergadi, E.; Vaporidi, K.; Theodorakis, E.E.; Doxaki, C.; Lagoudaki, E.; Ieronymaki, E.; Alexaki, V.I.; Helms, M.; Kondili, E.; Soennichsen, B.; et al. Akt2 Deficiency Protects from Acute Lung Injury via Alternative Macrophage Activation and miR-146a Induction in Mice. *J. Immunol.* **2013**, *192*, 394–406. [CrossRef] [PubMed]
54. Jiang, P.; Liu, R.; Zheng, Y.; Liu, X.; Chang, L.; Xiong, S.; Chu, Y. MiR-34a inhibits lipopolysaccharide-induced inflammatory response through targeting Notch1 in murine macrophages. *Exp. Cell Res.* **2012**, *318*, 1175–1184. [CrossRef]
55. Lopatina, T.; Bruno, S.; Tetta, C.; Kalinina, N.; Porta, M.; Camussi, G. Platelet-derived growth factor regulates the secretion of extracellular vesicles by adipose mesenchymal stem cells and enhances their angiogenic potential. *Cell Commun. Signal.* **2014**, *12*, 26. [CrossRef]
56. Chang, P.-Y.; Qu, Y.-Q.; Wang, J.; Dong, L.-H. The potential of mesenchymal stem cells in the management of radiation enteropathy. *Cell Death Dis.* **2015**, *6*, e1840. [CrossRef]
57. Raker, V.K.; Domogalla, M.P.; Steinbrink, K. Tolerogenic Dendritic Cells for Regulatory T Cell Induction in Man. *Front. Immunol.* **2015**, *6*, 569. [CrossRef]
58. Budoni, M.; Fierabracci, A.; Luciano, R.; Petrini, S.; Di Ciommo, V.; Muraca, M. The immunosuppressive effect of mesenchymal stromal cells on B lymphocytes is mediated by membrane vesicles. *Cell Transplant.* **2013**, *22*, 369–379. [CrossRef]
59. Wang, L.; Gu, Z.; Zhao, X.; Yang, N.; Wang, F.; Deng, A.; Zhao, S.; Luo, L.; Wei, H.; Guan, L.; et al. Extracellular Vesicles Released from Human Umbilical Cord-Derived Mesenchymal Stromal Cells Prevent Life-Threatening Acute Graft-Versus-Host Disease in a Mouse Model of Allogeneic Hematopoietic Stem Cell Transplantation. *Stem Cells Dev.* **2016**, *25*, 1874–1883. [CrossRef]
60. Komaki, M.; Numata, Y.; Morioka, C.; Honda, I.; Tooi, M.; Yokoyama, N.; Ayame, H.; Iwasaki, K.; Taki, A.; Oshima, N.; et al. Exosomes of human placenta-derived mesenchymal stem cells stimulate angiogenesis. *Stem Cell Res. Ther.* **2017**, *8*, 1–12. [CrossRef]

61. Anderson, J.D.; Johansson, H.J.; Graham, C.S.; Vesterlund, M.; Pham, M.T.; Bramlett, C.S.; Montgomery, E.N.; Mellema, M.S.; Bardini, R.L.; Contreras, Z.; et al. Comprehensive Proteomic Analysis of Mesenchymal Stem Cell Exosomes Reveals Modulation of Angiogenesis via Nuclear Factor-KappaB Signaling. *Stem Cells* **2016**, *34*, 601–613. [CrossRef] [PubMed]
62. Chang, P.-Y.; Zhang, B.-Y.; Cui, S.; Qu, C.; Shao, L.-H.; Xu, T.-K.; Qu, Y.-Q.; Dong, L.-H.; Wang, J. MSC-derived cytokines repair radiation-induced intra-villi microvascular injury. *Oncotarget* **2017**, *8*, 87821–87836. [CrossRef] [PubMed]
63. Ulivi, V.; Tasso, R.; Cancedda, R.; Descalzi, F. Mesenchymal Stem Cell Paracrine Activity Is Modulated by Platelet Lysate: Induction of an Inflammatory Response and Secretion of Factors Maintaining Macrophages in a Proinflammatory Phenotype. *Stem Cells Dev.* **2014**, *23*, 1858–1869. [CrossRef] [PubMed]
64. Teng, X.; Chen, L.; Chen, W.; Yang, J.; Yang, Z.; Shen, Z. Mesenchymal Stem Cell-Derived Exosomes Improve the Microenvironment of Infarcted Myocardium Contributing to Angiogenesis and Anti-Inflammation. *Cell. Physiol. Biochem.* **2015**, *37*, 2415–2424. [CrossRef] [PubMed]
65. Liang, X.; Zhang, L.; Wang, S.; Han, Q.; Zhao, R.C. Exosomes secreted by mesenchymal stem cells promote endothelial cell angiogenesis by transferring miR-125a. *J. Cell Sci.* **2016**, *129*, 2182–2189. [CrossRef]
66. Yang, K.; Li, D.; Wang, M.; Xu, Z.; Chen, X.; Liu, Q.; Sun, W.; Li, J.; Gong, Y.; Liu, D.; et al. Exposure to blue light stimulates the proangiogenic capability of exosomes derived from human umbilical cord mesenchymal stem cells. *Stem Cell Res. Ther.* **2019**, *10*, 1–14. [CrossRef]
67. Gonzalez-King, H.; García, N.A.; Ontoria-Oviedo, I.; Ciria, M.; Montero, J.A.; Sepúlveda, P. Hypoxia Inducible Factor-1α Potentiates Jagged 1-Mediated Angiogenesis by Mesenchymal Stem Cell-Derived Exosomes. *Stem Cells* **2017**, *35*, 1747–1759. [CrossRef]
68. Xue, C.; Shen, Y.; Li, X.; Li, B.; Zhao, S.; Gu, J.; Chen, Y.; Ma, B.; Wei, J.; Han, Q.; et al. Exosomes Derived from Hypoxia-Treated Human Adipose Mesenchymal Stem Cells Enhance Angiogenesis Through the PKA Signaling Pathway. *Stem Cells Dev.* **2018**, *27*, 456–465. [CrossRef]
69. Han, Y.; Ren, J.; Bai, Y.; Pei, X.-T.; Han, Y. Exosomes from hypoxia-treated human adipose-derived mesenchymal stem cells enhance angiogenesis through VEGF/VEGF-R. *Int. J. Biochem. Cell Biol.* **2019**, *109*, 59–68. [CrossRef]
70. Liang, B.; Liang, J.-M.; Ding, J.-N.; Xu, J.; Xu, J.-G.; Chai, Y. Dimethyloxaloylglycine-stimulated human bone marrow mesenchymal stem cell-derived exosomes enhance bone regeneration through angiogenesis by targeting the AKT/mTOR pathway. *Stem Cell Res. Ther.* **2019**, *10*, 1–11. [CrossRef]
71. Ma, J.; Zhao, Y.; Sun, L.; Sun, X.; Zhao, X.; Sun, X.; Qian, H.; Xu, W.; Zhu, W. Exosomes Derived from Akt -Modified Human Umbilical Cord Mesenchymal Stem Cells Improve Cardiac Regeneration and Promote Angiogenesis via Activating Platelet-Derived Growth Factor, D. *Stem Cells Transl. Med.* **2016**, *6*, 51–59. [CrossRef] [PubMed]
72. Xu, H.-J.; Liao, W.; Liu, X.-Z.; Hu, J.; Zou, W.-Z.; Ning, Y.; Yang, Y.; Li, Z.-H. Down-regulation of exosomal microRNA-224-3p derived from bone marrow–derived mesenchymal stem cells potentiates angiogenesis in traumatic osteonecrosis of the femoral head. *FASEB J.* **2019**, *33*, 8055–8068. [CrossRef] [PubMed]
73. Zhang, L.; Jiao, G.; Ren, S.; Zhang, X.; Li, C.; Wu, W.; Wang, H.; Liu, H.; Zhou, H.; Chen, Y. Exosomes from bone marrow mesenchymal stem cells enhance fracture healing through the promotion of osteogenesis and angiogenesis in a rat model of nonunion. *Stem Cell Res. Ther.* **2020**, *11*, 1–15. [CrossRef]
74. Zhang, Y.; Hao, Z.; Wang, P.; Xia, Y.; Wu, J.; Xia, D.; Fang, S.; Xu, S. Exosomes from human umbilical cord mesenchymal stem cells enhance fracture healing through HIF-1α-mediated promotion of angiogenesis in a rat model of stabilized fracture. *Cell Prolif.* **2019**, *52*, e12570. [CrossRef] [PubMed]
75. Zhang, B.; Wu, X.; Zhang, X.; Sun, Y.; Yan, Y.; Shi, H.; Zhu, Y.; Wu, L.; Pan, Z.; Zhu, W.; et al. Human Umbilical Cord Mesenchymal Stem Cell Exosomes Enhance Angiogenesis Through the Wnt4/β-Catenin Pathway. *Stem Cells Transl. Med.* **2015**, *4*, 513–522. [CrossRef] [PubMed]
76. McBride, J.D.; Rodriguez-Menocal, L.; Guzman, W.; Candanedo, A.; Garcia-Contreras, M.; Van Badiavas, E. Bone Marrow Mesenchymal Stem Cell-Derived CD63+ Exosomes Transport Wnt3a Exteriorly and Enhance Dermal Fibroblast Proliferation, Migration, and Angiogenesis In Vitro. *Stem Cells Dev.* **2017**, *26*, 1384–1398. [CrossRef] [PubMed]
77. Choi, H.-J.; Pena, G.N.A.; Pradeep, S.; Cho, M.S.; Coleman, R.L.; Sood, A.K. Anti-vascular therapies in ovarian cancer: Moving beyond anti-VEGF approaches. *Cancer Metastasis Rev.* **2015**, *34*, 19–40. [CrossRef]
78. Li, J.W.; Li, J.; Han, Z.; Chen, Z. Mesenchymal stromal cells-derived exosomes alleviate ischemia/reperfusion injury in mouse lung by transporting anti-apoptotic miR-21-5p. *Eur. J. Pharmacol.* **2019**, *852*, 68–76. [CrossRef]
79. Huang, C.; Luo, W.; Ye, Y.; Lin, L.; Wang, Z.; Luo, M.-H.; Song, Q.-D.; He, X.-P.; Chen, H.; Kong, Y.; et al. Characterization of inflammatory factor-induced changes in mesenchymal stem cell exosomes and sequencing analysis of exosomal microRNAs. *World J. Stem Cells* **2019**, *11*, 859–890. [CrossRef]
80. Wang, T.; Gilkes, D.M.; Takano, N.; Xiang, L.; Luo, W.; Bishop, C.J.; Chaturvedi, P.; Green, J.J.; Semenza, G.L. Hypoxia-inducible factors and RAB22A mediate formation of microvesicles that stimulate breast cancer invasion and metastasis. *Proc. Natl. Acad. Sci. USA* **2014**, *111*, E3234–E3242. [CrossRef]
81. Gao, S.; Chen, T.; Hao, Y.; Zhang, F.; Tang, X.; Wang, D.; Wei, Z.; Qi, J. Exosomal miR-135a derived from human amnion mesenchymal stem cells promotes cutaneous wound healing in rats and fibroblast migration by directly inhibiting LATS2 expression. *Stem Cell Res. Ther.* **2020**, *11*, 1–11. [CrossRef] [PubMed]

82. Tao, S.-C.; Guo, S.-C.; Li, M.; Ke, Q.-F.; Guo, Y.-P.; Zhang, C.-Q. Chitosan Wound Dressings Incorporating Exosomes Derived from MicroRNA-126-Overexpressing Synovium Mesenchymal Stem Cells Provide Sustained Release of Exosomes and Heal Full-Thickness Skin Defects in a Diabetic Rat Model. *Stem Cells Transl. Med.* **2017**, *6*, 736–747. [CrossRef] [PubMed]
83. Zhu, Z.; Zhang, Y.; Zhang, Y.; Zhang, H.; Liu, W.; Zhang, N.; Zhang, X.; Zhou, G.; Wu, L.; Hua, K.; et al. Exosomes derived from human umbilical cord mesenchymal stem cells accelerate growth of VK2 vaginal epithelial cells through MicroRNAs in vitro. *Hum. Reprod.* **2018**, *34*, 248–260. [CrossRef] [PubMed]
84. Li, D.; Zhang, D.; Tang, B.; Zhou, Y.; Guo, W.; Kang, Q.; Wang, Z.; Shen, L.; Wei, G.; He, D. Exosomes from Human Umbilical Cord Mesenchymal Stem Cells Reduce Damage from Oxidative Stress and the Epithelial-Mesenchymal Transition in Renal Epithelial Cells Exposed to Oxalate and Calcium Oxalate Monohydrate. *Stem Cells Int.* **2019**, *2019*, 1–10. [CrossRef]
85. Cloonan, S.M.; Choi, A.M. Mitochondria in lung disease. *J. Clin. Investig.* **2016**, *126*, 809–820. [CrossRef]
86. Maremanda, K.P.; Sundar, I.K.; Rahman, I. Protective role of mesenchymal stem cells and mesenchymal stem cell-derived exosomes in cigarette smoke-induced mitochondrial dysfunction in mice. *Toxicol. Appl. Pharmacol.* **2019**, *385*, 114788. [CrossRef]
87. Hough, K.P.; Trevor, J.L.; Strenkowski, J.G.; Wang, Y.; Chacko, B.K.; Tousif, S.; Chanda, D.; Steele, C.; Antony, V.B.; Dokland, T.; et al. Exosomal transfer of mitochondria from airway myeloid-derived regulatory cells to T cells. *Redox Biol.* **2018**, *18*, 54–64. [CrossRef]
88. Jiang, X.; Jiang, X.; Qu, C.; Chang, P.; Zhang, C.; Qu, Y.; Liu, Y. Intravenous delivery of adipose-derived mesenchymal stromal cells attenuates acute radiation-induced lung injury in rats. *Cytotherapy* **2015**, *17*, 560–570. [CrossRef]
89. McCabe, A.; MacNamara, K.C. Macrophages: Key regulators of steady-state and demand-adapted hematopoiesis. *Exp. Hematol.* **2016**, *44*, 213–222. [CrossRef]
90. Zuo, R.; Liu, M.-H.; Wang, Y.; Li, J.; Wang, W.; Wu, J.; Sun, C.; Li, B.; Wang, Z.; Lan, W.-R.; et al. BM-MSC-derived exosomes alleviate radiation-induced bone loss by restoring the function of recipient BM-MSCs and activating Wnt/β-catenin signaling. *Stem Cell Res. Ther.* **2019**, *10*, 1–13. [CrossRef]
91. Yao, Y.; Chen, R.; Wang, G.; Zhang, Y.; Liu, F. Exosomes derived from mesenchymal stem cells reverse EMT via TGF-β1/Smad pathway and promote repair of damaged endometrium. *Stem Cell Res. Ther.* **2019**, *10*, 1–17. [CrossRef]
92. Nagaishi, K.; Mizue, Y.; Chikenji, T.; Otani, M.; Nakano, M.; Konari, N.; Fujimiya, M. Mesenchymal stem cell therapy ameliorates diabetic nephropathy via the paracrine effect of renal trophic factors including exosomes. *Sci. Rep.* **2016**, *6*, 34842. [CrossRef] [PubMed]
93. Zhang, C.; Zhu, Y.; Zhang, Y.; Gao, L.; Zhang, N.; Feng, H. Therapeutic Potential of Umbilical Cord Mesenchymal Stem Cells for Inhibiting Myofibroblastic Differentiation of Irradiated Human Lung Fibroblasts. *Tohoku J. Exp. Med.* **2015**, *236*, 209–217. [CrossRef]
94. Liu, D.; Kong, F.; Yuan, Y.; Seth, P.; Xu, W.; Wang, H.; Xiao, F.; Wang, L.-S.; Zhang, Q.; Yang, Y.; et al. Decorin-Modified Umbilical Cord Mesenchymal Stem Cells (MSCs) Attenuate Radiation-Induced Lung Injuries via Regulating Inflammation, Fibrotic Factors, and Immune Responses. *Int. J. Radiat. Oncol.* **2018**, *101*, 945–956. [CrossRef] [PubMed]
95. Gong, W.; Guo, M.; Han, Z.; Wang, Y.; Yang, P.; Xu, C.; Wang, Q.; Du, L.; Li, Q.; Zhao, H.; et al. Mesenchymal stem cells stimulate intestinal stem cells to repair radiation-induced intestinal injury. *Cell Death Dis.* **2016**, *7*, e2387. [CrossRef] [PubMed]
96. Rodgers, K.; Jadhav, S.S. The application of mesenchymal stem cells to treat thermal and radiation burns. *Adv. Drug Deliv. Rev.* **2018**, *123*, 75–81. [CrossRef]
97. Qian, L.; Cen, J. Hematopoietic Stem Cells and Mesenchymal Stromal Cells in Acute Radiation Syndrome. *Oxid. Med. Cell. Longev.* **2020**, *2020*, 8340756. [CrossRef]
98. Zahran, R.; Ghozy, A.; Elkholy, S.S.; El-Taweel, F.; El-Magd, M.A. Combination therapy with melatonin, stem cells and extracellular vesicles is effective in limiting renal ischemia-reperfusion injury in a rat model. *Int. J. Urol.* **2020**. [CrossRef]
99. Lin, K.-C.; Yip, H.-K.; Shao, P.-L.; Wu, S.-C.; Chen, K.-H.; Chen, Y.-T.; Yang, C.-C.; Sun, C.-K.; Kao, G.-S.; Chen, S.-Y.; et al. Combination of adipose-derived mesenchymal stem cells (ADMSC) and ADMSC-derived exosomes for protecting kidney from acute ischemia–reperfusion injury. *Int. J. Cardiol.* **2016**, *216*, 173–185. [CrossRef]
100. Wang, T.; Jian, Z.; Baskys, A.; Yang, J.; Li, J.; Guo, H.; Hei, Y.; Xian, P.; He, Z.; Li, Z.; et al. MSC-derived exosomes protect against oxidative stress-induced skin injury via adaptive regulation of the NRF2 defense system. *Biomaterials* **2020**, *257*, 120264. [CrossRef]
101. Li, Q.C.; Liang, Y.; Su, Z.B. Prophylactic treatment with MSC-derived exosomes attenuates traumatic acute lung injury in rats. *Am. J. Physiol. Lung Cell. Mol. Physiol.* **2019**, *316*, L1107–L1117. [CrossRef] [PubMed]
102. Rager, T.M.; Olson, J.K.; Zhou, Y.; Wang, Y.; Besner, G.E. Exosomes secreted from bone marrow-derived mesenchymal stem cells protect the intestines from experimental necrotizing enterocolitis. *J. Pediatr. Surg.* **2016**, *51*, 942–947. [CrossRef] [PubMed]
103. Ma, T.; Fu, B.; Yang, X.; Xiao, Y.; Pan, M. Adipose mesenchymal stem cell-derived exosomes promote cell proliferation, migration, and inhibit cell apoptosis via Wnt/β-catenin signaling in cutaneous wound healing. *J. Cell. Biochem.* **2019**, *120*, 10847–10854. [CrossRef] [PubMed]
104. Chang, H.H.Y.; Pannunzio, N.R.; Adachi, N.; Lieber, M.R. Non-homologous DNA end joining and alternative pathways to double-strand break repair. *Nat. Rev. Mol. Cell Biol.* **2017**, *18*, 495–506. [CrossRef]
105. Wen, S.; Dooner, M.; Papa, E.; Del Tatto, M.; Pereira, M.; Borgovan, T.; Cheng, Y.; Goldberg, L.; Liang, O.; Camussi, G.; et al. Biodistribution of Mesenchymal Stem Cell-Derived Extracellular Vesicles in a Radiation Injury Bone Marrow Murine Model. *Int. J. Mol. Sci.* **2019**, *20*, 5468. [CrossRef]

106. Roy, S.J.; Koval, O.M.; Sebag, S.C.; Ait-Aissa, K.; Allen, B.G.; Spitz, D.R.; Grumbach, I.M. Inhibition of CaMKII in mitochondria preserves endothelial barrier function after irradiation. *Free. Radic. Biol. Med.* **2020**, *146*, 287–298. [CrossRef]
107. Schultz, G.S.; Grant, M.B. Neovascular growth factors. *Eye* **1991**, 170–180. [CrossRef]
108. Patt, L.M.; Houck, J.C. Role of polypeptide growth factors in normal and abnormal growth. *Kidney Int.* **1983**, *23*, 603–610. [CrossRef]
109. Clayton, A.; Turkes, A.; DeWitt, S.; Steadman, R.; Mason, M.D.; Hallett, M.B. Adhesion and signaling by B cell-derived exosomes: The role of integrins. *FASEB J.* **2004**, *18*, 977–979. [CrossRef]
110. Rana, S.; Yue, S.; Stadel, D.; Zöller, M. Toward tailored exosomes: The exosomal tetraspanin web contributes to target cell selection. *Int. J. Biochem. Cell Biol.* **2012**, *44*, 1574–1584. [CrossRef]
111. Gong, X.; Liu, H.; Wang, S.; Liang, S.; Wang, G. Exosomes derived from SDF1-overexpressing mesenchymal stem cells inhibit ischemic myocardial cell apoptosis and promote cardiac endothelial microvascular regeneration in mice with myocardial infarction. *J. Cell. Physiol.* **2019**, *234*, 13878–13893. [CrossRef] [PubMed]
112. Stone, R.C.; Pastar, I.; Ojeh, N.; Chen, V.; Liu, S.; Garzon, K.I.; Tomic-Canic, M. Epithelial-mesenchymal transition in tissue repair and fibrosis. *Cell Tissue Res.* **2016**, *365*, 495–506. [CrossRef] [PubMed]
113. Abramowicz, A.; Łabaj, W.; Mika, J.; Szołtysek, K.; Ślęzak-Prochazka, I.; Mielańczyk, Ł.; Story, M.D.; Pietrowska, M.; Polański, A.; Widłak, P. MicroRNA Profile of Exosomes and Parental Cells is Differently Affected by Ionizing Radiation. *Radiat. Res.* **2020**, *194*, 133. [CrossRef] [PubMed]
114. Jeyaram, A.; Lamichhane, T.N.; Wang, S.; Zou, L.; Dahal, E.; Kronstadt, S.M.; Levy, D.; Parajuli, B.; Knudsen, D.R.; Chao, W.; et al. Enhanced Loading of Functional miRNA Cargo via pH Gradient Modification of Extracellular Vesicles. *Mol. Ther.* **2020**, *28*, 975–985. [CrossRef] [PubMed]
115. Zhu, W.; Huang, L.; Li, Y.; Zhang, X.; Gu, J.; Yan, Y.; Xu, X.; Wang, M.; Qian, H.; Xu, W. Exosomes derived from human bone marrow mesenchymal stem cells promote tumor growth in vivo. *Cancer Lett.* **2012**, *315*, 28–37. [CrossRef]
116. Farias, V.D.A.; O'Valle, F.; Serrano-Saenz, S.; Anderson, P.; Andrés-León, E.; López-Peñalver, J.J.; Tovar, I.; Nieto, A.; Santos, A.; Martin, F.; et al. Exosomes derived from mesenchymal stem cells enhance radiotherapy-induced cell death in tumor and metastatic tumor foci. *Mol. Cancer* **2018**, *17*, 1–12. [CrossRef]

 cells

MDPI

Review

Chronic Inflammation and Radiation-Induced Cystitis: Molecular Background and Therapeutic Perspectives

Carole Helissey [1,2], Sophie Cavallero [1], Clément Brossard [3], Marie Dusaud [4], Cyrus Chargari [1,5,6] and Sabine François [1,*]

1 Department of Radiation Biological Effects, French Armed Forces Biomedical Research Institute, 91220 Brétigny-sur-Orge, France; carole.helissey@intradef.gouv.fr (C.H.); sophie.cavallero@def.gouv.fr (S.C.); cyrus.chargari@gustaveroussy.fr (C.C.)
2 Clinical Unit Research, HIA Bégin, 94160 Saint-Mandé, France
3 Radiobiology of Medical Exposure Laboratory (LRMed), Institute for Radiological Protection and Nuclear Safety (IRSN), 92260 Fontenay-aux-Roses, France; clement.brossard@irsn.fr
4 Department of Urology, HIA Bégin, 94160 Saint-Mand, France; marie.dusaud@gmail.com
5 Gustave Roussy Comprehensive Cancer Center, Department of Radiation Oncology, 94805 Villejuif, France
6 French Military Health Academy, Ecole du Val-de-Grâce (EVDG), 75005 Paris, France
* Correspondence: sfm.francois@gmail.com

Abstract: Radiation cystitis is a potential complication following the therapeutic irradiation of pelvic cancers. Its clinical management remains unclear, and few preclinical data are available on its underlying pathophysiology. The therapeutic strategy is difficult to establish because few prospective and randomized trials are available. In this review, we report on the clinical presentation and pathophysiology of radiation cystitis. Then we discuss potential therapeutic approaches, with a focus on the immunopathological processes underlying the onset of radiation cystitis, including the fibrotic process. Potential therapeutic avenues for therapeutic modulation will be highlighted, with a focus on the interaction between mesenchymal stromal cells and macrophages for the prevention and treatment of radiation cystitis.

Keywords: radiation therapy; radiation cystitis; fibrosis; treatment; stem cells therapy; macrophages

Citation: Helissey, C.; Cavallero, S.; Brossard, C.; Dusaud, M.; Chargari, C.; François, S. Chronic Inflammation and Radiation-Induced Cystitis: Molecular Background and Therapeutic Perspectives. *Cells* **2021**, *10*, 21. https://dx.doi.org/10.3390/cells10010021

Received: 4 November 2020
Accepted: 22 December 2020
Published: 24 December 2020

1. Introduction

External pelvic radiation therapy is an important tool in the therapeutic arsenal for the treatment of pelvic cancers, such as prostate cancer, cervical cancer, rectal cancer or bladder cancer. Improvements in radiation techniques, such as intensity-modulated radiotherapy (IMRT), stereotactic radiotherapy and image-guided brachytherapy, have made it possible to deliver increasingly effective doses in smaller volumes with a clear improvement in treatment tolerance. However, the bladder is a critical organ thatmay be sensitive to low doses of radiation. Despite improved techniques, pelvic irradiation is still responsible for acute and/or late adverse events affecting the bladder. The term "radiation cystitis" therefore includes all lesions and symptoms of the bladder following the irradiation of the pelvic organs. Its severity is related to the volume of radiation exposure, the total dose delivered as well as the administration schedule and fractionation. This adverse event may have an impact on patients' quality of life. As cancer patient survival improves, long-term survivorship issues are of increasing importance, and an improved understanding of radiation-induced cystitis mechanisms is essential [1].

In this review, we review the available literature on clinical presentation and pathophysiology of acute and late radiation cystitis. Then, currently available treatments are examined. Due to the lack of long-term clinical benefit, other therapeutic avenues must be developed for the management of this adverse event. Finally, we highlight the place of immunity in the pathological processes of radiation cystitis and its potential as a therapeutic target, focusing on the interaction between Mesenchymal Stromal Cells (MSCs) and macrophages.

2. Background Information

The reported incidence of radiation cystitis ranges from 9.1% to 80% [2]. This variability is linked to methods of evaluation and monitoring. Indeed, symptoms of late radiation cystitis may occur very late (sometimes decades) after their therapeutic irradiation, and some patients may be lost to follow-up. Similarly, acute manifestations may be underestimated (and therefore not reported) while irreversible radiation-induced bladder lesions are developing. Late radiation cystitis is the result of an ongoing process of destruction of bladder tissue and histological changes, and a continuum between acute and late radiation lesions do exist.

2.1. Acute Radiation Cystitis

Acute radiation cystitis is defined as any adverse event occurring during or up to threemonths after the end of radiation therapy (the threshold of sixmonths was also proposed). Its incidence is estimated at nearly 50% following pelvic irradiation at full curative doses (e.g., prostate or locally advanced cervical cancer treatment). Clinical symptoms may include increased urinary urgency and frequency (pollakiuria), both during the day and at night, dysuria, but also cystalgia with bladder spasms, and hematuria, albeit rarely at this early stage. An international grade classification ranging from 1 to 5 can be used to assess the severity and impact on the quality of life (Figure 1) [3]. Acute radiation tissue injury to the bladder is caused primarily by damage to the bladder mucosa. It involves an acute inflammatory response and tissue edema. Urothelial regeneration thus comes to a halt, and the epithelium is desquamated with no regeneration, which results in urothelial lesions making the bladder vulnerable to trauma and infections [4]. These lesions are characterized by edema, hyperemia and inflammation of the mucous membrane. In most of the cases, the prognosis is favorable, as these reactions usually disappear spontaneously within fourto sixweeks after the completion of radiation therapy [4,5], but an interruption of radiation therapy may be considered in case of severe grade 3–4 symptoms. Such treatment disruptions may potentially lead to a decrease in tumor control because of an increase in overall treatment time and should, therefore, be discussed on an individual basis [6].

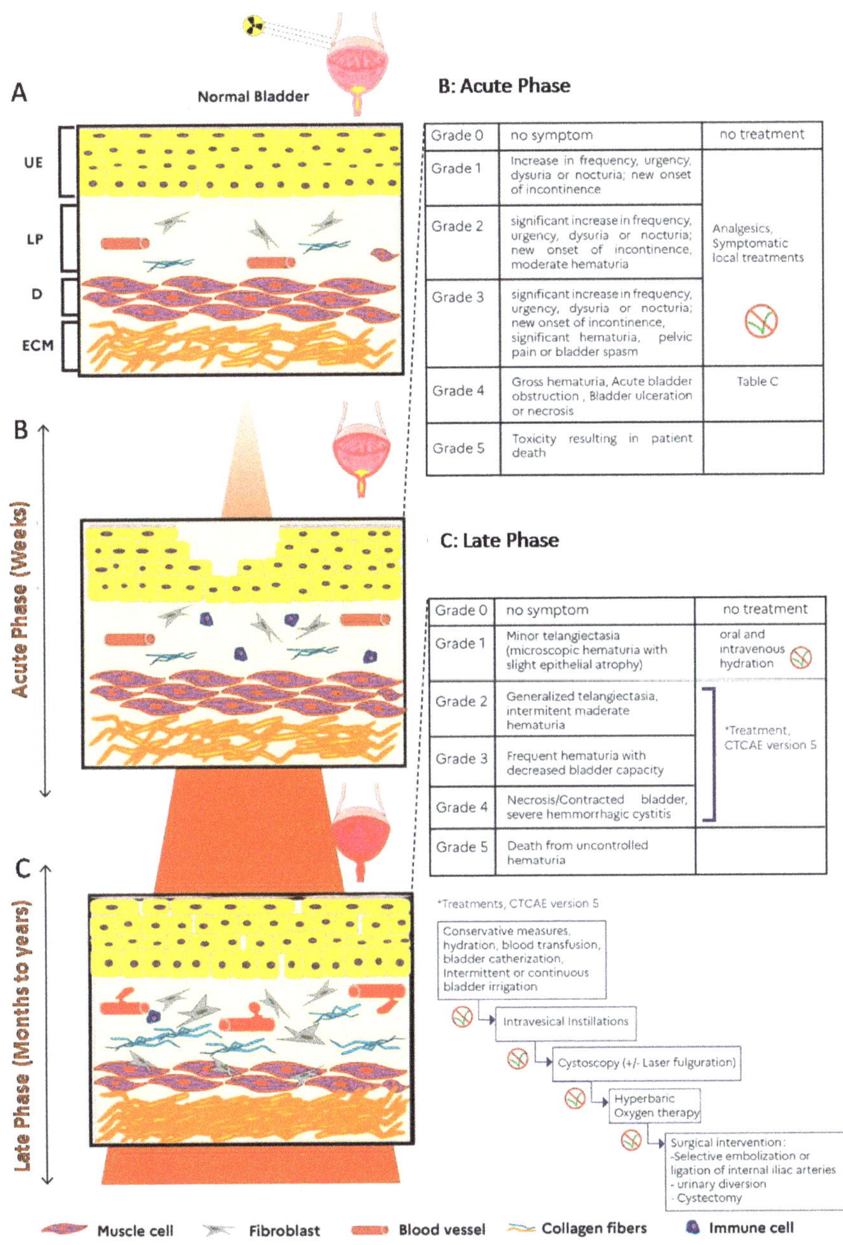

Figure 1. Illustration of radiation cystitis (RC) and clinical management adapted with "Modeling and treatment of radiation cystitis [7], development of RC after radiotherapy (**A**), in the acute phase (infiltration of immune cells into the lamina propria (LP) and depletion of the urothelium (UE)), in latent phase (proliferation of fibroblasts with hematuria, dilation of vessels, bleeding, decrease in the detrusor muscle layer (D) and production of collagen in LP and extracellular matrix(ECM). (F: fibroblast, V: vessel, M: muscle cells, C: collagen fibers, I: immune cells) (**B**) Clinical management during the acute phase of CR. (**C**) Clinical management during the latent phase of CR. * corresponds to cascade treatments of grades 2 to 4 according to CTCAE version 5 [3].

2.2. Late Radiation Cystitis

Late radiation cystitis is defined as an adverse event associated with pelvic irradiation that occurs after a minimum of threemonths and possibly even several years after completion of radiation therapy. Toxicities occurring between three and sixmonths are sometimes considered as "early delayed". On average, late radiation cystitis appears within the following 2–3 years. The incidence of late symptomatic radiation cystitis is stable over time at 5–10%, despite improved radiation techniques [8–10]. The clinical presentation can be variable, including bladder pain, urinary urgency, isolated urinary disorders and pollakiuria. Given that these symptoms are nonspecific and appear long after treatment, urine culture, or even cystoscopy may be useful to rule out other differential diagnoses. The most pathognomonic clinical feature is recurrent hematuria, with varying severity. In its most (and rare) severe forms, late radiation cystitis may be life-threatening. The incidence of late radiation cystitis was approximately 5% at 5 years and 10% at 20 years with conventional radiotherapy techniques [9]. It is important to eliminate any local recurrence or new cancer by performing cystoscopy. It should be highlighted that severe late symptoms (e.g., fistulas) may be worsened by inappropriate bladder biopsies, which should therefore be avoided in previously irradiated areas. Patients with a pelvic tumor extending to the bladder are also at high risk of fistulas [11]. A classification of this adverse event was developed (Figure 1). Severe late radiation cystitis is related to the volume and the dose of radiation exposure, the administration schedule and the technique used, but it is also important to identify patients with risk factors for developing a severe form. Marks et al. reported that patients with co-morbidities, such as hypertension, diabetes, a history of abdominal surgery, and patients receiving concomitant chemotherapy were at higher risk of developing radiation cystitis, especially in its late form [12]. Recent data suggested that after high dose exposures (such as after brachytherapy treatment), some anatomic subpart of the bladder may be at higher risk of complication, such as the bladder neck [13]. Although the pathophysiology of late radiation cystitis still remains unclear, endothelial cells appear to play an important role in this mechanism. Indeed, the submucosal vascularity is damaged by fibrosis of the vascular intima resulting in vessel obliteration and submucosal/muscular fibrosis. This is followed by urothelial atrophy, hypoxia with hypovascularization and ischemia of the bladder leading to the development of fibrosis and atrophy of the bladder tissue with the emergence of neovascularization in the form of telangiectasia that may easily bleed [14,15]. At the later stage, reduction in bladder capacity is observed linked to complete bladder fibrosis, mucosal ulcers with the risk of fistulization and spontaneous perforations of the bladder (or fistulae resulting from biopsies).

3. Current Treatments and Clinical Trials

3.1. Acute and Late Radiation Cystitis with Storage, Voiding Symptoms or Occasional Bleeding

The clinical management of storage symptoms for acute and late radiation cystitis is largely symptomatic with analgesics and anti-inflammatory drugs. Good hydration is recommended for patients in order to increase diuresis, cleanse the bladder, and avoid urinary obstruction resulting from blood clots [16].

Likewise, anticholinergics, like oxybutynin, trospium chloride, solifenacin, fesoterodine or flavoxate hydrochloride, can be prescribed to help alleviate urgency and increased daytime frequency. Their action is to decrease the contractility of the detrusor and improve symptoms [4].

In some cases, antibiotics may be proposed to prevent the condition from worsening in the event of infection.

Alpha-blockers, 5-reductase inhibitors or phosphodiesterase 5 inhibitors may be useful to alleviate voiding symptoms. Their action is to decrease the tone of the posterior urethra, bladder neck and the volume of the prostate [1]. In severe cases, it is sometimes necessary to hospitalize the patient for transfusions or clot evacuation [4]. In fact, bladder irrigations are performed in order to obtain a dilution of hematuria and drain the clots. It is a sterile technique with lubrication for standard catheter insertion with a large three-way catheter.

Blood clot evacuation is performed manually by using a large Toomey or catheter syringe until no further clots and output begin to clear. Then, we use normal saline (0.9%) for continuous irrigation [17].

If acute active bleeding does persist and is refractory to irrigations, electrocoagulation should be discussed, as described by Martinez and colleagues [18]. The procedure was performed with a rigid 22 French cystoscope. It was performed to identify the source of bleeding and rule out any other unidentified pathology. The Green Light laser was used to target any active source of bleeding. These areas were coagulated with the laser. Throughout the procedure, saline irrigation was used, and care was taken to ensure that the ureteral orifices were not injured. At the completion of the procedure, the bladder was drained under direct visualization to ensure adequate hemostasis. Very minimal bladder mucosal damage was reported. Then, a large three-way catheter was placed, and continuous irrigation was maintained overnight and stopped the next morning [18].

These treatments are tailored according to the severity of the symptoms (Figure 1).

3.2. *Late Radiation Cystitis with Persistent or Recurrent Hematuria*

3.2.1. Intravesical Instillations

Different molecules have been used for this indication, with different mechanisms of action. Their objectives are sterilization, cleansing and arrest of focal bleeding points.

Aluminum salt: Intravesical aluminous salts are considered astringent agents. They exert their action through protein precipitation on the cell surface and in interstitial spaces. They decrease blood vessel diameter and stiffness of capillary endothelium [9,19]. Aluminum salts are typically delivered as a 1% concentration of alum mixed with sterile water. Westerman et al. evaluated the benefit of alum instillations in 40 patients with hematuria, which was linked in 95% of patients to radiation cystitis [20]. These instillations led to a reduction in transfusion requirements (82% before instillation vs. 59% after instillation, $p = 0.05$). Moreover, 32.5% of patients did not require additional treatment after a median follow-up of 17 months. Tolerance was generally good. The main side effect reported was bladder spasm in 35% of patients [20].

Formalin: Formalin action consists of precipitating cellular proteins in the mucosa of the bladder. The consequence is to create occlusion within telangiectatic tissue. It appears to be the most effective intravesical agent with complete resolution rates ranging from 70 to 89%. However, the safety profile for this treatment is mediocre. First of all, its instillation is quite painful and must therefore be performed under general anesthesia. In addition, formalin has a high rate of morbidity and mortality (31%), with risks of vesicoureteral reflux complicated by severe bilateral pyelonephritis, ureteral stenosis and fibrosis of the bladder with reduced capacity and increased urinary frequency [21]. To date, its use remains very limited due to its poor safety profile.

Hyaluronic acid: Hyaluronic acid is a mucopolysaccharidethathelps to repair the normal glycosaminoglycan layer of the bladder when administrated through intravesical instillations. It has immunomodulatory properties that enhance connective tissue healing. Shao et al. evaluated the efficacy of intravesical hyaluronic acid (HA) instillation and hyperbaric oxygen (HBO) in the management of hemorrhagic radiation cystitis [22]. The clinical benefit was identical in the 2 groups but was maintained over time significantly in the HA arm. Indeed, complete resolution of hematuria was noted in 88%, 75%, and 50% of HA patients and in 75%, 50%, and 45% of patients in the HBO group, at 6-, 12- and 18-months following therapy, respectively. Hyaluronic acid appears to be an interesting therapeutic alternative, though this must be confirmed in a larger cohort.

Other agents have shown interesting results but have been studied only in small cohorts, like botulinum toxin, chondroitin sulfate, polydeoxyribonucleotides, early placental extract [23–26].

3.2.2. Hyperbaric Oxygen Therapy (HBOT)

This technique consists of placing the patient in a pressurized chamber (hyperbaric chamber) to administer pure or mixed oxygen at a pressure greater than atmospheric pressure, for 5–7 days a week, for a daily duration of 60–90 min up to approximately 30–45 sessions [27]. The effect of hyperbaric oxygen therapy is to allow better oxygen diffusion in tissues and to disrupt the continuum between hypoxia and fibrosis. Hyperoxia induces primary neovascularization, secondary growth of healthy granulation tissue, and induces short-term vasoconstriction, which may help control active bleeding [28,29]. It is the most widely reported therapeutic technique in the management of hemorrhagic radiation cystitis. Dellis et al. evaluated the benefit of HBOT in 38 patients with severe radiation cystitis. The complete response rate was 86.8%, and the partial response rate was 13.2%. The mean follow-up was 29.3 months. For the thirty-three patients with complete response who received HBO therapy within 6 months of the hematuria onset, the mean time interval was 4.9 months (range 1–6), while in the remaining five patients with partial response, the mean time interval was 22 months (range 8–48) ($p < 0.001$). Thirty-three patients were alive at the end of follow-up [29].

Recently, the randomized phase 2–3 RICH-ART evaluated the benefit of HBOT compared to standard of care for patients with late radiation cystitis and a value of fewerthan 80 points in the urinary domain of the expanded prostate index composite score (EPIC score). Forty-one patients were randomized in the HBOT arm and 38 in the standard of care arm. HBOT significantly alleviated patient-perceived symptoms of late radiation cystitis and improved HRQOL. The mean improvement in EPIC urinary total score was higher (17.8 [SD 18.4]) in the hyperbaric oxygen therapy group compared with patients in the control group (7.7 [SD 15.5]). Seventy patients in HBOT presented a grade 1–2 adverse events. The main adverse events grade 1–2 were ear pain (15%), myopia (12%) and barotrauma (10%). No grade 3–4 or 5 was reported in this group [30]. The HBOT's benefit was maintained in the time. In fact, Pereira et al. reviewed 105 patients diagnosed with RIHC whowere treated with HBOT between 2007 and 2016. After a median follow-up of 63 months, 76.3% had a complete response [31]. Cardinal et al. evaluated the benefits of HBOT through a meta-analysis of data from 602 patients treated with HBOT for hemorrhagic radiation cystitis. They determined that 84% of patients achieved partial or complete resolution, while 75% saw an improvement in hematuria. In their analysis of 499 patients with documented follow-up, authors observed a recurrence rate of 14%, with a median time to recurrence of 10 months (6 to 16.5 months). To summarize, this treatment is well-tolerated, the most common side effects being pressure-related, most notably ear and sinus barotrauma. HBOT is offered to patients for whom bladder washings and instillations are ineffective [32]. In a systemic review, Villeirs et al. emphasized HBOT benefit in radiation cystitis. In a cohort of 815 patients, an overall and complete response rate varied from 64.8% to 100% and 20% to 100%, respectively. Blood transfusion before HBOT, other treatment modalities before HBOT, use of anticoagulant therapy, along the interval between the onset of hematuria and start of HBOT were possible factors associated with lower efficacy of HBOT [27]. It is important to start HBOT in the onset of late radiation cystitis symptoms [33,34]. However, the availability and cost-effectiveness of high-pressure oxygen tanks is a critical factor in the success of HBOT [27].

3.3. *Late Radiation Cystitis with Refractory or Life-Threatening Hematuria*

In late radiation cystitis with refractory or life-threatening hematuria, the treatments aim at achieving volume expansion and at limiting the need for frequent transfusions due to active bleeding [35].

3.3.1. Arterial Embolization

Improvements in interventional radiological techniques have led to improvements in morbidity and mortality compared with surgery in patients with refractory hemorrhagic radiation cystitis. The technical success rate reported is 88–100%. The main adverse events were Brown–Sequard's syndrome, bladder necrosis, and gluteal paresis or skin necrosis. Thanks to improved techniques, the incidence of adverse events has decreased from 65% to 9–31% [36,37]. The follow-up of these studies is brief.

3.3.2. Cystectomy and Urinary Diversion

In some patients, treatment by means of cystectomy with urinary diversion is unfortunately inevitable when clot evacuation, bladder fulguration and bladder irrigation have failed. This therapeutic option should be reserved for patients for whom local and conservative treatments have proven unsuccessful, given its high rate of morbidity and mortality. Linder et al. reported a postoperative complication rate of 42% and a 90-day mortality rate of 16% [38].

4. Clinical Trials: Other Therapeutic Avenues, Antifibrotics

4.1. Antifibrotic or Antioxidant Pharmacological Agents

Vitamin E has protectiveeffects against oxidative stress and also plays an important role in preventing lipid peroxidation in the cellular membrane [39]. Between April 2003 and July 2009, 53 breast cancer patients were recruited to determine if a combination of Pentoxifylline (PTX) and Vitamin E could prevent the development of radiation fibrosis after radiotherapy for the definitive management (NCT00583700). This clinical study of post-irradiation cancer patients treated with PTX/vitamin E showed a significant difference in radiation-induced fibrosis. Of importance, the combination of PTX/vitamin E did not impact local control or survival within the first 2 years of follow-up (which is still quite a short follow-up). The oral PTX/vitamin E treatment was safe and well-tolerated. After pelvic irradiation in high-risk patients, the combination of Pentoxifylline (PTX) and Vitamin E can thus be considered clinically useful in preventing fibrosis [40,41]. Orgotein copper-zinc superoxide dismutase (SOD) is an enzyme found in various tissues throughout the body and has a fundamental role in the elimination of reactive oxidative species and free radicals that cause tissue damage and fibrosis [42]. SOD was found to be effective in reducing radiation-induced fibrosis by a reduced pain score and a decrease in the size of the fibrotic area in half of the cases after 6 months in 44 patients with clinical radiofibrosis following conservative treatment of breast cancer [43]. However, the role of antifibrotic agents in reducing or mitigating radiation cystitis remains unknown.

4.2. Angiotensin-Converting-Enzyme Inhibitors

An initial clinical study had revealed that angiotensin-converting-enzyme inhibitors mightdecrease the incidence of radiation pneumonitis in patients receiving thoracic radiation for lung cancer [44]. An ongoing clinical study (NCT01754909) is evaluating the efficacy of enalapril on radiation pneumonitis and fibrosis for patients undergoing radiation therapy for lung cancer or other intrathoracic cancers. A recently published phase 2 prospective study supported the notion of radio-induced fibrosis reversibility, showing that pravastatin (40 mg/d for 12 months) was an efficient antifibrotic agent in patients with grade ≥ 2 cutaneous and subcutaneous fibrosis following head and neck radiotherapy [45]. However, no data are available in the setting of radiation cystitis.

5. Impact of Macrophages in the Development of Radiation Fibrosis

Recent insights regarding the functional role of inflammatory cells suggest that inflammation could play a role beyond the classical "acute" phase. During the radiation wound repair process, recruitment of inflammatory cells occurs at the site of injury, which can contribute to late inflammatory tissue damage through a continuous mechanismbetween inflammation, hypoxia and fibrosis [46]. During normal healing, sequential activation of the classical, proinflammatory, M1 and alternately activated macrophages, M2a, M2b and M2c, is known to occur, which facilitates the transitions between the inflammatory, proliferative and remodeling phases of the repair process [47]. Thus, macrophage dysfunction or deficient generation can lead to the uncontrolled production of inflammatory mediators and growth factors. This can modify their communications with other cells (epithelial and/or endothelial cells, fibroblasts, progenitors and stem cells) and contribute to a state of persistent injury, which could, in turn, lead to the onset and maintenance of a pathological fibrotic process [48,49]. Macrophages are heterogeneous cells with various phenotypes and functions in part regulated by their micro-environment. Macrophages commonly exist in two (basically defined) distinct subsets, M1 and M2 macrophages, which have different functions and transcriptional profiles (schematically: M1 macrophages are associated with the production of proinflammatory cytokines, while M2 phenotype reprogramming enhances the production of anti-inflammatory cytokines) [50–52]. The representation of macrophage subtypes and their different actions are illustrated in Figure 2. The macrophage reprogramming processes, and steps have been reviewed elsewhere. Briefly, those include 1/an amplified macrophage response, which is a direct amplification following exposure to a reprogramming factor (e.g.,INFγ) but also a cross-amplification consecutive to another factor, such as lipopolysaccharide; 2/a reciprocal suppression of the alternate phenotype (M1/M2); 3/a cascade activation of the reprogramming mechanisms; 4/a feedback phenomenon [53].There are schematically two typical macrophage reprogramming signaling pathways: on the one hand, activation pathways such as JNK, Notch, TLR/NF-κB (p65/p50), PI3K/Akt2, JAK/STAT1, and HIF1αfavor the M1 phenotype; while on the other hand pathways such as PI3K/Akt1, JAK/STAT3/6, TGF-β/SMAD, TLR/NF-κB (p50/p50) and HIF2α are mainly involved in M2 phenotype programming [54]. After a fractionatedirradiation, an abnormal wound healing response occurs, which is characterized by the accumulation of M2 macrophages that promote fibrosis through the production of TGF-β1 [48,55]. Temporal and spatial coordination of myofibroblast activities with inflammatory macrophages is crucial for the controlled healing process and restored homeostasis in injured tissue, such as after irradiation. In this context, the M1 phenotype has also been associated with an antifibrotic effect by releasing MMPs (MMP-9, MMP-12 and MMP-13 that degrade ECM). However, if the injury process persists, fibrosis progresses with the proliferation of myofibroblasts and ECM deposition (such as collagen and fibronectin) in and around inflamed or damaged tissue [56,57]. In this context, M1 macrophages represent the starting point of the profibrotic process. Indeed, M1 macrophages release proinflammatory cytokines and chemokines that indirectly promote the proliferation of myofibroblasts. In this deregulated tissue microenvironment. There is a modulating of the macrophage phenotype, in which M2 macrophages phenotype can be generated by apoptotic bodies accumulated, ECM quality modified and Th2 cytokines stimulation [58]. M2 initially involves anti-inflammatory cells, with the release of IL10, arginase, TGFβ and HO-1 [59]. When the tissue microenvironment homeostasis is deregulated, M2 activation leads to fibrocytes recruitment and proliferation, myofibroblastactivation, and fibroblasts proliferation. In addition to collagen production, pro-fibro-fibrotic genes are transcripted, leading to the secretion of a large number of profibrotic factors such as TGF-B1, PDGF, IL6, IL-13 IL-7 and galactin-3and to an increase in tissue inhibitors of metalloproteinases (TIMPs) expression [55,57,60,61]. Recent data highlighted that the exacerbation of radiation-induced pulmonary fibrosis might depend on the mesenchymal transition of epithelial cells, promoted by the TGF-β-secreting M2 macrophages [62]. Therefore, any change in the M1/M2 balance will have a central role in terms of fibrosis control or worsening. In addition, exposure to irradiation may also activate macrophages indirectly. An abnormal wound healing response occurs, which is characterized by the accumulation of M2 macrophages, which promote fibrosis through the production of TGF-β1 [63]. Non-coding RNA appears to be involved in the initiation and progression of radiation-induced lung fibrosis by modulating the M2-mediated signaling pathway [64]. In animal models of lung fibrosis, it was reported a high expression of let-7i and low expression of miR-21 at 3 weeks post-irradiation. At a later stage (after 26 weeks), let-7i expression decreased, and miR-21 was upregulated. In addition, it has been described that let-7i targets TGFBR1, inhibiting TGF-β signaling, while miR-21 degrades the TGF-β inhibitor SMAD7. In support of this, the miR-21 expression is upregulated in several models of fibrosis [65,66]. Significant functional and temporal differences have been shown among several distinct miRNAs that are found both in the inflammatory phase (immediately post-irradiation) and in the later fibrotic stages [58]. Mukherji et al. proposed that macrophage activation could be a secondary effect of radiation exposure, which may result from cellular damage signals and clearance of radiation-induced apoptotic cells, rather than a direct effect of irradiation [67]. In Oncoimmunology, Meziani and colleagues et al. provide a thorough discussion of the immune system/macrophage responses to radiotherapy and their involvement in the development of radiation injury. Macrophages are described as a promising therapeutic target for the prevention or the treatment of radiation-induced toxicities [68]. This immunomodulatory approach aims not only to increase the antitumor efficacy of radiotherapy but also to limit its side effects. To limit the initiation of fibrosis in healthy irradiated tissue, the migration of type 2 macrophages or the reprogramming of M1/M2 must be controlled [68]. However, the dynamics of myeloid cells in the bladder after pelvic radiotherapy have not yet been elucidated and must be investigated.

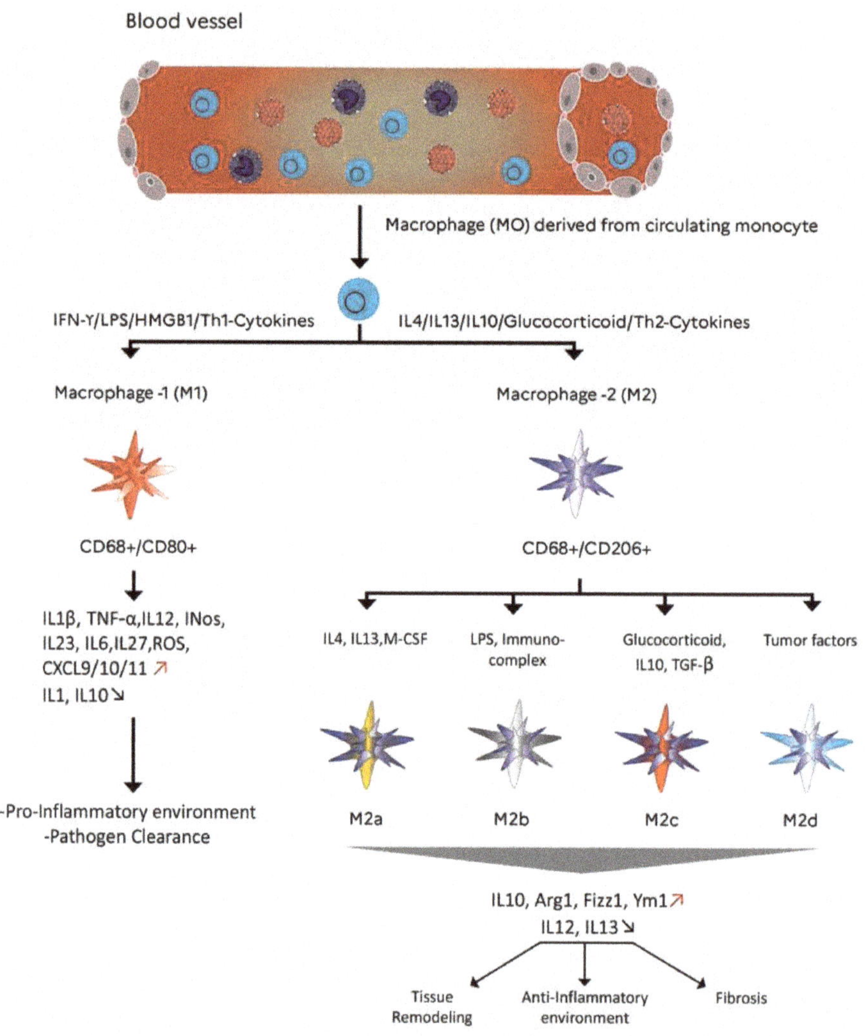

Figure 2. Role of macrophage polarization in inflammation, phenotype markers and signaling molecules involved in M1/M2, M2a, M2b, M2c, M2d. Importance of M1/M2 balance in tissue repair. The M1 CD68 +/CD80 + macrophages which are pro-inflammatory and the M2 CD68 +/CD206 + macrophages which are anti-inflammatory. Red arrows: elevation. Blue arrow: decrease

6. Preclinical Studies of Radiation Cystitis and Cell Therapy: A New Therapeutic Avenue

6.1. Preclinical Studies of Radiation-Induced Cystitis

Animal models of radiation cystitis are preferably performed in rodents. As illustrated in Table 1, the radiation exposure is mostly delivered as a single dose via an X-ray or gamma-ray irradiator. However, it appears more appropriate to use X-rays to study the effects of radiotherapy on tissues. For example, the SARRP (Small Animal Radiation Research Platform) is one of the X-ray irradiators used in preclinical research. These small animal radiotherapy devices enable state-of-the-art image-guided therapy (IGRT) research to be performed by combining high-resolution cone-beam computed tomography (CBCT) imaging with an isocentric irradiation system [69]. This radiation fractionation is, however, not clinically relevant, as most pelvic therapeutic irradiations are delivered through fractionated schemes in patients. Among the factors inducing fibrosis, Th-2 cytokines were among the first to be recognized to have strong profibrotic properties. Typical cytokines released from Th-2 cells are IL-4, IL-5, IL-10 and IL-13. Three of them, IL-4, IL-5 and IL-13, are linked to fibrosis development [70]. A few months after irradiation of the bladder, degenerative epithelial tissue, urothelial swelling, pseudo-carcinomatous epithelial hyperplasia, fibrous tissue in the lamina propria and between muscle cells, a mild increase in inflammatory cells, disruptions in tight junction formation, edema, loss of endothelial cells, urothelial hyperplasia, and bleedings (in the most severe cases) were detected [71]. As described in Table 1, three preclinical treatments were shown to reduce the development of radiation cystitis, including Hyperbaric oxygen therapy (HBOT), liposomal and tacrolimus instillations and also vasculogenic and angiogenic localized therapies [72–74]. For HBOT, 14 days after radiation, rats were treated in the chamber (95% oxygen and pressurized to 200 kPa for 90 min), twice daily, for a period of two weeks. This therapy reduced radiation oxidative stress and TGF-Beta and consequently lowered levels of IL-10. Using endothelial cells as a vasculogenic therapy and vascular endothelial growth factor (VEGF) as an angiogenic therapy wasbeneficial in the early chronic phase. But this angiogenic therapy using endothelial cells could promote tumor revascularization, although routine endothelial cells culture is still limited [75,76]. It is well documented that ionizing radiations activate the Nuclear Factor κB (NF-κB) signaling cascade directly or via induction of double-strand breaks and oxidative stress [77]. The NF-κB pathway is a link to the immune system in radiation response [78]. Thalidomide, an immunosuppressive drug thatinterferes with the activation of NF-κB, may be a valid treatment option for patients with inflammatory diseases refractory to other first- and second-line treatments. Considering the immunomodulatory effect of thalidomide, Kowaliuk et al. recently investigated the role of NF-κB and the functional effects of this treatment on radiogenic bladder dysfunction. Early thalidomide infusion after pelvic irradiation using a YXLON MG325 X-ray device showed beneficial and promising effects on the incidence and severity of bladder dysfunction [79]. The late administration of thalidomide showed no significant effect on functionality with possible neurological side effects, limiting its use [80]. Oral administration of clarithromycin or isoflavone before and after irradiation results in the anti-inflammatory macrophage subtype switch and reduction of macrophage infiltrate, respectively [81,82]. Intraperitoneal injection of Melatonin before radiation reduces lymphocytic and macrophagic infiltrates [83]. Moreover, Intraperitoneal injection over 8 weeks of Purified murine anti-IL-13 IgG antibody, starting 3 Week post radiation exposure participate ininhibition of polarization of alternatively M2-macrophages, also after Iterative IV infusion 5 × 10 Adipose -MSCs/infusion [84,85]. To increase the chances of finding a potential preclinical treatment for radiation cystitis, it is imperative to explore novel mitigators of radio-induced inflammatory reactions.

Table 1. Recent animal models for preclinical studies of radiation cystitis (RC) and preclinical studies targeting immune cells to limit the development of radio-induced fibrosis.

Animals	Method of Radiation Cystitis Induction	Treatments		Effect(s)	References
		Administration Type	Time Post-Radiation Exposure (PE)		
Adult female Sprague–Dawley rats	Single 20 Gy dose by a linear accelerator (6MV)	20 sessions of HBOT over a fortnight	14 days PE	Reduction of oxidative stress and proinflammatory factors	Oscarsson N et al., 2017
Adult female Sprague–Dawley rats	SARRP, singledose 40 Gy 6–8 weeks PE: histological tissue damage to the bladder	Liposomal tacrolimus instillation	6 weeks PE	Increase in inter-micturition intervals	Rajaganapathy BR et al., 2015
Adult female Lewis rats	A single 20 Gy dose of using a cesium isotope-based irradiator.	Injection into the bladder wall of a solution containing VEGF +/− endothelial cells	30 days PE	Revascularization of radiation-damaged urinary bladders	Soler R et al., 2011
Female BALB/c mice	Single 10 Gy dose by Siemens Stabilipan X-ray to the whole lung, Thickened alveolar septa, reflective of pneumonitis at 18 weeks PE	Isoflavone mixture gavage	Before and after radiation exposure	M1 subtype switched to an anti-inflammatory M2 subtype with increased levels of Arg-1 and decreased NOS2	Abernathy LM et al., 2015
Female C57BL/6J mice	Single 18 Gy dose by linear accelerator (21EX 3153 VARIAN) to the whole lung Interstitial edema and fibrosis sections at 16 weeks PE,	Oral clarithromycin	Before and after radiation exposure, and continuing until the day of sacrifice	Inhibition of fibrosis scoring, influx of macrophages and interstitial edema	Lee SJ et al., 2015
C57BL/6 female mice	5 × 6 Gy thoracic irradiation by X-RAD 320, Macrophage accumulation in the irradiated lung at 10 weeks PE	Purified murine anti-IL-13 IgG antibody by intraperitoneal (ip) injection	Weekly ip injection over 8 weeks, starting 3 weeksPE	Inhibition of recruitment and polarization of alternatively activated YM-1 positive macrophages	Chung SI et al., 2016
Adult male Wistar rats	^{60}Co source Single dose Gy 15 to the whole lung - Mild fibrosis at 17 weeks PE	- 1 mL of melatonin solution (100 mg/kg) - Intraperitoneal injection	30 min before irradiation	Increased levels of IL-4, DuoX1, Duox-2 and decreased lymphocyte and macrophage infiltration	Aliasgharzadeh A et al., 2019
Sprague-Dawley rats	Single 27 Gy dose by ^{60}Co irradiator inthe colorectal region, Anastomosis in the colon at 4 weeks PE	Iterative IV infusion 5 × 10 adipose-MSCs/infusion	3 weeks PE	The proportion of anti-inflammatory M2 macrophages grew, favoring the M2 phenotype and promoting wound healing	Van de Putte D et al., 2017

6.2. Stem Cell Therapy: A New Therapeutic Avenue

Because of their ability to migrate to the irradiated site and of their immunomodulatory and antioxidant properties in promoting tissue repair, mesenchymal stem (or stromal) cells (MSCs) are a potential antifibrotic therapeutic candidate [86–90]. Preclinical studies have described their beneficial effects, in particular their ability to limit the development of pulmonary and colorectal after irradiation by modulating the polarization of macrophages. From these investigations, it seems that MSCs could not only replace damaged epithelial cells but also promote tissue repair through the secretion of anti-inflammatory and antifibrotic factors [85,91,92]. However, it is important to note that these studies were performed on non-cancerous models. In a recent preclinical study of radiotherapy to treat colorectal cancer, it was shown that treatment with bone marrow (BM)-derived MSCs significantly reduced both cancer initiation and cancer progression by increasing the number of tumor-free animals as well as decreasing the number and the size of the tumors by half, thereby extending their lifespan. The attenuation of cancer progression was mediated by the capacity of the MSCs to modulate the immune component. The MSCs reprogrammed the macrophages to become regulatory cells involved in phagocytosis, thereby inhibiting the production of proinflammatory cytokines. Thus in the long term post-radiotherapy, this biotherapy allows the maintenance of tissue homeostasis and inhibits tumor progression [93]. MSCs inhibit fibrosis by reducing the expression of TGF-β1, modulating the inflammatory response, apoptosis, oxidative stress and remodeling of the extracellular matrix. In particular, preclinical studies have shown that MSCs could act on fibrosis by directing the polarization of macrophages and the differentiation of CD4+ T lymphocytes [94–96]. In response to signals derived from tissue damage, macrophages undergo reprogramming, which leads to the emergence of a spectrum of distinct functional phenotypes (Figure 2). A study by Chen et al. showed that MSCs couldpromote M2 macrophage polarization by secreting TGF-β3 and TSP1 [97]. Recent publications have shown that MSCs could induce M2 macrophages through the secretion of exosomes, and these effects could be due to the activation of transcription factors Stat6, MafB [98] and the secretion of miR-223 targeting PKNOX1 in macrophages [99]. These regulatory mechanisms are involved in acute inflammation. However, in the case of chronic radiation cystitis, fibrosis is triggered by chronic inflammation. MSCs could inhibit chronic inflammation by altering the polarization of macrophages to resolve chronic inflammation through the secretion of exosomes containing miR let-7b [100]. Moreover, HGF and TSG-6 have been shown to be major effectors of the antifibrotic activity of MSCs in several models (e.g., cutaneous and renal fibrosis [101–103]. HGF has been shown to be up the urine of prostate cancer survivors with a radiation history [104]. HGF could potentially play a dual role in radiation cystitis whereby it promotes angiogenesis and is protective against fibrosis [105]. TSG-6 is able to form hyaluronan polymers, which trigger the activation of NF-κB and the subsequent acquisition of the M1 phenotype [82]. Thus, TSG-6 could act as a negative regulator of M2 activity by promoting the availability of hyaluronan. As prolonged M2 activity has previously been associated with worsening fibrosis, newly secreted TSG-6 could be a major regulator of inflammation after MSC transplantation [106,107].

7. Discussion and Conclusions

Although irradiation techniques have improved over time, the incidence of radiation cystitis still poses a real problem for clinical management. Indeed, the management of radiation cystitis, especially in the late-stage, is based largely on symptomatic treatments. This was historically explained by the (theoretical) irreversibility of late-stage histological fibrotic lesions. Despite encouraging results, evidence that radiation cystitis can be modulated pharmacologically is insufficient and requires further confirmation as these findings are based only on small sample sizes or on retrospective analyses. The pathogenesis of fibrotic diseases remains a major challenge, due not only to the variety and multiplicity of initiating events but also to a large number of profibrotic mediators involved. Nevertheless, MSCs appear to be a promising therapeutic alternative for the treatment of fibrosis in chronic radiation cystitis. The pro- or antitumor effects of biotherapies using MSCs have been widely discussed in the literature and are one major parameter that must be better understood before clinical application [108–110]. Other stem cells may be promising treatments of hemorrhagic cystitis. The placenta is a potential source of stromal cells, with decidual stromal cells (DSCs). These stem cells are easily amplified in vitro and have greater immunosuppressive potential than BM-MSCs. DSCs inhibit alloreactive T cell proliferation better than stromal cells from other sources and induce coagulation more effectively than BM-MSCs. Iterative infusions can be considered in patients with inflammatory pathologies [111–113]. In preclinical studies, compared to bone marrow-derived MSCs, DSCs had better viability [114]. Their clinical use must, however, be optimized. It should be noted that stromal cell injections must be carried out only in a patient with a long complete remission to limit their potential implantation near dormant cancer cells. Such safety issues may limit the use of MSC in clinics. MSC paracrine action is widely described in the literature [115,116]. To limit the implantation of these cells, it is possible to use the extracellular microvesicles (Evs-MSCs) that they secrete to reduce radiation-induced lesions, including fibrosis, without exposing patients to the risk of cancer reactivation [117–119].

To increase the likelihood of finding a potential preclinical treatment for radiation cystitis, it is imperative to test novel mitigators of the radio-induced inflammatory reaction. For example, TCDO/WF10 is a chemically stabilized chlorite matrix that has previously been shown to have a positive effect in the context of chronic inflammatory conditions. It induces natural immunity and stimulates cellular defense mechanisms through its actions on natural killer cells, cytotoxic T lymphocytes, and modification of the monocyte-macrophage system. It reduces inflammation quickly so that healing can begin [120,121]. In an early-stage clinical study, 20 patients with grade 3 radiation cystitis received intravenous TCDO treatment for 5 consecutive days. From 1 to 9 months after TCDO treatment, patients had no recurrent bleeding, and no side effects from treatment with TCDO were observed [122]. In a phase-II study involving 100 patients, Veerasarn et al. evaluated the effectiveness of WF10 in combination with standard care compared to standard care alone. The complete resolution rate for hematuria was comparable in both arms (74% vs. 64% in the experimental arm and in the standard arm, respectively). However, a significant reduction in the relapse of hematuria was noted among responders in the experimental arm (47% vs. 77%, *p*-0.01). No severe adverse events were reported [123,124].

To date, no preclinical treatment without reconstructive surgery appears to fully restore the function and structure of the bladder after radiation exposure. A number of preclinical studies have described strategies for limiting fibrosis (cf. Table 1). The majority of studies have been carried out in mice on a model of radio-induced pulmonary fibrosis. Those were mitigating and continuous treatments (isoflavone mixture gavage, oral clarithromycin) that were evaluated with respect to the inflammation and migration of immune cells, including macrophages, and the development of fibrosis [81,125]. Clarithromycin (CLA) administration, before and after lung radiation exposure, reduced expression levels of TNF-α, TNFR1, TNFR2, TGF-β1, CTGF and type I collagen, and inhibits both the increased acetylation of NF-κB p65 and the elevated expression of COX-2 with reduction of both fibrosis and macrophage infiltration [81]. To reduce fibrosis, as described by Chung et al., it is necessary to target type 2 cytokines such as IL13, limiting recruitment and polarization of Ym1/Chi3L3-positive macrophages, which are alternatively activated in the lungs following thoracic irradiation. Intraperitoneal injection of IL-13 neutralizing antibody does not completely suppress radiation-induced TGF-β expression, suggesting that sustained IL-13 or TGF-β neutralization therapy may be necessary to durably mitigate fibrotic progression [84]. Rapamycin is a potent immunosuppressive drug used in solid organ transplantation for the prevention of allograft rejection. In oncology, mTOR (mammalian target of rapamycin) inhibitors are currently being evaluated in several types of cancers. Targeting mTOR signaling may provide a therapeutic option for radiation-induced lung injury. Indeed, mTORC2 activity has been described in the non-canonical signaling of TGF-β, and mTORC2 inhibitors could thus be of interest in the case of fibrosis [126].

In rats, melatonin or metformin administration before irradiation using a ^{60}Co source of gamma rays in the thoracic area (i.e., heart and lungs) helped prevent the infiltration of macrophages and lymphocytes, as well as the upregulation of IL-4, IL4ra1, Duox1, andDuox2 [83,127]. Furthermore, metformin treatment could stimulate the activity of antioxidant enzymes such as superoxide dismutase (SOD) and glutathione (GSH) [128]. Other antioxidants, such as fucoidan, have been orally administered preclinically after whole lung irradiation at 10Gy, mitigating different proteins (TIMP-1, CXCL1, MCP-1, MIP-2, and IL-1Ra) expression in pleural fluid, decreasing pleural fluid accumulation and reducing neutrophil and macrophage infiltration in lung tissues. Fucoidan changed the expression patterns of inflammatory cytokines, which may consequently attenuate lung fibrosis [129,130].

Many molecular mechanisms still need to be better understood in order to develop a targeted treatment for fibrosis and radiation cystitis. It is, therefore, necessary to broaden our knowledge of myeloid and lymphocytic dynamics in the development of this fibrosis induced after pelvic radiotherapy, based on data obtained in other models (ex: lung fibrosis). It is also necessary to follow a step-by-step translational development to ensure that patient outcome may be improved, through the integration of robust biomarkers of toxicity, as well as through the implementation of modern radiotherapy tools in clinical research to minimize the doses to organs at risk, including the bladder, and therefore improve the therapeutic index. The current knowledge on the biological processes involved in late radiation toxicity clearly suggests that it is indeed easier to prevent late toxicity than to reverse [131].

Funding: This research received no external funding.

Conflicts of Interest: The authors declare no conflict of interest.

References

1. Rehailia-Blanchard, A.; He, M.Y.; Rancoule, C.; Guillaume, É.; Guy, J.-B.; Vial, N.; Nivet, A.; Orliac, H.; Chargari, C.; Magné, N. Medical prevention and treatment of radiation-induced urological and nephrological complications. *Cancer Radiother. J. Soc. Fr. Radiother. Oncol.* **2019**, *23*, 151–160.
2. Martin, S.E.; Begun, E.M.; Samir, E.; Azaiza, M.T.; Allegro, S.; Abdelhady, M. Incidence and Morbidity of Radiation-Induced Hemorrhagic Cystitis in Prostate Cancer. *Urology* **2019**, *131*, 190–195. [CrossRef] [PubMed]
3. Freites-Martinez, A.; Santana, N.; Arias-Santiago, S.; Viera, A. CTCAE versión 5.0. Evaluación de la gravedad de los eventos adversos dermatológicos de las terapias antineoplásicas. *Actas Dermo-Sifiliográficas* **2020**. [CrossRef] [PubMed]
4. Rigaud, J.; Hetet, J.-F.; Bouchot, O. Management of radiation cystitis. *Progres. En. Urol. J. Assoc. Fr. Urol. Soc. Fr. Urol.* **2004**, *14*, 568–572.
5. Smit, S.G.; Heyns, C.F. Management of radiation cystitis. *Nat. Rev. Urol.* **2010**, *7*, 206–214. [CrossRef]
6. Mazeron, R.; Castelnau-Marchand, P.; Dumas, I.; Del Campo, E.R.; Kom, L.K.; Martinetti, F.; Farha, G.; Tailleur, A.; Morice, P.; Chargari, C.; et al. Impact of treatment time and dose escalation on local control in locally advanced cervical cancer treated by chemoradiation and image-guided pulsed-dose rate adaptive brachytherapy. *Radiother. Oncol.* **2015**, *114*, 257–263. [CrossRef]
7. Zwaans, B.M.M.; Chancellor, M.B.; Lamb, L.E. Modeling and Treatment of Radiation Cystitis. *Urology* **2016**, *88*, 14–21. [CrossRef]
8. Cox, J.D.; Stetz, J.; Pajak, T.F. Toxicity criteria of the Radiation Therapy Oncology Group (RTOG) and the European Organization for Research and Treatment of Cancer (EORTC). *Int. J. Radiat. Oncol. Biol. Phys.* **1995**, *31*, 1341–1346. [CrossRef]
9. Denton, A.S.; Clarke, N.W.; Maher, E.J. Non-surgical interventions for late radiation cystitis in patients who have received radical radiotherapy to the pelvis. *Cochrane Database Syst. Rev.* **2002**, CD001773. [CrossRef]
10. Pavlidakey, P.G.; MacLennan, G.T. Radiation cystitis. *J. Urol.* **2009**, *182*, 1172–1173. [CrossRef]
11. Sun, R.; Koubaa, I.; Limkin, E.J.; Dumas, I.; Bentivegna, E.; Castanon, E.; Gouy, S.; Baratiny, C.; Monnot, F.; Maroun, P.; et al. Locally advanced cervical cancer with bladder invasion: Clinical outcomes and predictive factors for vesicovaginal fistulae. *Oncotarget* **2018**, *9*, 9299–9310. [CrossRef] [PubMed]
12. Marks, L.B.; Carroll, P.R.; Dugan, T.C.; Anscher, M.S. The response of the urinary bladder, urethra, and ureter to radiation and chemotherapy. *Int. J. Radiat. Oncol. Biol. Phys.* **1995**, *31*, 1257–1280. [CrossRef]
13. Manea, E.; Escande, A.; Bockel, S.; Khettab, M.; Dumas, I.; Lazarescu, I.; Fumagalli, I.; Morice, P.; Deutsch, E.; Haie-Meder, C.; et al. Risk of Late Urinary Complications Following Image Guided Adaptive Brachytherapy for Locally Advanced Cervical Cancer: Refining Bladder Dose-Volume Parameters. *Int. J. Radiat. Oncol.* **2018**, *101*, 411–420. [CrossRef] [PubMed]
14. Mendenhall, W.M.; Henderson, R.H.; Costa, J.A.; Hoppe, B.S.; Dagan, R.; Bryant, C.M.; Nichols, R.C.; Williams, C.R.; Harris, S.E.; Mendenhall, N.P. Hemorrhagic Radiation Cystitis. *Am. J. Clin. Oncol.* **2015**, *38*, 331–336. [CrossRef] [PubMed]

15. Muruve, N.A. Radiation Cystitis. 2017. Available online: https://emedicine.medscape.com/article/2055124-overview (accessed on 23 December 2020).
16. D'Ancona, C.; Haylen, B.; Oelke, M.; Abranches-Monteiro, L.; Arnold, E.; Goldman, H.; Hamid, R.; Homma, Y.; Marcelissen, T.; Rademakers, K.; et al. The International Continence Society (ICS) report on the terminology for adult male lower urinary tract and pelvic floor symptoms and dysfunction. *Neurourol. Urodynam.* **2019**, *38*, 433–477. [CrossRef]
17. Pascoe, C.; Duncan, C.; Lamb, B.W.; Davis, N.F.; Lynch, T.H.; Murphy, D.G.; Lawrentschuk, N. Current management of radiation cystitis: A review and practical guide to clinical management. *BJU Int.* **2019**, *123*, 585–594. [CrossRef]
18. Martinez, D.R.; Ercole, C.E.; Lopez, J.G.; Parker, J.; Hall, M.K. A Novel Approach for the Treatment of Radiation-Induced Hemorrhagic Cystitis with the GreenLightTM XPS Laser. *Int. Braz. J. Urol.* **2015**, *41*, 584–587. [CrossRef]
19. Arrizabalaga, M.; Extramina, J.; Parra, J.L.; Ramos, C.; Gonzàlez, R.D.; Leiva, O. Treatment of Massive Haematuria with Aluminous Salts. *BJU Int.* **1987**, *60*, 223–226. [CrossRef]
20. Westerman, M.E.; Boorjian, S.A.; Linder, B.J. Safety and efficacy of intravesical alum for intractable hemorrhagic cystitis: A contemporary evaluation. *Int. Braz. J. Urol.* **2016**, *42*, 1144–1149. [CrossRef]
21. Donahue, L.A.; Frank, I.N. Intravesical formalin for hemorrhagic cystitis: Analysis of therapy. *J. Urol.* **1989**, *141*, 809–812. [CrossRef]
22. Shao, Y.; Lu, G.; Shen, Z. Comparison of intravesical hyaluronic acid instillation and hyperbaric oxygen in the treatment of radiation-induced hemorrhagic cystitis. *BJU Int.* **2012**, *109*, 691–694. [CrossRef] [PubMed]
23. Chuang, Y.-C.; Kim, D.K.; Chiang, P.-H.; Chancellor, M.B. Bladder botulinum toxin A injection can benefit patients with radiation and chemical cystitis. *BJU Int.* **2008**, *102*, 704–706. [CrossRef] [PubMed]
24. Hazewinkel, M.H.; Stalpers, L.J.A.; Dijkgraaf, M.G.; Roovers, J.-P.W.R. Prophylactic vesical instillations with 0.2% chondroitin sulfate may reduce symptoms of acute radiation cystitis in patients undergoing radiotherapy for gynecological malignancies. *Int. Urogynecol. J.* **2011**, *22*, 725–730. [CrossRef] [PubMed]
25. Bonfili, P.; Franzese, P.; Marampon, F.; La Verghetta, M.E.; Parente, S.; Cerasani, M.; Di Genova, D.; Mancini, M.; Vittorini, F.; Gravina, G.L.; et al. Intravesical instillations with polydeoxyribonucleotides reduce symptoms of radiation-induced cystitis in patients treated with radiotherapy for pelvic cancer: A pilot study. *Support. Care Cancer* **2013**, *22*, 1155–1159. [CrossRef]
26. Mićić, S.; Genbacev, O. Post-irradiation cystitis improved by instillation of early placental extract in saline. *Eur. Urol.* **1988**, *14*, 291–293.
27. Villeirs, L.; Tailly, T.; Ost, P.; Waterloos, M.; Decaestecker, K.; Fonteyne, V.; Van Praet, C.; Lumen, N. Hyperbaric oxygen therapy for radiation cystitis after pelvic radiotherapy: Systematic review of the recent literature. *Int. J. Urol.* **2019**, *27*, 98–107. [CrossRef]
28. Capelli-Schellpfeffer, M.; Gerber, G.S. The use of hyperbaric oxygen in urology. *J. Urol.* **1999**, *162*, 647–654. [CrossRef]
29. Dellis, A.; Deliveliotis, C.; Kalentzos, V.; Vavasis, P.; Skolarikos, A. Is there a role for hyberbaric oxygen as primary treatment for grade IV radiation-induced haemorrhagic cystitis? A prospective pilot-feasibility study and review of literature. *Int. Braz J. Urol.* **2014**, *40*, 296–305. [CrossRef]
30. Oscarsson, N.; Müller, B.; Rosén, A.; Lodding, P.; Mölne, J.; Giglio, D.; Hjelle, K.M.; Vaagbø, G.; Hyldegaard, O.; Vangedal, M.; et al. Radiation-induced cystitis treated with hyperbaric oxygen therapy (RICH-ART): A randomised, controlled, phase 2–3 trial. *Lancet Oncol.* **2019**, *20*, 1602–1614. [CrossRef]
31. Pereira, D.; Ferreira, C.; Catarino, R.; Correia, T.; Cardoso, A.; Reis, F.; Cerqueira, M.; Prisco, R.; Camacho, O. Hyperbaric oxygen for radiation-induced cystitis: A long-term follow-up. *Actas Urológicas Españolas (English Edition)* **2020**, *44*, 561–567. [CrossRef]
32. Cardinal, J.; Slade, A.; McFarland, M.; Keihani, S.; Hotaling, J.N.; Myers, J.B. Scoping Review and Meta-analysis of Hyperbaric Oxygen Therapy for Radiation-Induced Hemorrhagic Cystitis. *Curr. Urol. Rep.* **2018**, *19*, 38. [CrossRef]
33. Chong, K.T.; Hampson, N.B.; Corman, J.M. Early hyperbaric oxygen therapy improves outcome for radiation-induced hemorrhagic cystitis. *Urology* **2005**, *65*, 649–653. [CrossRef] [PubMed]
34. Nakada, T.; Nakada, H.; Yoshida, Y.; Nakashima, Y.; Banya, Y.; Fujihira, T.; Karasawa, K. Hyperbaric Oxygen Therapy for Radiation Cystitis in Patients with Prostate Cancer: A Long-Term Follow-Up Study. *Urol. Int.* **2012**, *89*, 208–214. [CrossRef] [PubMed]
35. Choong, S.K.; Walkden, M.; Kirby, R. The management of intractable haematuria. *BJU Int.* **2000**, *86*, 951–959. [CrossRef] [PubMed]
36. Loffroy, R.; Pottecher, P.; Cherblanc, V.; Favelier, S.; Estivalet, L.; Koutlidis, N.; Moulin, M.; Cercueil, J.; Cormier, L.; Krausé, D. Current role of transcatheter arterial embolization for bladder and prostate hemorrhage. *Diagn. Interv. Imaging* **2014**, *95*, 1027–1034. [CrossRef]
37. Gowda, G.G.; Vijayakumar, R.; Tigga, M.P. Endovascular Management of Radiation-Induced Hemorrhagic Cystitis. *Indian J. Palliat. Care* **2019**, *25*, 471–473.
38. Linder, B.J.; Tarrell, R.F.; Boorjian, S.A. Cystectomy for refractory hemorrhagic cystitis: Contemporary etiology, presentation and outcomes. *J. Urol.* **2014**, *192*, 1687–1692. [CrossRef]
39. Marquardt, D.; Williams, J.A.; Kučerka, N.; Atkinson, J.; Wassall, S.R.; Katsaras, J.; Harroun, T.A. Tocopherol Activity Correlates with Its Location in a Membrane: A New Perspective on the Antioxidant Vitamin E. *J. Am. Chem. Soc.* **2013**, *135*, 7523–7533. [CrossRef]
40. Jacobson, G.M.; Bhatia, S.; Smith, B.J.; Button, A.M.; Bodeker, K.; Buatti, J. Randomized Trial of Pentoxifylline and Vitamin E vs Standard Follow-up After Breast Irradiation to Prevent Breast Fibrosis, Evaluated by Tissue Compliance Meter. *Int. J. Radiat. Oncol.* **2013**, *85*, 604–608. [CrossRef]

41. Younus, H. Therapeutic potentials of superoxide dismutase. *Int. J. Health Sci.* **2018**, *12*, 88–93.
42. Campana, F.; Zervoudis, S.; Perdereau, B.; Gez, E.; Fourquet, A.; Badiu, C.; Tsakiris, G.; Koulaloglou, S. Topical superoxide dismutase reduces post-irradiation breast cancer fibrosis. *J. Cell. Mol. Med.* **2007**, *8*, 109–116. [CrossRef]
43. Kharofa, J.; Cohen, E.P.; Tomic, R.; Xiang, Q.; Gore, E. Decreased risk of radiation pneumonitis with incidental concurrent use of angiotensin-converting enzyme inhibitors and thoracic radiation therapy. *Int. J. Radiat. Oncol. Biol. Phys.* **2012**, *84*, 238–243. [CrossRef] [PubMed]
44. Gensel, J.C.; Zhang, B. Macrophage activation and its role in repair and pathology after spinal cord injury. *Brain Res.* **2015**, *1619*, 1–11. [CrossRef] [PubMed]
45. Bourgier, C.; Auperin, A.; Rivera, S.; Boisselier, P.; Petit, B.; Lang, P.; Lassau, N.; Taourel, P.; Tetreau, R.; Azria, D.; et al. Pravastatin Reverses Established Radiation-Induced Cutaneous and Subcutaneous Fibrosis in Patients With Head and Neck Cancer: Results of the Biology-Driven Phase 2 Clinical Trial Pravacur. *Int. J. Radiat. Oncol.* **2019**, *104*, 365–373. [CrossRef] [PubMed]
46. Chargari, C.; Supiot, S.; Hennequin, C.; Chapel, A.; Simon, J.-M. Treatment of radiation-induced late effects: What's new? *Cancer Radiother. J. Soc. Francaise Radiother. Oncol.* **2020**, *24*, 602–611.
47. Bentzen, S.M. Preventing or reducing late side effects of radiation therapy: Radiobiology meets molecular pathology. *Nat. Rev. Cancer* **2006**, *6*, 702–713. [CrossRef] [PubMed]
48. Wynn, T.A.; Vannella, K.M. Macrophages in Tissue Repair, Regeneration, and Fibrosis. *Immunity* **2016**, *44*, 450–462. [CrossRef]
49. Mantovani, A.; Sica, A.; Sozzani, S.; Allavena, P.; Vecchi, A.; Locati, M. The chemokine system in diverse forms of macrophage activation and polarization. *Trends Immunol.* **2004**, *25*, 677–686. [CrossRef]
50. Shapouri-Moghaddam, A.; Mohammadian, S.; Vazini, H.; Taghadosi, M.; Esmaeili, S.A.; Mardani, F.; Seifi, B.; Mohammadi, A.; Afshari, J.T.; Sahebkar, A. Macrophage plasticity, polarization, and function in health and disease. *J. Cell. Physiol.* **2018**, *233*, 6425–6440. [CrossRef]
51. Zhang, L.; Wang, Y.; Wu, G.R.; Xiong, W.N.; Gu, W.K.; Wang, C.Y. Macrophages: Friend or foe in idiopathic pulmonary fibrosis? *Respir. Res.* **2018**, *19*, 170. [CrossRef]
52. Malyshev, I.; Malyshev, Y. Current Concept and Update of the Macrophage Plasticity Concept: Intracellular Mechanisms of Reprogramming and M3 Macrophage 'Switch' Phenotype. *BioMed Res. Int.* **2015**, *2015*, 341308. [CrossRef]
53. Zhou, D.; Huang, C.; Lin, Z.; Zhan, S.; Kong, L.; Fang, C.; Li, J. Macrophage polarization and function with emphasis on the evolving roles of coordinated regulation of cellular signaling pathways. *Cell. Signal.* **2014**, *26*, 192–197. [CrossRef] [PubMed]
54. Lech, M.; Anders, H.-J. Macrophages and fibrosis: How resident and infiltrating mononuclear phagocytes orchestrate all phases of tissue injury and repair. *Biochim. Biophys. Acta* **2013**, *1832*, 989–997. [CrossRef] [PubMed]
55. Braga, T.T.; Agudelo, J.S.H.; Camara, N.O.S. Macrophages During the Fibrotic Process: M2 as Friend and Foe. *Front. Immunol.* **2015**, *6*, 602. [CrossRef] [PubMed]
56. Chou, J.; Chan, M.F.; Werb, Z. Metalloproteinases: A Functional Pathway for Myeloid Cells. *Microbiol. Spectr.* **2016**, *4*, 1–15.
57. Pakshir, P.; Hinz, B. The big five in fibrosis: Macrophages, myofibroblasts, matrix, mechanics, and miscommunication. *Matrix Biol. J. Int. Soc. Matrix Biol.* **2018**, *68–69*, 81–93. [CrossRef]
58. Wynn, T.A.; Barron, L. Macrophages: Master regulators of inflammation and fibrosis. *Semin. Liver Dis.* **2010**, *30*, 245–257. [CrossRef]
59. Nikolic-Paterson, D.J.; Wang, S.; Lan, H.Y. Macrophages promote renal fibrosis through direct and indirect mechanisms. *Kidney Int. Suppl.* **2014**, *4*, 34–38. [CrossRef]
60. Sunderkötter, C.; Steinbrink, K.; Goebeler, M.; Bhardwaj, R.; Sorg, C. Macrophages and angiogenesis. *J. Leukoc. Biol.* **1994**, *55*, 410–422. [CrossRef]
61. Vernon, M.A.; Mylonas, K.J.; Hughes, J. Macrophages and renal fibrosis. *Semin. Nephrol.* **2010**, *30*, 302–317. [CrossRef]
62. Park, H.-R.; Jo, S.-K.; Jung, U. Ionizing Radiation Promotes Epithelial-to-Mesenchymal Transition in Lung Epithelial Cells by TGF-β-producing M2 Macrophages. *Vivo Athens Greece* **2019**, *33*, 1773–1784. [CrossRef]
63. Huang, Y.; Zhang, W.; Yu, F.; Gao, F. The Cellular and Molecular Mechanism of Radiation-Induced Lung Injury. *Med. Sci. Monit. Int. Med. J. Exp. Clin. Res.* **2017**, *23*, 3446–3450. [CrossRef] [PubMed]
64. Duru, N.; Wolfson, B.; Zhou, Q. Mechanisms of the alternative activation of macrophages and non-coding RNAs in the development of radiation-induced lung fibrosis. *World J. Biol. Chem.* **2016**, *7*, 231–239. [CrossRef] [PubMed]
65. Patel, V.; Noureddine, L. MicroRNAs and fibrosis. *Curr. Opin. Nephrol. Hypertens.* **2012**, *21*, 410–416. [CrossRef] [PubMed]
66. Zhu, H.; Luo, H.; Li, Y.; Zhou, Y.; Jiang, Y.; Chai, J.; Xiao, X.; You, Y.; Zuo, X. MicroRNA-21 in Scleroderma Fibrosis and its Function in TGF-β- Regulated Fibrosis-Related Genes Expression. *J. Clin. Immunol.* **2013**, *33*, 1100–1109. [CrossRef] [PubMed]
67. Mukherjee, D.; Coates, P.J.; Lorimore, S.A.; Wright, E.G. Responses to ionizing radiation mediated by inflammatory mechanisms. *J. Pathol.* **2014**, *232*, 289–299. [CrossRef] [PubMed]
68. Meziani, L.; Deutsch, E.; Mondini, M. Macrophages in radiation injury: A new therapeutic target. *Oncoimmunology* **2018**, *7*, e1494488. [CrossRef]
69. Ghita, M.; McMahon, S.J.; Thompson, H.F.; McGarry, C.K.; King, R.B.; Osman, S.O.S.; Kane, J.L.; Tulk, A.; Schettino, G.; Butterworth, K.T.; et al. Small field dosimetry for the small animal radiotherapy research platform (SARRP). *Radiat. Oncol.* **2017**, *12*, 204. [CrossRef]
70. Gieseck, R.L.; Wilson, M.S.; Wynn, T.A. Type 2 immunity in tissue repair and fibrosis. *Nat. Rev. Immunol.* **2018**, *18*, 62–76. [CrossRef]

71. Zwaans, B.M.; Krueger, S.; Bartolone, S.N.; Chancellor, M.B.; Marples, B.; Lamb, L.E. Modeling of chronic radiation-induced cystitis in mice. *Adv. Radiat. Oncol.* **2016**, *1*, 333–343. [CrossRef]
72. Oscarsson, N.; Ny, L.; Mölne, J.; Lind, F.; Ricksten, S.-E.; Seeman-Lodding, H.; Giglio, D. Hyperbaric oxygen treatment reverses radiation induced pro-fibrotic and oxidative stress responses in a rat model. *Free. Radic. Biol. Med.* **2017**, *103*, 248–255. [CrossRef]
73. Rajaganapathy, B.R.; Jayabalan, N.; Tyagi, P.; Kaufman, J.; Chancellor, M.B. Advances in Therapeutic Development for Radiation Cystitis. *Low. Urin. Tract Symptoms* **2014**, *6*, 1–10. [CrossRef] [PubMed]
74. Soler, R.; Vianello, A.; Füllhase, C.; Wang, Z.; Atala, A.; Soker, S.; Yoo, J.J.; Williams, J.K. Vascular therapy for radiation cystitis. *Neurourol. Urodynam.* **2010**, *30*, 428–434. [CrossRef] [PubMed]
75. Ferrara, N. Role of vascular endothelial growth factor in physiologic and pathologic angiogenesis: Therapeutic implications. *Semin. Oncol.* **2002**, *29*, 10–14. [CrossRef] [PubMed]
76. Palchesko, R.N.; Lathrop, K.L.; Funderburgh, J.L.; Feinberg, A.W. In vitro expansion of corneal endothelial cells on biomimetic substrates. *Sci. Rep.* **2015**, *5*, 7955. [CrossRef]
77. Magné, N.; Toillon, R.-A.; Bottero, V.; Didelot, C.; Van Houtte, P.; Gérard, J.-P.; Peyron, J.-F. NF-kappaB modulation and ionizing radiation: Mechanisms and future directions for cancer treatment. *Cancer Lett.* **2006**, *231*, 158–168. [CrossRef]
78. Hellweg, C.E. The Nuclear Factor κB pathway: A link to the immune system in the radiation response. *Cancer Lett.* **2015**, *368*, 275–289. [CrossRef]
79. Kowaliuk, J.; Sarsarshahi, S.; Hlawatsch, J.; Kastsova, A.; Kowaliuk, M.; Krischak, A.; Kuess, P.; Duong, L.; Dörr, W. Translational Aspects of Nuclear Factor-Kappa B and Its Modulation by Thalidomide on Early and Late Radiation Sequelae in Urinary Bladder Dysfunction. *Int. J. Radiat. Oncol.* **2020**, *107*, 377–385. [CrossRef]
80. Bauditz, J. Effective treatment of gastrointestinal bleeding with thalidomide–Chances and limitations. *World J. Gastroenterol.* **2016**, *22*, 3158–3164. [CrossRef]
81. Lee, S.J.; Yi, C.O.; Song, D.H.; Cho, Y.J.; Jeong, Y.Y.; Kang, K.M.; Roh, G.S.; Lee, J.D. Clarithromycin Attenuates Radiation-Induced Lung Injury in Mice. *PLoS ONE* **2015**, *10*, e0131671. [CrossRef]
82. Abernathy, L.M.; Fountain, M.D.; Rothstein, S.E.; David, J.M.; Yunker, C.K.; Rakowski, J.; Lonardo, F.; Joiner, M.C.; Hillman, G.G. Soy Isoflavones Promote Radioprotection of Normal Lung Tissue by Inhibition of Radiation-Induced Activation of Macrophages and Neutrophils. *J. Thorac. Oncol.* **2015**, *10*, 1703–1712. [CrossRef]
83. Aliasgharzadeh, A.; Farhood, B.; Amini, P.; Saffar, H.; Motevaseli, E.; Rezapoor, S.; Nouruzi, F.; Shabeeb, D.H.; Musa, A.E.; Mohseni, M.; et al. Melatonin Attenuates Upregulation of Duox1 and Duox2 and Protects against Lung Injury following Chest Irradiation in Rats. *Cell J* **2019**, *21*, 236–242. [PubMed]
84. Chung, S.I.; Horton, J.A.; Ramalingam, T.R.; White, A.O.; Chung, E.J.; Hudak, K.E.; Scroggins, B.T.; Arron, J.R.; Wynn, T.A.; Citrin, D.E. IL-13 is a therapeutic target in radiation lung injury. *Sci. Rep.* **2016**, *6*, 39714. [CrossRef] [PubMed]
85. Van de Putte, D.; Demarquay, C.; Van Daele, E.; Moussa, L.; Vanhove, C.; Benderitter, M.; Ceelen, W.; Pattyn, P.; Mathieu, N. Adipose-Derived Mesenchymal Stromal Cells Improve the Healing of Colonic Anastomoses Following High Dose of Irradiation Through Anti-Inflammatory and Angiogenic Processes. *Cell Transplant.* **2017**, *26*, 1919–1930. [CrossRef] [PubMed]
86. Murphy, M.B.; Moncivais, K.; Caplan, A.I. Mesenchymal stem cells: Environmentally responsive therapeutics for regenerative medicine. *Exp. Mol. Med.* **2013**, *45*, e54. [CrossRef] [PubMed]
87. François, S.; Bensidhoum, M.; Mouiseddine, M.; Mazurier, C.; Allenet, B.; Semont, A.; Frick, J.; Saché, A.; Bouchet, S.; Thierry, D.; et al. Local Irradiation Not Only Induces Homing of Human Mesenchymal Stem Cells at Exposed Sites but Promotes Their Widespread Engraftment to Multiple Organs: A Study of Their Quantitative Distribution After Irradiation Damage. *Stem Cells* **2006**, *24*, 1020–1029. [CrossRef]
88. Sensebé, L.; Bourin, P. Mesenchymal stem cells for therapeutic purposes. *Transplantation* **2009**, *87*, S49–S53. [CrossRef]
89. Charbord, P. Bone marrow mesenchymal stem cells: Historical overview and concepts. *Hum. Gene Ther.* **2010**, *21*, 1045–1056. [CrossRef]
90. Khalifa, J.; François, S.; Rancoule, C.; Riccobono, D.; Magné, N.; Drouet, M.; Chargari, C. Gene therapy and cell therapy for the management of radiation damages to healthy tissues: Rationale and early results. *Cancer/Radiothérapie* **2019**, *23*, 449–465. [CrossRef]
91. Voswinkel, J.; François, S.; Simon, J.-M.; Benderitter, M.; Gorin, N.-C.; Mohty, M.; Fouillard, L.; Chapel, A.; Chapel, A. Use of Mesenchymal Stem Cells (MSC) in Chronic Inflammatory Fistulizing and Fibrotic Diseases: A Comprehensive Review. *Clin. Rev. Allergy Immunol.* **2013**, *45*, 180–192. [CrossRef]
92. Zanoni, M.; Cortesi, M.; Zamagni, A.; Tesei, A. The Role of Mesenchymal Stem Cells in Radiation-Induced Lung Fibrosis. *Int. J. Mol. Sci.* **2019**, *20*, 3876. [CrossRef]
93. François, S.; Usunier, B.; Forgue-Lafitte, M.-E.; L'Homme, B.; Benderitter, M.; Douay, L.; Gorin, N.-C.; Larsen, A.K.; Chapel, A. Mesenchymal Stem Cell Administration Attenuates Colon Cancer Progression by Modulating the Immune Component within the Colorectal Tumor Microenvironment. *Stem Cells Transl. Med.* **2019**, *8*, 285–300. [CrossRef]
94. Linard, C.; Busson, E.; Holler, V.; Strup-Perrot, C.; Lacave-Lapalun, J.-V.; Lhomme, B.; Prat, M.; Devauchelle, P.; Sabourin, J.-C.; Simon, J.-M.; et al. Repeated Autologous Bone Marrow-Derived Mesenchymal Stem Cell Injections Improve Radiation-Induced Proctitis in Pigs. *Stem Cells Transl. Med.* **2013**, *2*, 916–927. [CrossRef] [PubMed]

95. Bessout, R.; Semont, A.; Demarquay, C.; Charcosset, A.; Benderitter, M.; Mathieu, N. Mesenchymal stem cell therapy induces glucocorticoid synthesis in colonic mucosa and suppresses radiation-activated T cells: New insights into MSC immunomodulation. *Mucosal Immunol.* **2014**, *7*, 656–669. [CrossRef] [PubMed]
96. Bessout, R.; Demarquay, C.; Moussa, L.; René, A.; Doix, B.; Benderitter, M.; Sémont, A.; Mathieu, N. TH17 predominant T-cell responses in radiation-induced bowel disease are modulated by treatment with adipose-derived mesenchymal stromal cells. *J. Pathol.* **2015**, *237*, 435–446. [CrossRef] [PubMed]
97. Chen, X.; Yang, B.; Tian, J.; Hong, H.; Du, Y.; Li, K.; Li, X.; Wang, N.; Yu, X.; Wei, X. Dental Follicle Stem Cells Ameliorate Lipopolysaccharide-Induced Inflammation by Secreting TGF-β3 and TSP-1 to Elicit Macrophage M2 Polarization. *Cell. Physiol. Biochem.* **2018**, *51*, 2290–2308. [CrossRef] [PubMed]
98. Heo, J.S.; Choi, Y.; Kim, H.O. Adipose-Derived Mesenchymal Stem Cells Promote M2 Macrophage Phenotype through Exosomes. *Stem Cells Int.* **2019**, *2019*, 7921760. [CrossRef]
99. He, X.; Dong, Z.; Cao, Y.; Wang, H.; Liu, S.; Liao, L.; Jin, Y.; Yuan, L.; Li, B. MSC-Derived Exosome Promotes M2 Polarization and Enhances Cutaneous Wound Healing. *Stem Cells Int.* **2019**, *2019*, 1–16. [CrossRef]
100. Ti, D.; Hao, H.; Tong, C.; Liu, J.; Dong, L.; Zheng, J.; Zhao, Y.; Liu, H.; Fu, X.; Han, W. LPS-preconditioned mesenchymal stromal cells modify macrophage polarization for resolution of chronic inflammation via exosome-shuttled let-7b. *J. Transl. Med.* **2015**, *13*, 1–14. [CrossRef]
101. Qi, Y.; Jiang, D.; Sindrilaru, A.; Stegemann, A.; Schatz, S.; Treiber, N.; Rojewski, M.; Schrezenmeier, H.; Beken, S.V.; Wlaschek, M.; et al. TSG-6 Released from Intradermally Injected Mesenchymal Stem Cells Accelerates Wound Healing and Reduces Tissue Fibrosis in Murine Full-Thickness Skin Wounds. *J. Investig. Dermatol.* **2014**, *134*, 526–537. [CrossRef]
102. Wu, Y.; Huang, S.; Enhe, J.; Ma, K.; Yang, S.; Sun, T.; Fu, X. Bone marrow-derived mesenchymal stem cell attenuates skin fibrosis development in mice. *Int. Wound J.* **2013**, *11*, 701–710. [CrossRef]
103. Wang, B.; Yao, K.; Huuskes, B.M.; Shen, H.H.; Zhuang, J.L.; Godson, C.; Brennan, E.P.; Wilkinson-Berka, J.L.; Wise, A.F.; Ricardo, S.D. Mesenchymal Stem Cells Deliver Exogenous MicroRNA-let7c via Exosomes to Attenuate Renal Fibrosis. *Mol. Ther. J. Am. Soc. Gene Ther.* **2016**, *24*, 1290–1301. [CrossRef] [PubMed]
104. Zwaans, B.M.M.; Bartolone, S.N.; Chancellor, M.B.; Nicolai, H.E.; Lamb, L.E. Altered Angiogenic Growth Factors in Urine of Prostate Cancer Survivors With Radiation History and Radiation Cystitis. *Urology* **2018**, *120*, 180–186. [CrossRef] [PubMed]
105. Zwaans, B.M.M.; Nicolai, H.E.; Chancellor, M.B.; Lamb, L.E. Prostate cancer survivors with symptoms of radiation cystitis have elevated fibrotic and vascular proteins in urine. *PLoS ONE* **2020**, *15*, e0241388. [CrossRef]
106. Wojtan, P.; Mierzejewski, M.; Osińska, I.; Domagała-Kulawik, J. Macrophage polarization in interstitial lung diseases. *Cent. Eur. J. Immunol.* **2016**, *41*, 159–164. [CrossRef]
107. Song, W.-J.; Li, Q.; Ryu, M.-O.; Ahn, J.-O.; Bhang, D.H.; Jung, Y.C.; Youn, H.-Y. TSG-6 Secreted by Human Adipose Tissue-derived Mesenchymal Stem Cells Ameliorates DSS-induced colitis by Inducing M2 Macrophage Polarization in Mice. *Sci. Rep.* **2017**, *7*, 1–14. [CrossRef]
108. Hong, I.-S.; Lee, H.-Y.; Kang, K.-S. Mesenchymal stem cells and cancer: Friends or enemies? *Mutat. Res.* **2014**, *768*, 98–106. [CrossRef]
109. Cortes-Dericks, L.; Galetta, D. The therapeutic potential of mesenchymal stem cells in lung cancer: Benefits, risks and challenges. *Cell. Oncol. Dordr.* **2019**, *42*, 727–738. [CrossRef]
110. Galland, S.; Stamenkovic, I. Mesenchymal stromal cells in cancer: A review of their immunomodulatory functions and dual effects on tumor progression. *J. Pathol.* **2020**, *250*, 555–572. [CrossRef]
111. Karlsson, H.; Erkers, T.; Nava, S.; Ruhm, S.; Westgren, M.; Ringdén, O. Stromal cells from term fetal membrane are highly suppressive in allogeneic settings in vitro. *Clin. Exp. Immunol.* **2012**, *167*, 543–555. [CrossRef]
112. Moll, G.; Ignatowicz, L.; Catar, R.; Luecht, C.; Sadeghi, B.; Hamad, O.; Jungebluth, P.; Dragun, D.; Schmidtchen, A.; Ringden, O. Different Procoagulant Activity of Therapeutic Mesenchymal Stromal Cells Derived from Bone Marrow and Placental Decidua. *Stem Cells Dev.* **2015**, *24*, 2269–2279. [CrossRef]
113. Aronsson-Kurttila, W.; Baygan, A.; Moretti, G.; Remberger, M.; Khoein, B.; Moll, G.; Sadeghi, B.; Ringdén, O. Placenta-Derived Decidua Stromal Cells for Hemorrhagic Cystitis after Stem Cell Transplantation. *Acta Haematol.* **2018**, *139*, 106–114. [CrossRef] [PubMed]
114. Sadeghi, B.; Moretti, G.; Arnberg, F.; Samén, E.; Kohein, B.; Catar, R.; Kamhieh-Milz, J.; Geissler, S.; Moll, G.; Holmin, S.; et al. Preclinical Toxicity Evaluation of Clinical Grade Placenta-Derived Decidua Stromal Cells. *Front. Immunol.* **2019**, *10*, 2685. [CrossRef] [PubMed]
115. Caplan, A.I.; Dennis, J.E. Mesenchymal stem cells as trophic mediators. *J. Cell. Biochem.* **2006**, *98*, 1076–1084. [CrossRef] [PubMed]
116. Gnecchi, M.; Danieli, P.; Malpasso, G.; Ciuffreda, M.C. Paracrine Mechanisms of Mesenchymal Stem Cells in Tissue Repair. *Methods Mol. Biol.* **2016**, *1416*, 123–146. [PubMed]
117. Xu, S.; Liu, C.; Ji, H.-L. Concise Review: Therapeutic Potential of the Mesenchymal Stem Cell Derived Secretome and Extracellular Vesicles for Radiation-Induced Lung Injury: Progress and Hypotheses. *Stem Cells Transl. Med.* **2019**, *8*, 344–354. [CrossRef]
118. Jafarinia, M.; Alsahebfosoul, F.; Salehi, H.; Eskandari, N.; Ganjalikhani-Hakemi, M. Mesenchymal Stem Cell-Derived Extracellular Vesicles: A Novel Cell-Free Therapy. *Immunol. Invest.* **2020**, *49*, 758–780. [CrossRef]
119. Cavallero, S.; Riccobono, D.; Drouet, M.; François, S. MSC-Derived Extracellular Vesicles: New Emergency Treatment to Limit the Development of Radiation-Induced Hematopoietic Syndrome? *Health Phys.* **2020**, *119*, 21–36. [CrossRef]

120. McGrath, M.S.; Kahn, J.O.; Herndier, B.G. Development of WF10, a novel macrophage-regulating agent. *Curr. Opin. Investig. Drugs Lond. Engl.* **2002**, *3*, 365–373.
121. Giese, T.; McGrath, M.S.; Stumm, S.; Schempp, H.; Elstner, E.; Meuer, S.C. Differential effects on innate versus adaptive immune responses by WF10. *Cell. Immunol.* **2004**, *229*, 149–158. [CrossRef]
122. Srisupundit, S.; Kraiphibul, P.; Sangruchi, S.; Linasmita, V.; Chingskol, K.; Veerasarn, V. The efficacy of chemically-stabilized chlorite-matrix (TCDO) in the management of late postradiation cystitis. *J. Med Assoc. Thail.* **1999**, *82*, 798–802.
123. Veerasarn, V.; Khorprasert, C.; Lorvidhaya, V.; Sangruchi, S.; Tantivatana, T.; Narkwong, L.; Kongthanarat, Y.; Chitapanarux, I.; Tesavibul, C.; Panichevaluk, A. Reduced recurrence of late hemorrhagic radiation cystitis by WF10 therapy in cervical cancer patientsA multicenter, randomized, two-arm, open-label trial. *Radiother. Oncol.* **2004**, *73*, 179–185. [CrossRef]
124. Veerasarn, V.; Boonnuch, W.; Kakanaporn, C. A phase II study to evaluate WF10 in patients with late hemorrhagic radiation cystitis and proctitis. *Gynecol. Oncol.* **2006**, *100*, 179–184. [CrossRef] [PubMed]
125. Fountain, M.D.; Abernathy, L.M.; Lonardo, F.; Rothstein, S.E.; Dominello, M.M.; Yunker, C.K.; Chen, W.; Gadgeel, S.; Joiner, M.C.; Hillman, G.G. Radiation-Induced Esophagitis is Mitigated by Soy Isoflavones. *Front. Oncol.* **2015**, *5*, 238. [CrossRef] [PubMed]
126. Chung, E.J.; Sowers, A.; Thetford, A.; McKay-Corkum, G.; Chung, S.I.; Mitchell, J.B.; Citrin, D.E. Mammalian Target of Rapamycin Inhibition With Rapamycin Mitigates Radiation-Induced Pulmonary Fibrosis in a Murine Model. *Int. J. Radiat. Oncol.* **2016**, *96*, 857–866. [CrossRef]
127. Farhood, B.; Aliasgharzadeh, A.; Amini, P.; Saffar, H.; Motevaseli, E.; Rezapour, S.; Nouruzi, F.; Shabeeb, D.; Musa, A.E.; Ashabi, G.; et al. Radiation-Induced Dual Oxidase Upregulation in Rat Heart Tissues: Protective Effect of Melatonin. *Medicina* **2019**, *55*, 317. [CrossRef]
128. Xu, G.; Wu, H.; Zhang, J.; Li, D.; Wang, Y.; Wang, Y.; Zhang, H.; Lu, L.; Li, C.; Huang, S.; et al. Metformin ameliorates ionizing irradiation-induced long-term hematopoietic stem cell injury in mice. *Free. Radic. Biol. Med.* **2015**, *87*, 15–25. [CrossRef]
129. Senthilkumar, K.; Manivasagan, P.; Venkatesan, J.; Kim, S.-K. Brown seaweed fucoidan: Biological activity and apoptosis, growth signaling mechanism in cancer. *Int. J. Biol. Macromol.* **2013**, *60*, 366–374. [CrossRef]
130. Yu, H.-H.; Ko, E.C.; Chang, C.-L.; Yuan, K.S.-P.; Wu, A.T.; Shan, Y.; Wu, S.-Y. Fucoidan Inhibits Radiation-Induced Pneumonitis and Lung Fibrosis by Reducing Inflammatory Cytokine Expression in Lung Tissues. *Mar. Drugs* **2018**, *16*, 392. [CrossRef]
131. Chargari, C.; Levy, A.; Paoletti, X.; Soria, J.-C.; Massard, C.; Weichselbaum, R.R.; Deutsch, E. Methodological Development of Combination Drug and Radiotherapy in Basic and Clinical Research. *Clin. Cancer Res.* **2020**, *26*, 4723–4736. [CrossRef]

Review

Current and Future Perspectives of the Use of Organoids in Radiobiology

Peter W. Nagle [1,*] **and Robert P. Coppes** [2,3]

1 Cancer Research UK Edinburgh Centre, MRC Institute of Genetics and Molecular Medicine, University of Edinburgh, Edinburgh EH4 2XR, UK
2 Department of Biomedical Sciences of Cells & Systems, University Medical Center Groningen, University of Groningen, Groningen, 9700 RB Groningen, The Netherlands; r.p.coppes@umcg.nl
3 Department of Radiation Oncology, University Medical Center Groningen, University of Groningen, Groningen, 9700 RB Groningen, The Netherlands
* Correspondence: peter.nagle@ed.ac.uk

Received: 23 November 2020; Accepted: 8 December 2020; Published: 9 December 2020

Abstract: The majority of cancer patients will be treated with radiotherapy, either alone or together with chemotherapy and/or surgery. Optimising the balance between tumour control and the probability of normal tissue side effects is the primary goal of radiation treatment. Therefore, it is imperative to understand the effects that irradiation will have on both normal and cancer tissue. The more classical lab models of immortal cell lines and in vivo animal models have been fundamental to radiobiological studies to date. However, each of these comes with their own limitations and new complementary models are required to fill the gaps left by these traditional models. In this review, we discuss how organoids, three-dimensional tissue-resembling structures derived from tissue-resident, embryonic or induced pluripotent stem cells, overcome the limitations of these models and thus have a growing importance in the field of radiation biology research. The roles of organoids in understanding radiation-induced tissue responses and in moving towards precision medicine are examined. Finally, the limitations of organoids in radiobiology and the steps being made to overcome these limitations are considered.

Keywords: Radiation; radiobiology; stem/progenitor cells; organoids

1. Introduction—Optimising the Therapeutic Window of Radiation Treatment

With an ever-aging population the number of people diagnosed with cancer is constantly growing [1]. Therefore, there is an even greater onus on the need to develop both current and new methods to enhance the efficacy of cancer treatments. Traditional cancer treatments, such as radiotherapy, chemotherapy and surgery, are still the most common modalities, but newer treatments such as immunotherapy are becoming more and more prevalent. Radiotherapy (either alone or in combination with surgery and/or chemotherapy) is used to treat over half of all cancer patients, with a curative intent in the majority of these cases [2,3]. Furthermore, the number of patients undergoing radiotherapy is predicted to increase even further due to an aging and growing population, as well as rapid technological advances in radiotherapy delivery practices [4]. The primary goal of radiotherapy, as with all other forms of cancer treatment, is to maximise the therapeutic window. The therapeutic window describes the balance between the probability of increasing tumour cell kill while minimising the probability of normal tissue complications. This can be achieved by using drugs which target the intrinsic vulnerabilities of a tumour to make it more susceptible than healthy tissue, or alternatively by physically targeting the tumour with greater accuracy and minimising the co-irradiated normal healthy tissue (Figure 1).

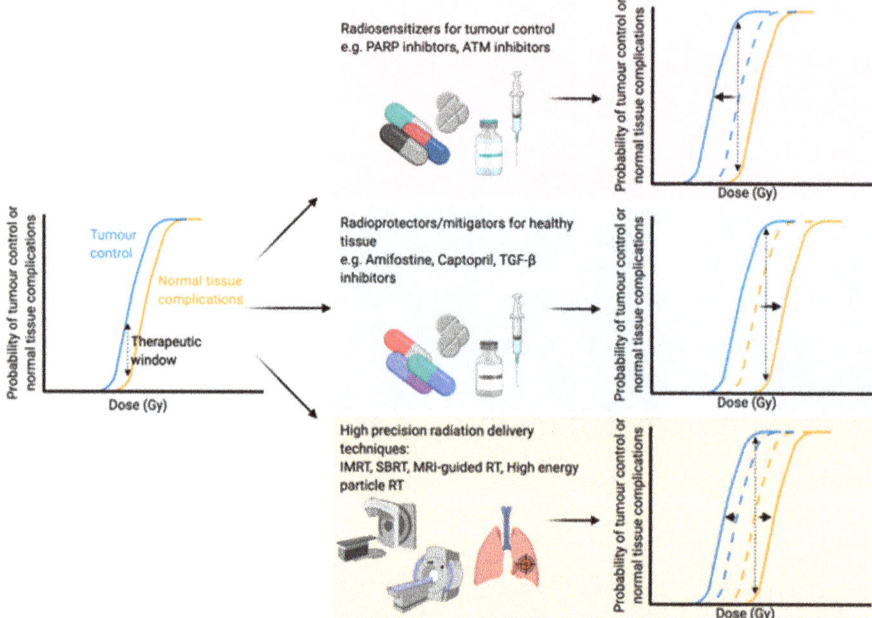

Figure 1. Optimising the therapeutic window of radiotherapy. The therapeutic window describes the balance between the probability of tumour control (blue line) and normal tissue complications (yellow line). There are three main rationales behind broadening the therapeutic window in radiation treatment: (1) Increasing tumour sensitivity using radiosensitisers, reducing the dose required for tumour kill (blue line shifts to the left), (2) protecting normal tissue using radioprotectors or mitigators, thus increasing the tolerable dose of normal tissue (shifting the yellow line to the right) or (3) high precision dose delivery which can reduce the volume of co-irradiated normal tissue (effectively shifting the yellow line to the right) while in the case of charged particles an increased relative biological effectiveness reduces the dose required for tumour control. Abbreviations: PARP; poly-ADP ribose polymerase, ATM; ataxia telangiectasia mutated, TGF-β; transforming growth factor beta, IMRT; intensity-modulated radiation therapy, SBRT; stereotactic body radiation therapy, MRI; magnetic resonance imaging. Created with BioRender.com.

The development of high precision means of dose delivery, such as intensity modulated radiation therapy [5], stereotactic radiation therapy [6] and charged particle radiotherapy [7], have allowed for substantial reductions in co-irradiated normal tissue during therapy. These strategies enable better sparing of crucial organs [8] or sub-regions [9] within organs during treatment or dose escalation to the tumour. Furthermore, real-time advanced imaging, such as magnetic resonance imaging (MRI), during radiation therapy has been suggested as a means to further optimise the delivery of radiation to the target tumour with an increased sparing of the surrounding healthy tissue. Initial in vitro studies showed no changes in survival in response to X-rays when a magnetic field of 1.5 T was applied [10]. Indeed, combining X-ray therapy with MRI-guidance has been successfully applied in clinical practice to increase the accuracy of dose delivery and thus spare a greater proportion of healthy tissue [11]. Particle therapies, such as proton therapies, can modulate the dose to encompass the whole tumour in a so-called "spread-out Bragg peak" with a minimised entrance dose and negligible exit dose, sparing healthy tissue [12,13]. Furthermore, MRI-guided proton therapy has also been proposed [14,15] and early in vitro findings suggest that a magnetic field perpendicular to the radiation beam has no effect on the radiobiological effectiveness of the dose [16], while a magnetic field longitudinal to the beam slightly changes the effectiveness [17], emphasising the potential of such advances in a clinical setting.

Further advances in radiation delivery include FLASH radiotherapy, which delivers ultra-high dose rates of ionising radiation which are believed to reduce normal tissue complications compared to conventional dose rates [18], although the therapeutic window of FLASH therapy still needs to be addressed [19].

All of these technological advances in the field of radiation beam delivery have significantly reduced the amount of co-irradiated healthy tissue during radiation treatment; however, none of these developments can completely eliminate dose to the surrounding tissue. Therefore, it is still necessary to develop in vivo and in vitro models to improve understanding of the mechanisms involved to better protect and/or regenerate normal tissue or to target intrinsic vulnerabilities of a tumour to enhance radiotherapy efficacy. These models should also take the therapeutic window into account as there is often an overlap between these mechanisms in both normal tissue and tumours, a feature which is regretfully often overlooked. Here we discuss one such in vitro model which could potentially allow the comparison of normal tissue and tumour responses at a patient-specific level, organoids, and the ever-growing role it has in radiobiological studies. We examine the strength of organoids in mechanistic studies in both normal and diseased tissue, but also examine the prospects of organoids in a more personalised medicine approach for patients. Finally, we discuss (potential) developments within the field of organoid research that could further benefit the radiobiology world.

2. The Need for New Models in Radiobiology

Since the beginning of the use of radiation treatment for cancer, radiobiology has made use of many different models to understand the molecular pathways triggered by radiation and to determine the consequences of radiation at a cellular, organ and system level in order to maximise the therapeutic window (Figure 2). Traditional cell cultures have been fundamental to radiobiology research, with many different techniques and findings crucial to other fields of biology, such as the development of the clonogenic survival assay [20,21]. Moreover, cell lines are highly amenable to high throughput drug screens, which in the field of radiobiology facilitates the efficient screening of large panels of potential radiosensitising agents over a radiation dose range [22]. However, cell lines cultured in two dimensions lack many features that are crucial to the overall response and survival of organisms following irradiation, such as cellular heterogeneity, cell–matrix interactions, "real" cell–cell interactions, a correct morphology and polarity, and functional relevance such as cytokine secretion [23,24]. Therefore, while they are invaluable, findings in cell lines often overstate findings, such as survival, compared to in vivo [25,26] and must therefore be treated with caution when translating to a more clinical patient setting.

Another model which has always been considered as a cornerstone for radiobiological research are in vivo animal models. Obviously animal models overcome many of the limitations of cell lines mentioned earlier, but they come with their own drawbacks, such as translatability to human settings, they are time consuming and expensive. Animal models are the most complete model available to researchers with the complete diversity of cell types and molecular interactions on an organismal level, as opposed to being constricted simply to a single cell type of a particular tissue when working with cell lines. Animal models are amenable to genetic manipulation and genetically modified animals offer the opportunity to study the impact of disease specific mutations, which in radiobiology allows researchers to study the effects on radio-resistance or -sensitivity of particular cancer associated mutations, such as *p53* [27,28] or *Atm* [29,30]. Furthermore, in vitro and in silico findings should always be confirmed in vivo as the final step prior to human translation, and therefore animal models will remain crucial to biomedical and radiobiological research. However, with an increasing growing pressure on researchers to limit (or even eradicate) the use of animals in research [31], it is necessary to find and implement alternative models in the search for treatments to a wide variety of diseases, not just cancer.

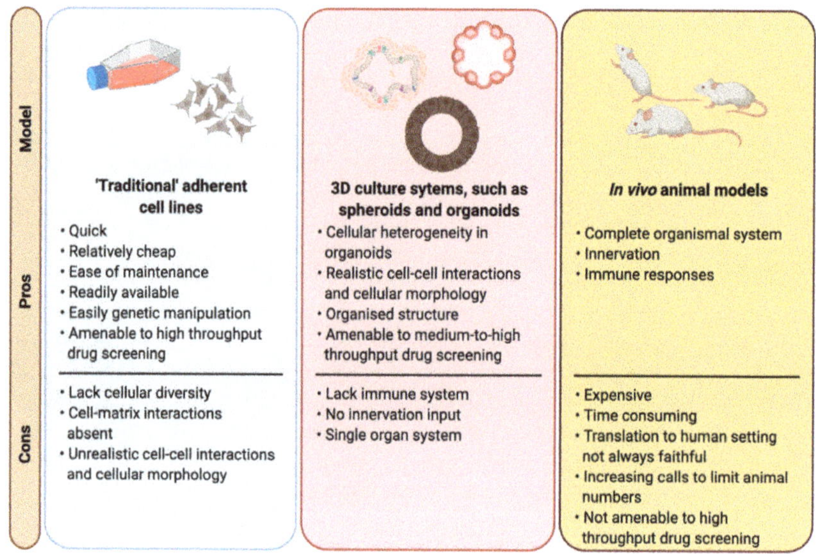

Figure 2. Laboratory models used in radiation biology. A comparison between the main laboratory-based models used in radiobiological studies highlighting the pros and cons of each model. Created with BioRender.com.

Organoids, three-dimensional in vitro structures derived from induced pluripotent stem cells, embryonic stem cells or tissue specific resident stem/progenitor cells [32,33], offer a "steppingstone" between more traditional in vitro cell lines and in vivo animal models. Organoids are self-assembling structures which resemble the tissue of origin [32–34]. They contain multiple cell types [32], overcoming the lack of cellular diversity of cell lines, although vasculature and endothelial cells are generally absent from these cultures. Distinct nomenclature has been proposed in some fields to distinguish between different 3D in vitro cultures, such as the suggested nomenclature differences between "enteroids", "colonoids" and "organoids" in the gastrointestinal field [35]. Furthermore, the term "tumouroids" is frequently used for tumour-derived organoids (or tumour-like organoids). Therefore, it should be noted that here we use the term "organoids" to encompass all self-organising 3D cellular structures derived from embryonic stem cells, induced pluripotent stem cells or tissue-resident stem/progenitor cells which contain multiple different cell types found within the tissue of origin. This is based on the definitions proposed by Lancaster and Knoblich (2014) [32] and Clevers (2016) [33].

As they are cultured in three dimensions, the cellular interactions and morphology become more "realistic" allowing for endpoint readouts which more closely resemble clinical observations. Furthermore, many organoid cultures have been shown to secrete functional enzymes under the right conditions [36], while transplantation of cultured organoids into murine models has been shown to rescue injured phenotypes [37]. Following radiation treatment, normal tissue stem cells are crucial to tissue regeneration. Conversely, cancer stem cells have increased radioresistance, repopulate tumours and are more prone to metastasize [38]. Therefore, it is important to be able to assess stem cell responses and the dynamics of those responses within the cellular heterogeneity (consisting of stem cells, progenitors and differentiated cells) of the tissue of origin. As they are derived from stem/progenitor cells, organoids can be used as a readout for such cells in an environment encompassing such heterogeneity [26]. Organoids are crucial to the studies of the mechanistic sequalae to irradiation, but also have an increasing role and potential in a more personalised approach to determining individual patient treatments. However, when designing experiments using organoids, researchers should always consider the question on hand when deciding which model (tissue-derived organoids, embryonic stem cell-derived or induced pluripotent stem cell-derived) to be used. For example, in cancer studies using organoids, pluripotent stem cell-derived and CRISPR-edited normal tissue-derived organoids can mimic germline mutations and thus allow accurate assessment of specific mutations in oncogenesis [39]. However, for treatment response studies, patient-derived organoids may represent a more suitable model, as they can encompass the true complexity of the disease, such microsatellite and chromosomal instabilities [40,41].

3. Organoids and Regeneration of Radiation-Induced Damaged Tissue

Since the identification of Lgr5 as a marker for intestinal stem cells [42], one of the most studied and established organoid models are the gastrointestinal "mini-gut" organoids. Originally established from mouse small intestinal stem cells [43], organoid "mini-gut" models have subsequently been established from human stem cells [44], as well as from various different locations along the gastrointestinal tract, including stomach [45], colon [44] and oesophagus [46]. Furthermore, pluripotent stem cells have been utilised to successfully generate intestinal [47] and oesophageal [48] organoid cultures. These models have opened novel avenues of study for intestinal development, cancer progression [49] and other diseases, such cystic fibrosis [50].

While there have been only a limited number of studies using organoids to investigate radiation-induced gastrointestinal injury, some recent studies have used organoids to complement and reinforce important insights from in vivo mouse studies [51–53]. Wang et al. [51] demonstrated using intestinal crypt organoids that selective inhibition of radiation-induced p53-mediated apoptosis using CHIR99021, an inhibitor of glycogen synthase kinase-3 (GSK-3), can protect intestinal stem cells against radiation due to an increased survival of Lgr5+ cells. This was recapitulated in vivo, indicating a pivotal role for p53 post-translational modifications in intestinal stem cell responses to irradiation [51]. More recently, using intestinal organoids from mouse jejunum and human colon, Bhanja et al. [52] revealed the potential of BCN057, an anti-neoplastic small molecular agent, to mitigate radiation-induced gastrointestinal syndrome in normal tissue. Interestingly, BCN057 did not have a radiomitigative effect in tumour-derived organoids with these findings again mimicking in vivo findings. The same group also investigated the potential of repurposing auranofin, an anti-rheumatoid drug containing gold, as a radioprotective agent against intestinal injury [53]. In both in vivo mice and ex vivo human colon organoids treatment with auranofin significantly reduced the toxicity of radiation [53]. Furthermore, Martin et al. [54] recently demonstrated that the profile of the Lgr5+ stem cell population of the large and small intestines following irradiation of organoids could act as a marker for predicting the sensitivity of these organs to radiation. The authors validated their approach using organoids with a well-established in vivo microcolony assay which quantifies the number of regenerating crypts per small intestinal circumference [54,55]. This assay is regarded as a benchmark assay for establishing the radiosensitivity of intestinal stem cell survival and highlights the potential of intestinal organoids to predict radiation responses [54]. These studies demonstrate the strength of "mini-gut" organoids as a model for radiation studies of the gastrointestinal tract and also the opportunities for radiobiological studies in other organoid systems, particularly in tissues which lack accurate in vitro models for radiobiological studies.

Radiotherapy is used to treat the majority of head and neck cancer patients, either alone or in combination with surgery and/or chemotherapy [56]. Frequently, irradiation of head and neck tumours leads to the unavoidable co-irradiation of salivary glands, with almost half of head and neck cancer patients subsequently suffering from radiation-induced xerostomia due to hyposalivation. This drastically impacts on the quality of life of patients due to impaired chewing, swallowing, speaking and an increased risk of oral infections [57]. In vivo studies using rats have shown that sparing a region of the salivary gland which contains a high density of tissue specific stem/progenitor cells has been shown to reduce the effects of salivary gland irradiation [9]. Therapeutic options are available to stimulate salivary gland flow post-irradiation but are limited in their effectiveness [57]. Therefore, a need for a more long-term strategy for salivary gland regeneration following radiotherapy remains [58]. While in vivo animal models have provided a wealth of knowledge as to the mechanisms behind salivary gland regeneration following injury, including radiation-induced damage [59–64], there is a limited number of in vitro systems to accurately study salivary glands following irradiation. Thus there is a growing niche for new models such as organotypic slice cultures [65] and organoids in the area of salivary gland radiation research.

Recently, our group has established protocols for the isolation and expansion of both murine [66] and human [67] submandibular salivary gland stem/progenitor cells. Using these protocols, we have shown that transplantation of enriched murine or human stem/progenitor cell populations improved functional readouts of irradiated mice salivary glands [37,67,68]. However, this effect may not only be directly from the expansion of the stem/progenitor cells in the transplanted tissue, but also due to paracrine effects of the transplanted cells acting on the recipient tissue [67]. Another recent study by Tanaka et al. has demonstrated the ability to derive salivary gland stem cells from embryonic stem cells [69]. Upon transplantation into parotid gland-defective mice, the induced salivary gland cells (transplanted either alone or together with mesenchymal cells) were capable of generating mature salivary gland tissue. The newly generated tissue was also shown to be functional as demonstrated by an increased saliva secretion in transplanted mice [69]. Combined, these studies hold significant preclinical promise for studying the mechanisms behind salivary gland regeneration and amelioration of salivary gland damage, both irradiation and non-irradiation induced damage [67,69]. However, the translation of any embryonic stem cell derived treatment [69] to a clinical application is always likely to be hindered by ethical concerns [70] and safety concerns regarding tumorigenicity [71].

Our models have been successfully utilised to study the survival responses of salivary gland stem/progenitor cells [26]. The salivary gland stem/progenitor organoids demonstrated a disproportionate sensitivity to low dose of radiation which was recapitulated in a functional low dose sensitivity in vivo [72]. While low dose hypersensitivity is not a new phenomenon [73,74], this was the first study to show the relevance of this phenomenon in stem/progenitor cells, with a potential clinical relevance. Furthermore, we have recently developed a protocol for the culturing of parotid salivary gland organoids and demonstrated that parotid gland stem cells display a similar radiosensitivity as those of submandibular salivary glands [75]. Importantly, as organoids are derived from stem/progenitor cell populations, they allow for the study of a more stem/progenitor specific response. As stem/progenitor cells play a prominent role in tissue regeneration following irradiation, models which allow for the understanding of these cells are crucial to protecting these tissues.

Another tissue in which the use of radiation is highly limited due to radiation-induced toxicity is the liver. Along with lung, breast, colorectal and pancreatic cancers, liver cancer deaths are one of the highest of all cancer-related deaths each year [76], while the prognosis is extremely poor due to limited treatment options [77]. The use of radiation treatment for liver cancer is severely hindered by the development of radiation-induced liver disease [78], a consequence which can also impede the utilisation of radiotherapy for other abdominal tumours in proximity to the liver, such as gastrointestinal cancers [79]. Much of what is known regarding radiation-induced liver disease is from retrospective clinical studies [79], as current lab models for studying it are limited with in vitro studies generally limited to cell lines lacking cellular heterogeneity and functionality. The recently developed models of both mouse [80] and human [81] derived liver organoid cultures from tissue resident stem cells, as well as pluripotent stem cell-derived liver cultures [82–85], may represent an ideal model for studying radiation-induced liver disease in the future. These models display cellular, functional activity and have structural organisation, while they have been successfully utilised to study genetic liver disorders mimicking the clinical pathology [81] and drug-induced liver injury [86]. Understanding the mechanisms of radiation-induced liver disease may eventually allow for increased treatment options for liver cancers.

4. A Platform for Treatment Response Studies; Moving towards Personalised Treatment?

The concept of precision treatments has been of growing interest in many fields of research in recent years, particularly oncology, as there is a wide variability of patient responses to standard "one size fits all" treatment regimens. In some cases, genetic factors which can be specifically targeted in a "personalised" manner are already known, for example non-small cell lung cancer patients with an activating mutation in tyrosine kinase are particularly sensitive to treatment with tyrosine kinases inhibitors such as gefitinib [87]. However, for other cancers, such as oesophageal cancers and locally advanced rectal cancers, there are currently no accurate predictors of patient responses to treatment. The standard of care for oesophageal cancer consists of neo-adjuvant chemoradiotherapy followed by surgery, with a complete pathological response observed in approximately a quarter at the time of surgery but no response in approximately one fifth of patients [88,89]. Similarly, for neoadjuvant chemoradiotherapy treatment of colorectal cancer while approximately one fifth of patients show a complete pathological response, almost 40% of patients show no benefit to the treatment [90]. In both cancers, patients would clearly benefit from more robust pre-treatment predictive models.

Therefore, there has been a concentrated effort in the field of organoids to establish reliable predictors of colorectal cancer treatment response to both chemotherapy alone [91,92] and neoadjuvant chemoradiotherapy [93,94]. Van de Wetering et al. [91] established colorectal cancer organoids, alongside paired healthy tissue, and demonstrated that the organoids recapitulated the genetic profiles and mutational spectra of the tumours of origin. Furthermore, by performing screening of 83 compounds, including both clinically used drugs and experimental compounds, the authors showed that the organoids facilitated the high-content drug screening [91], which could facilitate precision treatments in the future. Interestingly, a later study by Ooft et al. [92] investigating treatment response of metastatic colorectal cancer using organoids, was able to predict accuracy of irinotecan monotherapy and 5-flurouracil/irinotecan dual therapy, with 80% and 83.3% respectively. While greater accuracy is required to implement predictive models in a clinical setting, these studies show the developing potential of organoids in precision medicine. Furthermore, in recent years, there has been an increasing number of studies aimed at identifying and repurposing already available drugs as radiosensitisers [95–98]. Drugs which can be repurposed offer cheaper and quicker alternatives to developing new drugs from scratch, while many of the adverse side effects are already known [99]. The possibilities to quickly and accurately screen drugs, as shown in the studies of van de Wetering et al. [91] and Ooft et al. [92], in cancer organoids will greatly increase the possibilities in precision medicine and further benefit the search for potentiators of radiation therapy.

Indeed, recent studies by Ganesh et al. [93] and Yao et al. [94] have focussed on rectal cancer organoids for predicting patient responses to neoadjuvant chemoradiotherapy (Table 1 summarises the different cancer organoids that have been used in studies of radiation responses). Both studies further consolidated other evidence that rectal cancer organoids faithfully recapitulate the tumours of origin, performing histopathological and mutational comparisons between the two [93,94]. Moreover, Ganesh et al. showed that upon xenotransplantation of the organoids into mice they were found to metastasise to the same locations as the original tumours. Importantly, upon treating the organoids with chemotherapeutic drugs (such as 5-Flurouracil and oxaliplatin) heterogeneous treatment responses correlated with the clinical progression-free survival of patients. Interestingly, organoids which displayed resistance to radiation were derived from patients who either were resistant to therapy or showed disease recurrence following treatment [93]. Yao et al. [94] also correlated the therapeutic clinical outcomes to the standard neoadjuvant chemoradiotherapy with the organoid outcomes following treatment 5-Flurouracil, irinotecan or radiation. In sixty-eight out of the 80 patient-derived organoid lines generated, at least one of the three treatment courses was found to be predictive of the patient's tumour regression score after surgery [94]. Furthermore, in a recent study, Pasch et al. established patient-derived cancer organoids and were prospectively able to predict the treatment response of a patient with metastatic colon cancer [100]. These studies combined with the works of van de Wetering et al. and Ooft et al. provide a significant step towards a model for patient-specific response prediction.

Table 1. Radiation response studies using different cancer organoid models.

Tumour Type	Organoids Radiation Treatment	Key Findings	Ref.
Rectal cancer	5-Fluorouracil (5-FU), FOLFOX (5-FU, leucovorin and oxaliplatin) or radiation	Tumour organoids displayed clinically relevant chemo- and radiation responses. Established an orthotopic endoluminal rectal cancer mouse model which reflected patient-specific responses.	[93]
Rectal cancer	Irradiation, 5-FU, or Irinotecan	Colorectal cancer organoids could predict patient outcome in 68 out of 80 patients, based on at least on organoid treatment course.	[94]
Multiple cancers, including lung, colorectal and pancreatic adenocarinomas	5-FU and/or radiation	Colorectal cancer patient-derived organoids displayed differential responses to 5-FU chemotherapy and/or radiation. Prospectively predicted treatment outcome of patient with metastatic colon cancer.	[100]
Head and neck squamous cell carcinoma	Doses ranging from 0–10 Gy	Differential responses which could potentially indicate clinical correlations. However, no resistance mechanisms could be identified via differential gene expression patterns.	[101]
Glioblastoma	Radiation (3 Gy)	Edges of organoids displayed increased apoptosis in Sox2- cells. However, Sox2+ cells (considered as the glioblastoma cancer stem cells) showed an increased resistance.	[102]
Cerebral organoid glioma	Radiation (5 or 10 Gy)	Established organoids combining glioblastoma and healthy cerebral tissue (GLICO). Glioblastoma stem cells showed increased radioresistance in GLICOs compared to when cultured in 2D.	[103]

A recent study also established an organoid model for metastatic gastrointestinal cancers which were histologically, genetically and molecularly similar to the tumour of origin [104]. Following drug treatment of the organoids, the outcomes were compared with the clinical outcomes of the patients enrolled in Phase I/II clinical trials and were found to closely mimic the clinical outcomes of the patients [104]. Moreover, the study successfully identified differential inter- and intra-patient responses to common chemotherapeutic agents for gastrointestinal cancer treatment [104]. This study represents an important advance for organoids in the field of personalised precision medicine.

As mentioned above, currently the ability to predict patient responses to chemoradiotherapy for oesophageal cancer is also extremely limited. Great strides are being made towards the optimisation of imaging techniques for predicting treatment outcomes for oesophageal cancer treatment [89,105–107]; however, there is still no means to accurately predict patient outcomes. Recent advances in the culturing of oesophageal adenocarcinoma organoids have established new models to study the development and heterogeneity of the disease [108]. The established patient-derived oesophageal adenocarcinoma organoids shared histopathological features with patient-matched tumour samples and genetic mutations were conserved at a patient-specific level [108]. They further showed a loss of cellular polarity, which is often considered a hallmark of cancer. Drug screening in the organoids revealed a highly diverse range of responses, which tallies with the difficulties in predicting patient responses. However, the diversity of the responses remained throughout passaging, indicating the stability of the model through time [108]. Unfortunately, the findings of this study were somewhat limited due to a low success rate of establishing organoids (organoids were established from only 10 out of 32 patients). Reasons for a low success yield included failure to initiate culture, infection, fibroblast overgrowth, and arrested growth [108], while others also working on developing oesophageal adenocarcinoma organoids have recently identified the presence of Barrett's epithelium as another potential contamination source in culture [109]. These new models will be essential to opening new avenues for testing new drugs and treatment regimens for oesophageal adenocarcinomas. Furthermore, as mentioned above, radiotherapy is an also important arm of treatment for other cancers in the head and neck region. Recently established protocols for generating organoids from oral mucosa and head and neck squamous cell carcinomas may facilitate a more personalised treatment planning for more tumours in this region [101]. Comparisons of the responses of tumour organoids with matched normal tissue organoids may even allow for studies of the therapeutic window on a personalised scale.

Glioblastoma is a highly aggressive brain tumour with an extremely poor prognosis for patients for whom radiotherapy is an integral arm of treatment [110]. This remains the case even with significant advances in the understanding of glioblastoma development, cellular heterogeneity within the tumour, and the role of cancer stem cells play in this [111–113]. Many of the models used for studying glioblastoma utilise adherent monolayers which, although they have been highly revealing of the mechanisms of glioma stem cell resistance [114], have thus far not been representative of the tumour microenvironment or levels of therapeutic resistance of glioblastoma seen in vivo. However, recently new organoid models have been established that could shed light on the initiation, development, tumour invasion, and treatment of glioblastoma. In two independent studies, Bian et al. [115] and Ogawa et al. [116] utilised CRISPR/Cas9 genome editing technology to manipulate cerebral organoids towards tumorigenesis. In both studies, cells derived from the generated tumour organoids exhibited epithelial-mesenchymal properties, indicative that they are representative of the invasive mesenchymal subtype of glioblastoma. Indeed, the cells were invasive when seeded with normal cerebral organoids [115,116] and were capable of forming tumours when xeno-transplanted into mouse recipients [116]. While neither group determined radiation responses of the glioblastoma organoids, Bian et al. demonstrated that CRISPR/Cas9 generated glioblastoma organoid models are appropriate for preclinical in vitro drug screening [115].

Indeed, studies which have investigated the radiosensitivity of glioblastoma organoids have demonstrated that they more closely resemble in vivo tumour sensitivity than monolayer cultures [102,103]. Furthermore, importantly, particularly from a radiobiology point-of-view, Hubert et al. showed that although the non-stem cells of the organoids were radiosensitive, the tumour-initiating cancer stem cells were indeed resistant [102], recapitulating important in vivo findings from previous studies [114]. Although these glioblastoma models offer excellent platforms to study glioblastoma development and biology, and to test new treatments, the duration of culturing generally does not facilitate rapid screening for a more personalised approach to treatment. However, recently a robust and rapid (within 1-2 weeks) protocol for establishing glioblastoma organoids capable of facilitating moderate to high throughput screening for a potentially more personalised response prediction [117].

5. The Future Directions of Organoid Models in Radiation Biology

Despite a growing role for organoids in radiobiology (as well as other fields of biology) and continuous advances of the models to faithfully simulate the tissue of origin, organoids still have limitations. However, these drawbacks may represent opportunities. Opportunities for researchers to optimise and improve current organoid systems, and opportunities to complement their research with other techniques, such as clinical imaging techniques for enhancing patient treatment response predictions.

While organoids consist of heterogeneous cell types and are cultured in three dimensions, they still lack important microenvironmental cues, such as sympathetic and parasympathetic innervation and immune cells (such as macrophages and cytokines). These are crucial factors in both development and regeneration of tissue. There is growing evidence for the role of parasympathetic innervation in salivary gland development [118] and regeneration [119], including following radiation-induced damage [120]. Finding means to accurately mimic autonomic innervation in organoids may be important to fully utilising them as models for regeneration in tissues with a similar architecture to the salivary glands. Similarly, the lack of stroma and immune cells in organoids in response to both injury and treatment are important factors which still need addressing especially considering the rising number of applications of immunotherapy. In the aforementioned study of Ooft et al. [92], while the patient-derived organoids were predictive of patient response to irinotecan-based treatments, they were not predictive of 5-FU–oxaliplatin combination therapy, which the authors suggest may, at least in part, be done to the lack of crucial stroma and immune system interactions. Recent advances have been made to overcome these issues, with Neal et al. [121] successfully developing patient-derived organoids with the T-cell spectra of the original tumours capable of modelling the immune checkpoint blockade. Alternatively, co-culturing of organoids with immune cells will offer a theoretically more realistic tissue response. Indeed, in co-culture experiments with macrophages and mammary organoids, macrophages were shown to migrate to organoids with an increased migration rate towards irradiated organoids [122].

Furthermore, stroma also plays an important role in radiation responses, of both normal and tumour tissue. In organoid cultures derived from whole tissue biopsies (without stem cell selection) stromal cells and effects can be found within the culture system [123]. However, in organoid cultures from selected stem cells stroma is absent, and therefore stromal co-culturing is necessary to recapitulate the effects of the tissue's stroma. In prostate organoids, an increased viability and maintained branching was induced upon co-culture with prostate stroma [124]. Furthermore, the generation of organoids derived from prostate cancer was also improved upon stromal co-culture. These effects were suggested to be primarily due to direct contact with stromal cells and the expression of factors, such as TGF-β, by the stromal cells [124]. Besides the advance that this model represents in development and disease studies, the co-culture of organoids with tissue-specific stromal cells could have important implications for treatment responses, due to the important role of stromal cells [125] and the effects of signalling factors, such as TGF-β [126], in tissue responses.

Radiation-induced bystander effects have been suggested to act both proximally [127,128] and distally [129,130] to the site of irradiation; however, organoids derived from a single tissue currently do not recapitulate such interactions. Various anti-cancer therapies, including radiation, are known to induce senescence and an induction of a senescent associated secretory phenotype [131,132] which, it has been suggested, can in turn contribute to therapy-induced normal tissue side effects [133]. Studies using cultured media from irradiated cells has long been shown to induce paracrine bystander effects in non-irradiated cells [134] and such techniques may be insightful into the effects of secreted SASP proteins on untreated cells or organoids. Indeed, our group recently demonstrated that cultured media from irradiation-induced senescent organoids inhibits organoid forming efficiency in freshly passaged salivary gland-derived organoids [135]; however, these models still lack a true interaction between treated and non-treated organoids and the potential paracrine effects of other tissues in their vicinity in vivo.

Furthermore, both organ–organ, tumour–organ and vasculature interactions are generally absent in organoid cultures. Some of the glioblastoma organoid studies mentioned above elegantly show that cancer cells and healthy cells can be cultured together as organoids allowing for the study of tumour invasion [103,115,116]. Moreover, these models may be useful in revealing new therapeutic targets for tumour radiosensitisation or normal tissue radioprotection. Implementing organoid models alongside newly-established microfluidic devices which allow for the study of metabolic gradients [136] in radiation studies has the potential to reveal valuable insights of how such as signalling gradients can influence both irradiated and non-irradiated cells in perhaps a physiological relevant setting than organoids alone. Indeed, gut-on-a-chip models have recently been utilised in studies of radiation-induced intestinal injury and faithfully mimicked epithelial cell loss due to reactive oxygen production as seen in vivo [137] and may represent an excellent model for complementary studies to the abovementioned "mini-gut" organoid models in radiation studies. Organ-on-a-chip devices have been established for various other tissues, including lung [138], kidney [139] and liver [140]. The capacity of these platforms to mimic functional mechanics, such as breathing movements in lung-on-a-chip [138], could potentially offer more physiologically relevant models to complement and add a translational element to findings from organoid radiation studies. Furthermore, multiple chamber "on-a-chip" devices [141] could overcome the limitation of organoids of studying organs in isolation, in which each chamber potentially could contain cells from different tissues, vessels, stroma or nerves.

Radiation-induced endothelial cell loss and vascular damage are known to be major contributors to the response of both normal tissue and tumours [142,143]. Vasculopathy significantly increases the chances of ischemic stroke following radiation treatment [144], while preclinical models have been used to demonstrate that vascular remodelling is a major contributor to radiation-induced lung toxicity [145]. The vasculature of a tissue is essential for nutrient availability and regeneration following damage, as well as effective engraftment after tissue transplantation [146]. Furthermore, the response of tumour vasculature, particularly vasculogenesis, has also been shown to play a key role in tumour recurrence following radiation treatment [147]. While radiation can initially control the tumour, a reduced flow through tumour blood vessels and increased hypoxia can induce the hypoxia inducible factor-1 pathway. This in turn can activate pathways to re-promote vasculature and can subsequently cause tumour regrowth [143]. Therefore, it is important that in vitro models, particularly tumour models, can recapitulate such vasculature features. Recently, many techniques have been established to engineer vascularisation of organoids, including bioprinting, implantation into highly vascularised tissue and growing organoids in the vicinity of endothelial cell monolayers [148]. Vascularised organoids, such as recently-established vascularised cortical organoids [149,150] and tumour organoids [151], offer new opportunities to study disease pathology but also to study the impact vascularisation can have on treatment (including radiation) responses.

Finally, although there are many different protocols and technical considerations for the isolation and propagation of organoids, they are often arduous and time consuming. In order to have enough cells or organoids to test still often requires weeks to months of culturing. This is of particular importance for the development of organoids as a model for predicting patient responses in proposed precision therapies, where it is frequently necessary to treat patients as soon as possible. Protocols are being established to reduce culture time of organoids while maintaining fidelity of the systems of various different tissue origins (such as the aforementioned glioblastoma organoid model [117]); however, it is important that organoid models are further optimised for rapid and accurate screening of responses before implementation in a personalised medicine.

Despite their limitations, the future of organoid models in the field of radiobiology remains bright. As highlighted, many valuable studies are already overcoming the shortcomings of organoids, and as our knowledge and availability of organoid models grow, so too will their place in radiobiology. New organoid models can potentially shed some much-needed light on tissues which are perhaps less studied or highly limiting to the clinical application of radiation treatment, such as the liver. Moreover, while it could be questioned if a response prediction accuracy of approximately 80–85% is good enough, this will surely only improve as the models themselves are further optimised. Combining clinical patient imaging techniques currently used to predict patient responses, such as PET/CT, with the in vitro predictions from organoids may in the future bring around more accurate means to forecast treatment outcomes. Organoids could also potentially be used in discovery and validation of radiation biomarker and in radiomics. Understanding the mechanisms behind tissue regeneration are key to mitigating radiation-induced side effects, whether it is by stem cell therapy or through druggable targets to protect against damage, and organoids have already proven themselves as excellent models for such studies.

Author Contributions: Writing—original draft preparation, P.W.N.; writing—editing and reviewing, P.W.N. and R.P.C.; All authors have read and agreed to the published version of the manuscript.

Funding: This work was funded by the Dutch Cancer Society (KWF-grant number 10417).

Acknowledgments: While we have tried to give a broad overview of the state of organoids in radiation biology, there are many other important works in this area of the field that unfortunately we could not include due to manuscript constraints.

Conflicts of Interest: The authors declare no conflict of interest.

References

1. Fitzmaurice, C.; Allen, C.; Barber, R.M.; Barregard, L.; Bhutta, Z.A.; Brenner, H.; Dicker, D.J.; Chimed-Orchir, O.; Dandona, R.; Dandona, L.; et al. Global, regional, and national cancer incidence, mortality, years of life lost, years lived with disability, and disability-adjusted life-years for 32 cancer groups, 1990 to 2015: A Systematic Analysis for the Global Burden of Disease Study Global Burden. *JAMA Oncol.* **2017**, *3*, 524–548. [CrossRef] [PubMed]
2. Moding, E.J.; Kastan, M.B.; Kirsch, D.G. Strategies for optimizing the response of cancer and normal tissues to radiation. *Nat. Rev. Drug Discov.* **2013**, *12*, 526–542. [CrossRef] [PubMed]
3. Barker, H.E.; Paget, J.T.E.; Khan, A.A.; Harrington, K.J. The tumour microenvironment after radiotherapy: Mechanisms of resistance and recurrence. *Nat. Rev. Cancer* **2015**, *15*, 409–425. [CrossRef] [PubMed]
4. Borras, J.M.; Lievens, Y.; Barton, M.; Corral, J.; Ferlay, J.; Bray, F.; Grau, C. How many new cancer patients in Europe will require radiotherapy by 2025? An ESTRO-HERO analysis. *Radiother. Oncol.* **2016**, *119*, 5–11. [CrossRef]
5. Wang, X.; Eisbruch, A. IMRT for head and neck cancer: Reducing xerostomia and dysphagia. *J. Radiat. Res.* **2016**, *57*, i69–i75. [CrossRef] [PubMed]
6. Folkert, M.R.; Timmerman, R.D. Stereotactic ablative body radiosurgery (SABR) or Stereotactic body radiation therapy (SBRT). *Adv. Drug Deliv. Rev.* **2017**, *109*, 3–14. [CrossRef]
7. Durante, M.; Loeffler, J.S. Charged particles in radiation oncology. *Nat. Rev. Clin. Oncol.* **2010**, *7*, 37–43. [CrossRef]
8. Nutting, C.M.; Morden, J.P.; Harrington, K.J.; Urbano, T.G.; Bhide, S.A.; Clark, C.; Miles, E.A.; Miah, A.B.; Newbold, K.; Tanay, M.A.; et al. Parotid-sparing intensity modulated versus conventional radiotherapy in head and neck cancer (PARSPORT): A phase 3 multicentre randomised controlled trial. *Lancet Oncol.* **2011**, *12*, 127–136. [CrossRef]
9. Van Luijk, P.; Pringle, S.; Deasy, J.O.; Moiseenko, V.V.; Faber, H.; Hovan, A.; Baanstra, M.; Van Der Laan, H.P.; Kierkels, R.G.J.; Van Der Schaaf, A.; et al. Sparing the region of the salivary gland containing stem cells preserves saliva production after radiotherapy for head and neck cancer. *Sci. Transl. Med.* **2015**, *7*, 305ra147. [CrossRef]
10. Wang, L.; Hoogcarspel, S.J.; Wen, Z.; van Vulpen, M.; Molkentine, D.P.; Kok, J.; Lin, S.H.; Broekhuizen, R.; Ang, K.K.; Bovenschen, N.; et al. Biological responses of human solid tumor cells to X-ray irradiation within a 1.5-Tesla magnetic field generated by a magnetic resonance imaging–linear accelerator. *Bioelectromagnetics* **2016**, *37*, 471–480. [CrossRef]
11. Raaymakers, B.W.; Jürgenliemk-Schulz, I.M.; Bol, G.H.; Glitzner, M.; Kotte, A.N.T.J.; Van Asselen, B.; De Boer, J.C.J.; Bluemink, J.J.; Hackett, S.L.; Moerland, M.A.; et al. First patients treated with a 1.5 T MRI-Linac: Clinical proof of concept of a high-precision, high-field MRI guided radiotherapy treatment. *Phys. Med. Biol.* **2017**, *62*, L41–L50. [CrossRef] [PubMed]
12. Lomax, A.J. Charged particle therapy: The physics of interaction. *Cancer J.* **2009**, *15*, 285–291. [CrossRef]
13. Newhauser, W.D.; Zhang, R. The physics of proton therapy. *Phys. Med. Biol.* **2015**, *60*, R155–R209. [CrossRef] [PubMed]
14. Oborn, B.M.; Dowdell, S.; Metcalfe, P.E.; Crozier, S.; Mohan, R.; Keall, P.J. Proton beam deflection in MRI fields: Implications for MRI-guided proton therapy. *Med. Phys.* **2015**, *42*, 2113–2124. [CrossRef]
15. Oborn, B.M.; Dowdell, S.; Metcalfe, P.E.; Crozier, S.; Mohan, R.; Keall, P.J. Future of medical physics: Real-time MRI-guided proton therapy: Real-time. *Med. Phys.* **2017**, *44*, e77–e90. [CrossRef]
16. Nagle, P.W.; van Goethem, M.J.; Kempers, M.; Kiewit, H.; Knopf, A.; Langendijk, J.A.; Brandenburg, S.; van Luijk, P.; Coppes, R.P. In vitro biological response of cancer and normal tissue cells to proton irradiation not affected by an added magnetic field. *Radiother. Oncol.* **2019**, *137*, 125–129. [CrossRef] [PubMed]

17. Inaniwa, T.; Suzuki, M.; Sato, S.; Muramatsu, M.; Noda, A.; Iwata, Y.; Kanematsu, N.; Shirai, T.; Noda, K. Effect of External Magnetic Fields on Biological Effectiveness of Proton Beams. *Int. J. Radiat. Oncol. Biol. Phys.* **2020**, *106*, 597–603. [CrossRef] [PubMed]
18. Wilson, J.D.; Hammond, E.M.; Higgins, G.S.; Petersson, K. Ultra-High Dose Rate (FLASH) Radiotherapy: Silver Bullet or Fool's Gold? *Front. Oncol.* **2020**, *9*, 1563. [CrossRef]
19. Vozenin, M.-C.; Baumann, M.; Coppes, R.P.; Bourhis, J. FLASH radiotherapy International Workshop. *Radiother. Oncol.* **2019**, *139*, 1–3. [CrossRef]
20. Barendsen, G.W. Dose-survival curves of human cells in tissue culture irradiated with alpha-, beta-, 20-kV. X- and 200-kV. X-radiation. *Nature* **1962**, *193*, 1153–1155. [CrossRef]
21. Franken, N.A.P.; Rodermond, H.M.; Stap, J.; Haveman, J.; van Bree, C. Clonogenic assay of cells in vitro. *Nat. Protoc.* **2006**, *1*, 2315–2319. [CrossRef] [PubMed]
22. Liu, Q.; Wang, M.; Kern, A.M.; Khaled, S.; Han, J.; Yeap, B.Y.; Hong, T.S.; Settleman, J.; Benes, C.H.; Held, K.D.; et al. Adapting a drug screening platform to discover associations of molecular targeted radiosensitizers with genomic biomarkers. *Mol. Cancer Res.* **2015**, *13*, 713–720. [CrossRef] [PubMed]
23. Pampaloni, F.; Reynaud, E.G.; Stelzer, E.H.K. The third dimension bridges the gap between cell culture and live tissue. *Nat. Rev. Mol. Cell Biol.* **2007**, *8*, 839–845. [CrossRef] [PubMed]
24. Langhans, S.A. Three-dimensional in vitro cell culture models in drug discovery and drug repositioning. *Front. Pharmacol.* **2018**, *9*, 6. [CrossRef] [PubMed]
25. Eke, I.; Cordes, N. Radiobiology goes 3D: How ECM and cell morphology impact on cell survival after irradiation. *Radiother. Oncol.* **2011**, *99*, 271–278. [CrossRef] [PubMed]
26. Nagle, P.W.; Hosper, N.A.; Ploeg, E.M.; Van Goethem, M.J.; Brandenburg, S.; Langendijk, J.A.; Chiu, R.K.; Coppes, R.P. The in Vitro Response of Tissue Stem Cells to Irradiation with Different Linear Energy Transfers. *Int. J. Radiat. Oncol. Biol. Phys.* **2016**, *95*, 103–111. [CrossRef]
27. Lee, J.M.; Abrahamson, J.L.A.; Kandel, R.; Donehower, L.A.; Bernstein, A. Susceptibility to radiation-carcinogenesis and accumulation of chromosomal breakage in p53 deficient mice. *Oncogene* **1994**, *9*, 3731–3736.
28. Kemp, C.J.; Wheldon, T.; Balmain, A. P53-deficient mice are extremely susceptible to radiation-induced tumorigenesis. *Nat. Genet.* **1994**, *8*, 66–69. [CrossRef]
29. Hammond, E.M.; Muschel, R.J. Radiation and ATM inhibition: The heart of the matter. *J. Clin. Investig.* **2014**, *124*, 3289–3291. [CrossRef]
30. Moding, E.J.; Lee, C.L.; Castle, K.D.; Oh, P.; Mao, L.; Zha, S.; Min, H.D.; Ma, Y.; Das, S.; Kirsch, D.G. ATM deletion with dual recombinase technology preferentially radiosensitizes tumor endothelium. *J. Clin. Investig.* **2014**, *124*, 3325–3338. [CrossRef]
31. De Boo, J.; Hendriksen, C. Reduction strategies in animal research: A review of scientific approaches at the intra-experimental, supra-experimental and extra-experimental levels. *ATLA Altern. Lab. Anim.* **2005**, *33*, 369–377. [CrossRef] [PubMed]
32. Lancaster, M.A.; Knoblich, J.A. Organogenesisin a dish: Modeling development and disease using organoid technologies. *Science* **2014**, *345*, 1247125. [CrossRef]
33. Clevers, H. Modeling Development and Disease with Organoids. *Cell* **2016**, *165*, 1586–1597. [CrossRef] [PubMed]
34. Dutta, D.; Heo, I.; Clevers, H. Disease Modeling in Stem Cell-Derived 3D Organoid Systems. *Trends Mol. Med.* **2017**, *23*, 393–410. [CrossRef] [PubMed]
35. Stelzner, M.; Helmrath, M.; Dunn, J.C.Y.; Henning, S.J.; Houchen, C.W.; Kuo, C.; Lynch, J.; Li, L.; Magness, S.T.; Martin, M.G.; et al. A nomenclature for intestinal in vitro cultures. *Am. J. Physiol.-Gastrointest. Liver Physiol.* **2012**, *302*, G1359–G1363. [CrossRef] [PubMed]
36. Shankar, A.S.; Du, Z.; Mora, H.T.; van den Bosch, T.P.P.; Korevaar, S.S.; Van den Berg-Garrelds, I.M.; Bindels, E.; Lopez-Iglesias, C.; Groningen, M.C.; Gribnau, J.; et al. Human kidney organoids produce functional renin. *Kidney Int.* **2020**, in press. [CrossRef]
37. Maimets, M.; Rocchi, C.; Bron, R.; Pringle, S.; Kuipers, J.; Giepmans, B.N.G.; Vries, R.G.J.; Clevers, H.; De Haan, G.; Van Os, R.; et al. Long-Term in Vitro Expansion of Salivary Gland Stem Cells Driven by Wnt Signals. *Stem Cell Rep.* **2016**, *6*, 150–162. [CrossRef]
38. Wang, D.; Plukker, J.T.M.; Coppes, R.P. Cancer stem cells with increased metastatic potential as a therapeutic target for esophageal cancer. *Semin. Cancer Biol.* **2017**, *44*, 60–66. [CrossRef]

39. Matano, M.; Date, S.; Shimokawa, M.; Takano, A.; Fujii, M.; Ohta, Y.; Watanabe, T.; Kanai, T.; Sato, T. Modeling colorectal cancer using CRISPR-Cas9-mediated engineering of human intestinal organoids. *Nat. Med.* **2015**, *21*, 256–262. [CrossRef]
40. Seidlitz, T.; Merker, S.R.; Rothe, A.; Zakrzewski, F.; Von Neubeck, C.; Grützmann, K.; Sommer, U.; Schweitzer, C.; Schölch, S.; Uhlemann, H.; et al. Human gastric cancer modelling using organoids. *Gut* **2019**, *68*, 207–217. [CrossRef]
41. Li, X.; Larsson, P.; Ljuslinder, I.; Öhlund, D.; Myte, R.; Löfgren-Burström, A.; Zingmark, C.; Ling, A.; Edin, S.; Palmqvist, R. Ex vivo organoid cultures reveal the importance of the tumor microenvironment for maintenance of colorectal cancer stem cells. *Cancers* **2020**, *12*, 923. [CrossRef] [PubMed]
42. Barker, N.; Van Es, J.H.; Kuipers, J.; Kujala, P.; Van Den Born, M.; Cozijnsen, M.; Haegebarth, A.; Korving, J.; Begthel, H.; Peters, P.J.; et al. Identification of stem cells in small intestine and colon by marker gene Lgr5. *Nature* **2007**, *449*, 1003–1007. [CrossRef] [PubMed]
43. Sato, T.; Vries, R.G.; Snippert, H.J.; Van De Wetering, M.; Barker, N.; Stange, D.E.; Van Es, J.H.; Abo, A.; Kujala, P.; Peters, P.J.; et al. Single Lgr5 stem cells build crypt-villus structures in vitro without a mesenchymal niche. *Nature* **2009**, *459*, 262–265. [CrossRef] [PubMed]
44. Sato, T.; Stange, D.E.; Ferrante, M.; Vries, R.G.J.; Van Es, J.H.; Van Den Brink, S.; Van Houdt, W.J.; Pronk, A.; Van Gorp, J.; Siersema, P.D.; et al. Long-term expansion of epithelial organoids from human colon, adenoma, adenocarcinoma, and Barrett's epithelium. *Gastroenterology* **2011**, *141*, 1762–1772. [CrossRef] [PubMed]
45. Barker, N.; Huch, M.; Kujala, P.; van de Wetering, M.; Snippert, H.J.; van Es, J.H.; Sato, T.; Stange, D.E.; Begthel, H.; van den Born, M.; et al. Lgr5+ve Stem Cells Drive Self-Renewal in the Stomach and Build Long-Lived Gastric Units In Vitro. *Cell Stem Cell* **2010**, *6*, 25–36. [CrossRef]
46. DeWard, A.D.; Cramer, J.; Lagasse, E. Cellular heterogeneity in the mouse esophagus implicates the presence of a nonquiescent epithelial stem cell population. *Cell Rep.* **2014**, *9*, 701–711. [CrossRef]
47. Spence, J.R.; Mayhew, C.N.; Rankin, S.A.; Kuhar, M.F.; Vallance, J.E.; Tolle, K.; Hoskins, E.E.; Kalinichenko, V.V.; Wells, S.I.; Zorn, A.M.; et al. Directed differentiation of human pluripotent stem cells into intestinal tissue in vitro. *Nature* **2011**, *470*, 105–110. [CrossRef]
48. Trisno, S.L.; Philo, K.E.D.; McCracken, K.W.; Catá, E.M.; Ruiz-Torres, S.; Rankin, S.A.; Han, L.; Nasr, T.; Chaturvedi, P.; Rothenberg, M.E.; et al. Esophageal Organoids from Human Pluripotent Stem Cells Delineate Sox2 Functions during Esophageal Specification. *Cell Stem Cell* **2018**, *23*, 501–515. [CrossRef]
49. Drost, J.; Van Boxtel, R.; Blokzijl, F.; Mizutani, T.; Sasaki, N.; Sasselli, V.; De Ligt, J.; Behjati, S.; Grolleman, J.E.; Van Wezel, T.; et al. Use of CRISPR-modified human stem cell organoids to study the origin of mutational signatures in cancer. *Science* **2017**, *358*, 234–238. [CrossRef]
50. Dekkers, J.F.; Wiegerinck, C.L.; De Jonge, H.R.; Bronsveld, I.; Janssens, H.M.; De Winter-De Groot, K.M.; Brandsma, A.M.; De Jong, N.W.M.; Bijvelds, M.J.C.; Scholte, B.J.; et al. A functional CFTR assay using primary cystic fibrosis intestinal organoids. *Nat. Med.* **2013**, *19*, 939–945. [CrossRef]
51. Wang, X.; Wei, L.; Cramer, J.M.; Leibowitz, B.J.; Judge, C.; Epperly, M.; Greenberger, J.; Wang, F.; Li, L.; Stelzner, M.G.; et al. Pharmacologically blocking p53-dependent apoptosis protects intestinal stem cells and mice from radiation. *Sci. Rep.* **2015**, *5*, 8566. [CrossRef] [PubMed]
52. Bhanja, P.; Norris, A.; Gupta-Saraf, P.; Hoover, A.; Saha, S. BCN057 induces intestinal stem cell repair and mitigates radiation-induced intestinal injury. *Stem Cell Res. Ther.* **2018**, *9*, 26. [CrossRef] [PubMed]
53. Nag, D.; Bhanja, P.; Riha, R.; Sanchez-Guerrero, G.; Kimler, B.F.; Tsue, T.T.; Lominska, C.; Saha, S. Auranofin protects intestine against radiation injury by modulating p53/p21 pathway and radiosensitizes human colon tumor. *Clin. Cancer Res.* **2019**, *25*, 4791–4807. [CrossRef] [PubMed]
54. Martin, M.L.; Adileh, M.; Hsu, K.S.; Hua, G.; Lee, S.G.; Li, C.; Fuller, J.D.; Rotolo, J.A.; Bodo, S.; Klingler, S.; et al. Organoids reveal that inherent radiosensitivity of small and large intestinal stem cells determines organ sensitivity. *Cancer Res.* **2020**, *256*, 1219–1227. [CrossRef] [PubMed]
55. Withers, H.R.; Elkind, M.M. Microcolony survival assay for cells of mouse intestinal mucosa exposed to radiation. *Int. J. Radiat. Biol.* **1970**, *17*, 261–267. [CrossRef] [PubMed]
56. Nutting, C. Radiotherapy in head and neck cancer management: United Kingdom National Multidisciplinary Guidelines. *J. Laryngol. Otol.* **2016**, *130*, S66–S67. [CrossRef]
57. Vissink, A.; Mitchell, J.B.; Baum, B.J.; Limesand, K.H.; Jensen, S.B.; Fox, P.C.; Elting, L.S.; Langendijk, J.A.; Coppes, R.P.; Reyland, M.E. Clinical management of salivary gland hypofunction and xerostomia in

head-and-neck cancer patients: Successes and barriers. *Int. J. Radiat. Oncol. Biol. Phys.* **2010**, *78*, 983–991. [CrossRef]

58. Rocchi, C.; Emmerson, E. Mouth-Watering Results: Clinical Need, Current Approaches, and Future Directions for Salivary Gland Regeneration. *Trends Mol. Med.* **2020**, *26*, 649–669. [CrossRef]
59. Zeilstra, L.J.W.; Vissink, A.; Konings, A.W.T.; Coppes, R.P. Radiation induced cell loss in rat submandibular gland and its relation to gland function. *Int. J. Radiat. Biol.* **2000**, *76*, 419–429. [CrossRef]
60. Coppes, R.P.; Vissink, A.; Konings, A.W.T. Comparison of radiosensitivity of rat parotid and submandibular glands after different radiation schedules. *Radiother. Oncol.* **2002**, *63*, 321–328. [CrossRef]
61. Marmary, Y.; Adar, R.; Gaska, S.; Wygoda, A.; Maly, A.; Cohen, J.; Eliashar, R.; Mizrachi, L.; Orfaig-Geva, C.; Baum, B.J.; et al. Radiation-induced loss of salivary gland function is driven by cellular senescence and prevented by IL6 modulation. *Cancer Res.* **2016**, *76*, 1170–1180. [CrossRef] [PubMed]
62. Emmerson, E.; May, A.J.; Nathan, S.; Cruz-Pacheco, N.; Lizama, C.O.; Maliskova, L.; Zovein, A.C.; Shen, Y.; Muench, M.O.; Knox, S.M. SOX2 regulates acinar cell development in the salivary gland. *eLife* **2017**, *6*, e26620. [CrossRef] [PubMed]
63. Ninche, N.; Kwak, M.; Ghazizadeh, S. Diverse epithelial cell populations contribute to the regeneration of secretory units in injured salivary glands. *Development* **2020**, *147*, dev192807. [CrossRef] [PubMed]
64. Lombaert, I.M.A.; Patel, V.N.; Jones, C.E.; Villier, D.C.; Canada, A.E.; Moore, M.R.; Berenstein, E.; Zheng, C.; Goldsmith, C.M.; Chorini, J.A.; et al. CERE-120 Prevents Irradiation-Induced Hypofunction and Restores Immune Homeostasis in Porcine Salivary Glands. *Mol. Ther.-Methods Clin. Dev.* **2020**, *18*, 839–855. [CrossRef] [PubMed]
65. Meyer, R.; Wong, W.Y.; Guzman, R.; Burd, R.; Limesand, K. Radiation treatment of organotypic cultures from submandibular and parotid salivary glands models key in vivo characteristics. *J. Vis. Exp.* **2019**, *2019*, 59484. [CrossRef]
66. Nanduri, L.S.Y.; Baanstra, M.; Faber, H.; Rocchi, C.; Zwart, E.; De Haan, G.; Van Os, R.; Coppes, R.P. Purification and Ex vivo expansion of fully functional salivary gland stem cells. *Stem Cell Rep.* **2014**, *3*, 957–964. [CrossRef]
67. Pringle, S.; Maimets, M.; Van Der Zwaag, M.; Stokman, M.A.; Van Gosliga, D.; Zwart, E.; Witjes, M.J.H.; De Haan, G.; Van Os, R.; Coppes, R.P. Human salivary gland stem cells functionally restore radiation damaged salivary glands. *Stem Cells* **2016**, *34*, 640–652. [CrossRef]
68. Nanduri, L.S.Y.; Maimets, M.; Pringle, S.A.; Van Der Zwaag, M.; Van Os, R.P.; Coppes, R.P. Regeneration of irradiated salivary glands with stem cell marker expressing cells. *Radiother. Oncol.* **2011**, *99*, 367–372. [CrossRef]
69. Tanaka, J.; Ogawa, M.; Hojo, H.; Kawashima, Y.; Mabuchi, Y.; Hata, K.; Nakamura, S.; Yasuhara, R.; Takamatsu, K.; Irié, T.; et al. Generation of orthotopically functional salivary gland from embryonic stem cells. *Nat. Commun.* **2018**, *9*, 4216. [CrossRef]
70. Lo, B.; Parham, L. Ethical issues in stem cell research. *Endocr. Rev.* **2009**, *30*, 204–213. [CrossRef]
71. Nori, S.; Okada, Y.; Nishimura, S.; Sasaki, T.; Itakura, G.; Kobayashi, Y.; Renault-Mihara, F.; Shimizu, A.; Koya, I.; Yoshida, R.; et al. Long-term safety issues of iPSC-based cell therapy in a spinal cord injury model: Oncogenic transformation with epithelial-mesenchymal transition. *Stem Cell Rep.* **2015**, *4*, 360–373. [CrossRef] [PubMed]
72. Nagle, P.W.; Hosper, N.A.; Barazzuol, L.; Jellema, A.L.; Baanstra, M.; Van Goethem, M.J.; Brandenburg, S.; Giesen, U.; Langendijk, J.A.; Van Luijk, P.; et al. Lack of DNA damage response at low radiation doses in adult stem cells contributes to organ dysfunction. *Clin. Cancer Res.* **2018**, *24*, 6583–6593. [CrossRef] [PubMed]
73. Marples, B.; Joiner, M.C. The response of Chinese hamster V79 cells to low radiation doses: Evidence of enhanced sensitivity of the whole cell population. *Radiat. Res.* **1993**, *133*, 41–51. [CrossRef] [PubMed]
74. Martin, L.M.; Marples, B.; Lynch, T.H.; Hollywood, D.; Marignol, L. Exposure to low dose ionising radiation: Molecular and clinical consequences. *Cancer Lett.* **2014**, *349*, 98–106. [CrossRef] [PubMed]
75. Serrano Martinez, P.; Cinat, D.; van Luijk, P.; Baanstra, M.; de Haan, G.; Pringle, S.; Coppes, R.P. Mouse parotid salivary gland organoids for the in vitro study of stem cell radiation response. *Oral Dis.* **2020**. [CrossRef]
76. Siegel, R.L.; Miller, K.D.; Jemal, A. Cancer statistics, 2020. *CA Cancer J. Clin.* **2020**, *70*, 7–30. [CrossRef]
77. Llovet, J.M.; Villanueva, A.; Lachenmayer, A.; Finn, R.S. Advances in targeted therapies for hepatocellular carcinoma in the genomic era. *Nat. Rev. Clin. Oncol.* **2015**, *12*, 408–424. [CrossRef]

78. Guha, C.; Kavanagh, B.D. Hepatic Radiation Toxicity: Avoidance and Amelioration. *Semin. Radiat. Oncol.* **2011**, *21*, 256–263. [CrossRef]
79. Kim, J.; Jung, Y. Radiation-induced liver disease: Current understanding and future perspectives. *Exp. Mol. Med.* **2017**, *49*, e359. [CrossRef]
80. Huch, M.; Dorrell, C.; Boj, S.F.; Van Es, J.H.; Li, V.S.W.; Van De Wetering, M.; Sato, T.; Hamer, K.; Sasaki, N.; Finegold, M.J.; et al. In vitro expansion of single Lgr5 + liver stem cells induced by Wnt-driven regeneration. *Nature* **2013**, *494*, 247–250. [CrossRef]
81. Huch, M.; Gehart, H.; Van Boxtel, R.; Hamer, K.; Blokzijl, F.; Verstegen, M.M.A.; Ellis, E.; Van Wenum, M.; Fuchs, S.A.; De Ligt, J.; et al. Long-term culture of genome-stable bipotent stem cells from adult human liver. *Cell* **2015**, *160*, 299–312. [CrossRef]
82. Hay, D.C.; Zhao, D.; Fletcher, J.; Hewitt, Z.A.; McLean, D.; Urruticoechea-Uriguen, A.; Black, J.R.; Elcombe, C.; Ross, J.A.; Wolf, R.; et al. Efficient Differentiation of Hepatocytes from Human Embryonic Stem Cells Exhibiting Markers Recapitulating Liver Development In Vivo. *Stem Cells* **2008**, *26*, 894–902. [CrossRef] [PubMed]
83. Sullivan, G.J.; Hay, D.C.; Park, I.H.; Fletcher, J.; Hannoun, Z.; Payne, C.M.; Dalgetty, D.; Black, J.R.; Ross, J.A.; Samuel, K.; et al. Generation of functional human hepatic endoderm from human induced pluripotent stem cells. *Hepatology* **2010**, *51*, 329–335. [CrossRef] [PubMed]
84. Takebe, T.; Sekine, K.; Enomura, M.; Koike, H.; Kimura, M.; Ogaeri, T.; Zhang, R.R.; Ueno, Y.; Zheng, Y.W.; Koike, N.; et al. Vascularized and functional human liver from an iPSC-derived organ bud transplant. *Nature* **2013**, *499*, 481–484. [CrossRef] [PubMed]
85. Takebe, T.; Sekine, K.; Kimura, M.; Yoshizawa, E.; Ayano, S.; Koido, M.; Funayama, S.; Nakanishi, N.; Hisai, T.; Kobayashi, T.; et al. Massive and Reproducible Production of Liver Buds Entirely from Human Pluripotent Stem Cells. *Cell Rep.* **2017**, *21*, 2661–2670. [CrossRef] [PubMed]
86. Shinozawa, T.; Kimura, M.; Cai, Y.; Saiki, N.; Yoneyama, Y.; Ouchi, R.; Koike, H.; Maezawa, M.; Zhang, R.-R.; Dunn, A.; et al. High-Fidelity Drug Induced Liver Injury Screen Using Human PSC-derived Organoids. *Gastroenterology* **2020**, in press. [CrossRef]
87. Lynch, T.J.; Bell, D.W.; Sordella, R.; Gurubhagavatula, S.; Okimoto, R.A.; Brannigan, B.W.; Harris, P.L.; Haserlat, S.M.; Supko, J.G.; Haluska, F.G.; et al. Activating Mutations in the Epidermal Growth Factor Receptor Underlying Responsiveness of Non–Small-Cell Lung Cancer to Gefitinib. *N. Engl. J. Med.* **2004**, *350*, 2129–2139. [CrossRef]
88. van Hagen, P.; Hulshof, M.C.C.M.; van Lanschot, J.J.B.; Steyerberg, E.W.; Henegouwen, M.V.; Wijnhoven, B.P.L.; Richel, D.J.; Nieuwenhuijzen, G.A.P.; Hospers, G.A.P.; Bonenkamp, J.J.; et al. Preoperative Chemoradiotherapy for Esophageal or Junctional Cancer. *N. Engl. J. Med.* **2012**, *366*, 2074–2084. [CrossRef]
89. Sjoquist, K.M.; Burmeister, B.H.; Smithers, B.M.; Zalcberg, J.R.; Simes, R.J.; Barbour, A.; Gebski, V. Survival after neoadjuvant chemotherapy or chemoradiotherapy for resectable oesophageal carcinoma: An updated meta-analysis. *Lancet Oncol.* **2011**, *12*, 681–692. [CrossRef]
90. Janakiraman, H.; Zhu, Y.; Becker, S.A.; Wang, C.; Cross, A.; Curl, E.; Lewin, D.; Hoffman, B.J.; Warren, G.W.; Hill, E.G.; et al. Modeling rectal cancer to advance neoadjuvant precision therapy. *Int. J. Cancer* **2020**, *147*, 1405–1418. [CrossRef]
91. Van De Wetering, M.; Francies, H.E.; Francis, J.M.; Bounova, G.; Iorio, F.; Pronk, A.; Van Houdt, W.; Van Gorp, J.; Taylor-Weiner, A.; Kester, L.; et al. Prospective derivation of a living organoid biobank of colorectal cancer patients. *Cell* **2015**, *161*, 933–945. [CrossRef] [PubMed]
92. Ooft, S.N.; Weeber, F.; Dijkstra, K.K.; McLean, C.M.; Kaing, S.; van Werkhoven, E.; Schipper, L.; Hoes, L.; Vis, D.J.; van de Haar, J.; et al. Patient-derived organoids can predict response to chemotherapy in metastatic colorectal cancer patients. *Sci. Transl. Med.* **2019**, *11*, eaay2574. [CrossRef] [PubMed]
93. Ganesh, K.; Wu, C.; O'Rourke, K.P.; Szeglin, B.C.; Zheng, Y.; Sauvé, C.E.G.; Adileh, M.; Wasserman, I.; Marco, M.R.; Kim, A.S.; et al. A rectal cancer organoid platform to study individual responses to chemoradiation. *Nat. Med.* **2019**, *25*, 1607–1614. [CrossRef] [PubMed]
94. Yao, Y.; Xu, X.; Yang, L.; Zhu, J.; Wan, J.; Shen, L.; Xia, F.; Fu, G.; Deng, Y.; Pan, M.; et al. Patient-Derived Organoids Predict Chemoradiation Responses of Locally Advanced Rectal Cancer. *Cell Stem Cell* **2020**, *26*, 17–26. [CrossRef]

95. Brown, S.L.; Kolozsvary, A.; Isrow, D.M.; Al Feghali, K.; Lapanowski, K.; Jenrow, K.A.; Kim, J.H. A novel mechanism of high dose radiation sensitization by metformin. *Front. Oncol.* **2019**, *9*, 247. [CrossRef] [PubMed]
96. Spiegelberg, L.; van Hoof, S.J.; Biemans, R.; Lieuwes, N.G.; Marcus, D.; Niemans, R.; Theys, J.; Yaromina, A.; Lambin, P.; Verhaegen, F.; et al. Evofosfamide sensitizes esophageal carcinomas to radiation without increasing normal tissue toxicity. *Radiother. Oncol.* **2019**, *141*, 247–255. [CrossRef]
97. Zhang, L.; Bochkur Dratver, M.; Yazal, T.; Dong, K.; Nguyen, A.; Yu, G.; Dao, A.; Bochkur Dratver, M.; Duhachek-Muggy, S.; Bhat, K.; et al. Mebendazole Potentiates Radiation Therapy in Triple-Negative Breast Cancer. *Int. J. Radiat. Oncol. Biol. Phys.* **2019**, *103*, 195–207. [CrossRef]
98. Broadfield, L.A.; Marcinko, K.; Tsakiridis, E.; Zacharidis, P.G.; Villani, L.; Lally, J.S.V.; Menjolian, G.; Maharaj, D.; Mathurin, T.; Smoke, M.; et al. Salicylate enhances the response of prostate cancer to radiotherapy. *Prostate* **2019**, *79*, 489–497. [CrossRef]
99. Pushpakom, S.; Iorio, F.; Eyers, P.A.; Escott, K.J.; Hopper, S.; Wells, A.; Doig, A.; Guilliams, T.; Latimer, J.; McNamee, C.; et al. Drug repurposing: Progress, challenges and recommendations. *Nat. Rev. Drug Discov.* **2018**, *18*, 41–58. [CrossRef]
100. Pasch, C.A.; Favreau, P.F.; Yueh, A.E.; Babiarz, C.P.; Gillette, A.A.; Sharick, J.T.; Karim, M.R.; Nickel, K.P.; DeZeeuw, A.K.; Sprackling, C.M.; et al. Patient-derived cancer organoid cultures to predict sensitivity to chemotherapy and radiation. *Clin. Cancer Res.* **2019**, *25*, 5376–5387. [CrossRef]
101. Driehuis, E.; Kolders, S.; Spelier, S.; Lõhmussaar, K.; Willems, S.M.; Devriese, L.A.; de Bree, R.; de Ruiter, E.J.; Korving, J.; Begthel, H.; et al. Oral mucosal organoids as a potential platform for personalized cancer therapy. *Cancer Discov.* **2019**, *9*, 852–871. [CrossRef] [PubMed]
102. Hubert, C.G.; Rivera, M.; Spangler, L.C.; Wu, Q.; Mack, S.C.; Prager, B.C.; Couce, M.; McLendon, R.E.; Sloan, A.E.; Rich, J.N. A three-dimensional organoid culture system derived from human glioblastomas recapitulates the hypoxic gradients and cancer stem cell heterogeneity of tumors found in vivo. *Cancer Res.* **2016**, *76*, 2465–2477. [CrossRef] [PubMed]
103. Linkous, A.; Balamatsias, D.; Snuderl, M.; Edwards, L.; Miyaguchi, K.; Milner, T.; Reich, B.; Cohen-Gould, L.; Storaska, A.; Nakayama, Y.; et al. Modeling Patient-Derived Glioblastoma with Cerebral Organoids. *Cell Rep.* **2019**, *26*, 3203–3211. [CrossRef] [PubMed]
104. Vlachogiannis, G.; Hedayat, S.; Vatsiou, A.; Jamin, Y.; Fernández-Mateos, J.; Khan, K.; Lampis, A.; Eason, K.; Huntingford, I.; Burke, R.; et al. Patient-derived organoids model treatment response of metastatic gastrointestinal cancers. *Science* **2018**, *359*, 920–926. [CrossRef]
105. Roedl, J.B.; Halpern, E.F.; Colen, R.R.; Sahani, D.V.; Fischman, A.J.; Blake, M.A. Metabolic tumor width parameters as determined on PET/CT predict disease-free survival and treatment response in squamous cell carcinoma of the esophagus. *Mol. Imaging Biol.* **2009**, *11*, 54–60. [CrossRef] [PubMed]
106. Monjazeb, A.M.; Riedlinger, G.; Aklilu, M.; Geisinger, K.R.; Mishra, G.; Isom, S.; Clark, P.; Levine, E.A.; Blackstock, A.W. Outcomes of patients with esophageal cancer staged with [18F] fluorodeoxyglucose positron emission tomography (FDG-PET): Can postchemoradiotherapy FDG-PET predict the utility of resection? *J. Clin. Oncol.* **2010**, *28*, 4714–4721. [CrossRef]
107. Omloo, J.M.T.; Van Heijl, M.; Hoekstra, O.S.; Van Berge Henegouwen, M.I.; Van Lanschot, J.J.B.; Sloof, G.W. FDG-PET parameters as prognostic factor in esophageal cancer patients: A review. *Ann. Surg. Oncol.* **2011**, *18*, 3338–3352. [CrossRef]
108. Li, X.; Francies, H.E.; Secrier, M.; Perner, J.; Miremadi, A.; Galeano-Dalmau, N.; Barendt, W.J.; Letchford, L.; Leyden, G.M.; Goffin, E.K.; et al. Organoid cultures recapitulate esophageal adenocarcinoma heterogeneity providing a model for clonality studies and precision therapeutics. *Nat. Commun.* **2018**, *9*, 2983. [CrossRef]
109. Derouet, M.F.; Allen, J.; Wilson, G.W.; Ng, C.; Radulovich, N.; Kalimuthu, S.; Tsao, M.S.; Darling, G.E.; Yeung, J.C. Towards personalized induction therapy for esophageal adenocarcinoma: Organoids derived from endoscopic biopsy recapitulate the pre-treatment tumor. *Sci. Rep.* **2020**, *10*, 14514. [CrossRef]
110. Stupp, R.; Mason, W.P.; van den Bent, M.J.; Weller, M.; Fisher, B.; Taphoorn, M.J.B.; Belanger, K.; Brandes, A.A.; Marosi, C.; Bogdahn, U.; et al. Radiotherapy plus Concomitant and Adjuvant Temozolomide for Glioblastoma. *N. Engl. J. Med.* **2005**, *352*, 987–996. [CrossRef]

111. Pollard, S.M.; Yoshikawa, K.; Clarke, I.D.; Danovi, D.; Stricker, S.; Russell, R.; Bayani, J.; Head, R.; Lee, M.; Bernstein, M.; et al. Glioma Stem Cell Lines Expanded in Adherent Culture Have Tumor-Specific Phenotypes and Are Suitable for Chemical and Genetic Screens. *Cell Stem Cell* **2009**, *4*, 568–580. [CrossRef] [PubMed]
112. Patel, A.P.; Tirosh, I.; Trombetta, J.J.; Shalek, A.K.; Gillespie, S.M.; Wakimoto, H.; Cahill, D.P.; Nahed, B.V.; Curry, W.T.; Martuza, R.L.; et al. Single-cell RNA-seq highlights intratumoral heterogeneity in primary glioblastoma. *Science* **2014**, *344*, 1396–1401. [CrossRef] [PubMed]
113. Lan, X.; Jörg, D.J.; Cavalli, F.M.G.; Richards, L.M.; Nguyen, L.V.; Vanner, R.J.; Guilhamon, P.; Lee, L.; Kushida, M.M.; Pellacani, D.; et al. Fate mapping of human glioblastoma reveals an invariant stem cell hierarchy. *Nature* **2017**, *549*, 227–232. [CrossRef] [PubMed]
114. Bao, S.; Wu, Q.; McLendon, R.E.; Hao, Y.; Shi, Q.; Hjelmeland, A.B.; Dewhirst, M.W.; Bigner, D.D.; Rich, J.N. Glioma stem cells promote radioresistance by preferential activation of the DNA damage response. *Nature* **2006**, *444*, 756–760. [CrossRef] [PubMed]
115. Bian, S.; Repic, M.; Guo, Z.; Kavirayani, A.; Burkard, T.; Bagley, J.A.; Krauditsch, C.; Knoblich, J.A. Genetically engineered cerebral organoids model brain tumor formation. *Nat. Methods* **2018**, *15*, 631–639. [CrossRef] [PubMed]
116. Ogawa, J.; Pao, G.M.; Shokhirev, M.N.; Verma, I.M. Glioblastoma Model Using Human Cerebral Organoids. *Cell Rep.* **2018**, *23*, 1220–1229. [CrossRef] [PubMed]
117. Jacob, F.; Salinas, R.D.; Zhang, D.Y.; Nguyen, P.T.T.; Schnoll, J.G.; Wong, S.Z.H.; Thokala, R.; Sheikh, S.; Saxena, D.; Prokop, S.; et al. A Patient-Derived Glioblastoma Organoid Model and Biobank Recapitulates Inter- and Intra-tumoral Heterogeneity. *Cell* **2020**, *180*, 188–204. [CrossRef]
118. Ferreira, J.N.; Hoffman, M.P. Interactions between developing nerves and salivary glands. *Organogenesis* **2013**, *9*, 152–158. [CrossRef]
119. Knox, S.M.; Lombaert, I.M.A.; Haddox, C.L.; Abrams, S.R.; Cotrim, A.; Wilson, A.J.; Hoffman, M.P. Parasympathetic stimulation improves epithelial organ regeneration. *Nat. Commun.* **2013**, *4*, 1494. [CrossRef]
120. Emmerson, E.; May, A.J.; Berthoin, L.; Cruz-Pacheco, N.; Nathan, S.; Mattingly, A.J.; Chang, J.L.; Ryan, W.R.; Tward, A.D.; Knox, S.M. Salivary glands regenerate after radiation injury through SOX2-mediated secretory cell replacement. *EMBO Mol. Med.* **2018**, *10*, e8051. [CrossRef]
121. Neal, J.T.; Li, X.; Zhu, J.; Giangarra, V.; Grzeskowiak, C.L.; Ju, J.; Liu, I.H.; Chiou, S.H.; Salahudeen, A.A.; Smith, A.R.; et al. Organoid Modeling of the Tumor Immune Microenvironment. *Cell* **2018**, *175*, 1972–1988. [CrossRef] [PubMed]
122. Hacker, B.C.; Gomez, J.D.; Silvera Batista, C.A.; Rafat, M. Growth and characterization of irradiated organoids from mammary glands. *J. Vis. Exp.* **2019**, *2019*, 59293. [CrossRef] [PubMed]
123. Kretzschmar, K.; Clevers, H. Organoids: Modeling Development and the Stem Cell Niche in a Dish. *Dev. Cell* **2016**, *38*, 590–600. [CrossRef] [PubMed]
124. Richards, Z.; McCray, T.; Marsili, J.; Zenner, M.L.; Manlucu, J.T.; Garcia, J.; Kajdacsy-Balla, A.; Murray, M.; Voisine, C.; Murphy, A.B.; et al. Prostate Stroma Increases the Viability and Maintains the Branching Phenotype of Human Prostate Organoids. *iScience* **2019**, *12*, 304–317. [CrossRef]
125. Hellevik, T.; Martinez-Zubiaurre, I. Radiotherapy and the tumor stroma: The importance of dose and fractionation. *Front. Oncol.* **2014**, *4*, 1. [CrossRef]
126. Farhood, B.; Khodamoradi, E.; Hoseini-Ghahfarokhi, M.; Motevaseli, E.; Mirtavoos-Mahyari, H.; Eleojo Musa, A.; Najafi, M. TGF-β in radiotherapy: Mechanisms of tumor resistance and normal tissues injury. *Pharmacol. Res.* **2020**, *155*, 104745. [CrossRef]
127. Khan, M.A.; Hill, R.P.; Van Dyk, J. Partial volume rat lung irradiation: An evaluation of early DNA damage. *Int. J. Radiat. Oncol. Biol. Phys.* **1998**, *40*, 467–476. [CrossRef]
128. Khan, M.A.; Van Dyk, J.; Yeung, I.W.T.; Hill, R.P. Partial volume rat lung irradiation; Assessment of early DNA damage in different lung regions and effect of radical scavengers. *Radiother. Oncol.* **2003**, *66*, 95–102. [CrossRef]
129. Koturbash, I.; Loree, J.; Kutanzi, K.; Koganow, C.; Pogribny, I.; Kovalchuk, O. In Vivo Bystander Effect: Cranial X-Irradiation Leads to Elevated DNA Damage, Altered Cellular Proliferation and Apoptosis, and Increased p53 Levels in Shielded Spleen. *Int. J. Radiat. Oncol. Biol. Phys.* **2008**, *70*, 554–562. [CrossRef]

130. Mancuso, M.; Pasquali, E.; Leonardi, S.; Tanori, M.; Rebessi, S.; Di Majo, V.; Pazzaglia, S.; Toni, M.P.; Pimpinella, M.; Covelli, V.; et al. Oncogenic bystander radiation effects in Patched heterozygous mouse cerebellum. *Proc. Natl. Acad. Sci. USA* **2008**, *105*, 12445–12450. [CrossRef]
131. Igarashi, K.; Sakimoto, I.; Kataoka, K.; Ohta, K.; Miura, M. Radiation-induced senescence-like phenotype in proliferating and plateau-phase vascular endothelial cells. *Exp. Cell Res.* **2007**, *313*, 3326–3336. [CrossRef] [PubMed]
132. Chiche, A.; Le Roux, I.; von Joest, M.; Sakai, H.; Aguín, S.B.; Cazin, C.; Salam, R.; Fiette, L.; Alegria, O.; Flamant, P.; et al. Injury-Induced Senescence Enables In Vivo Reprogramming in Skeletal Muscle. *Cell Stem Cell* **2017**, *20*, 407–414. [CrossRef]
133. Demaria, M.; O'Leary, M.N.; Chang, J.; Shao, L.; Liu, S.; Alimirah, F.; Koenig, K.; Le, C.; Mitin, N.; Deal, A.M.; et al. Cellular senescence promotes adverse effects of chemotherapy and cancer relapse. *Cancer Discov.* **2017**, *7*, 165–176. [CrossRef]
134. Ryan, L.A.; Smith, R.W.; Seymour, C.B.; Mothersill, C.E. Dilution of irradiated cell conditioned medium and the bystander effect. *Radiat. Res.* **2008**, *169*, 188–196. [CrossRef]
135. Peng, X.; Wu, Y.; Brouwer, U.; van Vliet, T.; Wang, B.; Demaria, M.; Barazzuol, L.; Coppes, R.P. Cellular senescence contributes to radiation-induced hyposalivation by affecting the stem/progenitor cell niche. *Cell Death Dis.* **2020**, *11*, 854. [CrossRef]
136. Manfrin, A.; Tabata, Y.; Paquet, E.R.; Vuaridel, A.R.; Rivest, F.R.; Naef, F.; Lutolf, M.P. Engineered signaling centers for the spatially controlled patterning of human pluripotent stem cells. *Nat. Methods* **2019**, *16*, 640–648. [CrossRef] [PubMed]
137. Jalili-Firoozinezhad, S.; Prantil-Baun, R.; Jiang, A.; Potla, R.; Mammoto, T.; Weaver, J.C.; Ferrante, T.C.; Kim, H.J.; Cabral, J.M.S.; Levy, O.; et al. Modeling radiation injury-induced cell death and countermeasure drug responses in a human Gut-on-a-Chip article. *Cell Death Dis.* **2018**, *9*, 223. [CrossRef] [PubMed]
138. Huh, D.; Matthews, B.D.; Mammoto, A.; Montoya-Zavala, M.; Yuan Hsin, H.; Ingber, D.E. Reconstituting organ-level lung functions on a chip. *Science* **2010**, *328*, 1662–1668. [CrossRef]
139. Jang, K.J.; Suh, K.Y. A multi-layer microfluidic device for efficient culture and analysis of renal tubular cells. *Lab Chip* **2010**, *10*, 36–42. [CrossRef]
140. Carraro, A.; Hsu, W.M.; Kulig, K.M.; Cheung, W.S.; Miller, M.L.; Weinberg, E.J.; Swart, E.F.; Kaazempur-Mofrad, M.; Borenstein, J.T.; Vacanti, J.P.; et al. In vitro analysis of a hepatic device with intrinsic microvascular-based channels. *Biomed. Microdevices* **2008**, *10*, 795–805. [CrossRef]
141. Sung, J.H.; Kam, C.; Shuler, M.L. A microfluidic device for a pharmacokinetic-pharmacodynamic (PK-PD) model on a chip. *Lab Chip* **2010**, *10*, 446–455. [CrossRef] [PubMed]
142. Milliat, F.; François, A.; Isoir, M.; Deutsch, E.; Tamarat, R.; Tarlet, G.; Atfi, A.; Validire, P.; Bourhis, J.; Sabourin, J.C.; et al. Influence of endothelial cells on vascular smooth muscle cells phenotype after irradiation: Implication in radiation-induced vascular damages. *Am. J. Pathol.* **2006**, *169*, 1484–1495. [CrossRef] [PubMed]
143. Brown, J.M. Radiation Damage to Tumor Vasculature Initiates a Program That Promotes Tumor Recurrences. *Int. J. Radiat. Oncol. Biol. Phys.* **2020**, *108*, 734–744. [CrossRef] [PubMed]
144. Plummer, C.; Henderson, R.D.; O'Sullivan, J.D.; Read, S.J. Ischemic stroke and transient ischemic attack after head and neck radiotherapy: A review. *Stroke* **2011**, *42*, 2410–2418. [CrossRef]
145. Ghobadi, G.; Bartelds, B.; Van Der Veen, S.J.; Dickinson, M.G.; Brandenburg, S.; Berger, R.M.F.; Langendijk, J.A.; Coppes, R.P.; Van Luijk, P. Lung irradiation induces pulmonary vascular remodelling resembling pulmonary arterial hypertension. *Thorax* **2012**, *67*, 334–341. [CrossRef]
146. Pellegata, A.F.; Tedeschi, A.M.; De Coppi, P. Whole organ tissue vascularization: Engineering the tree to develop the fruits. *Front. Bioeng. Biotechnol.* **2018**, *6*, 56. [CrossRef]
147. Kioi, M.; Vogel, H.; Schultz, G.; Hoffman, R.M.; Harsh, G.R.; Brown, J.M. Inhibition of vasculogenesis, but not angiogenesis, prevents the recurrence of glioblastoma after irradiation in mice. *J. Clin. Investig.* **2010**, *120*, 694–705. [CrossRef]
148. Grebenyuk, S.; Ranga, A. Engineering organoid vascularization. *Front. Bioeng. Biotechnol.* **2019**, *7*, 39. [CrossRef]
149. Cakir, B.; Xiang, Y.; Tanaka, Y.; Kural, M.H.; Parent, M.; Kang, Y.J.; Chapeton, K.; Patterson, B.; Yuan, Y.; He, C.S.; et al. Engineering of human brain organoids with a functional vascular-like system. *Nat. Methods* **2019**, *16*, 1169–1175. [CrossRef]

150. Shi, Y.; Sun, L.; Wang, M.; Liu, J.; Zhong, S.; Li, R.; Li, P.; Guo, L.; Fang, A.; Chen, R.; et al. Vascularized human cortical organoids (vOrganoids) model cortical development in vivo. *PLoS Biol.* **2020**, *18*, e3000705. [CrossRef]
151. Wörsdörfer, P.; Dalda, N.; Kern, A.; Krüger, S.; Wagner, N.; Kwok, C.K.; Henke, E.; Ergün, S. Generation of complex human organoid models including vascular networks by incorporation of mesodermal progenitor cells. *Sci. Rep.* **2019**, *9*, 15663. [CrossRef] [PubMed]

Publisher's Note: MDPI stays neutral with regard to jurisdictional claims in published maps and institutional affiliations.

Article

Evaluation of the Effectiveness of Mesenchymal Stem Cells of the Placenta and Their Conditioned Medium in Local Radiation Injuries

Vitaliy Brunchukov [1], Tatiana Astrelina [1,*], Daria Usupzhanova [1], Anna Rastorgueva [1], Irina Kobzeva [1], Victoria Nikitina [1], Sergei Lishchuk [1], Elena Dubova [1], Konstantin Pavlov [1], Valentin Brumberg [1], Marc Benderitter [2] and Alexander Samoylov [1]

1 State Research Center Burnasyan Federal Medical Biophysical Center of Federal Medical Biological Agency, 123098 Moscow, Russia; brunya2008@yandex.ru (V.B.); usupzhanova94@mail.ru (D.U.); rastorgueva.ann@gmail.com (A.R.); irina-kobzeva@yandex.ru (I.K.); nikitinava@yandex.ru (V.N.); leycom@mail.ru (S.L.); dubovaea@gmail.com (E.D.); kpavlov@fmbcfmba.ru (K.P.); brumb1225@gmail.com (V.B.); asamoilov@fmbcfmba.ru (A.S.)

2 Institute for Radiological Protection and Nuclear Safety (IRSN), 92260 Fontenay-aux-Roses, France; marc.benderitter@irsn.fr

* Correspondence: t_astrelina@mail.ru; Tel.: +7-916-532-56-77

Received: 21 September 2020; Accepted: 23 November 2020; Published: 29 November 2020

Abstract: Background: The search for an effective therapy for local radiation injuries (LRI) is urgent; one option is mesenchymal stem cells (MSC) derived from the placenta and their conditioned medium for the regenerative processes of the skin. Methods: We used 80 animals, randomly assigned to four groups: control (C) animals that did not receive therapy; control with the introduction of culture medium concentrate (CM); introduction of MSCs (PL); introduction of CMPL. LRI modeling was performed on an X-ray machine at a dose of 110 Gy. Histological and immunohistochemical tests were performed. Results: On the 112th day, the area of the open wound surface in the CMPL group was 6.7 times less than in the control group. Complete healing of the open wound surface of the skin in the CM group was observed in 40%, in CMPL 60%, in the PL group 20%, and in the C group there were no animals with a prolonged wound defect. A decrease in inflammatory processes was observed in the CMPL group. Conclusions: the use of a concentrate of conditioned MSCs (CMPL group) in severe LRI in laboratory animals accelerates the transition of the wound process to the stage of regeneration and epithelization.

Keywords: mesenchymal stem cells; local radiation injuries; conditioned medium; cell technologies; X-ray radiation; skin; placenta

1. Introduction

Today, ionizing radiation sources are widely used in various fields of human activity, and their scope is constantly expanding, which increases the risk of radiation damage [1,2]. Studies have shown that radiation damage to the skin causes damage to the stem and proliferating cells of the epidermis, as well as in the vessels of the microcirculatory bed [3–5], so the final effect of ionizing radiation is determined by the balance between damage to the cells and recovery processes in the affected area and adjacent tissues [4,6].

Radiation to human skin in doses exceeding 8 Gy may lead to the development of local radiation injuries (LRI) [4]. In radiotherapy of oncological diseases, LRI is registered in 20–40% of cases [7]. Skin LRIs are characterized by the development of recurrent ulcers with pain syndrome, which significantly lengthens the treatment process due to persistent damage to blood and lymphatic vessels with the progression of tissue fibrosis, which worsens the results of treatment and the quality-of-life of patients [7]. Currently, there are no effective treatments for LRI.

Taking into account the pathogenetic mechanisms of radiation-induced lesions, the use of cellular technologies using mesenchymal stem cells (MSCs) and their waste products (paracrine factors) may become a promising method of treating skin LRI [5]. MSCs are capable of self-renewal and various types of differentiation in the adipogenic, osteogenic, chondrogenic, and myogenic directions [1]. The use of MSCs leads to the healing of the wound surface of the skin and its appendages, diabetic ulcers, damage to skeletal muscles and cartilage, and the heart. Intravenous, local administration of MSCs helps to reduce necrotic changes, reduce inflammation, and significantly improve the processes of granulation, reepithelialization, neoangiogenesis, and hair restoration [1,3,4].

The main effect of MSCs may be due to their secretory activity, associated with the production of a wide range of cytokines and growth and angiogenic factors [8]. Paracrine factors initiate the stimulation of host MSCs, triggering the regeneration of damaged tissues. Thus, the cytokines involved in the regulation of the inflammatory process include IL-1β, 4, 6, 10, 12, and 17, TNF-α, TGF-β1, PGE-2, PDGF, HGF and SDF-1. VEGF, FGF-2, EGF, TGF-α, HGF, IGF-1, etc., are responsible for neoangiogenesis, and the regulation of tissue fibrosis involves the participation of IL-4, 16, TGF-β1, HGF, bFGF, etc. MSC secretome injections in the form of conditioned media containing extracellular vesicles also have positive effects, as well as the MSCs themselves. Further assessment of the paracrine potential of MSCs may open up new ways of treating acute and chronic forms of MLP of the skin [1,3,4].

One of the most important advantages of MSCs is their low immunogenicity, which allows the use of allogeneic MSCs without the risk of rejection reactions. Sources of MSCs are various human tissues (bone marrow, adipose tissue, skin, placenta, synovial membrane, cartilage, etc.) [1,3,4,6,7]. The main sources of MSCs are bone marrow, mucosal and placental tissues, etc.

Placental tissue is of great interest due to the simplicity of sampling, the absence of ethical problems, and the ability to quickly obtain and accumulate the necessary amount of cellular material. MSCs derived from the placenta are known to have a higher regenerative potential compared to cells from other sources, but there are no data on the use of these cells in LRI [9].

Thus, the presented data indicate that MSCs derived from the placenta and the paracrine factors produced by them can be used to produce drugs intended for the treatment of LRI, which undoubtedly deserves further study.

The aim of this study was to study the effect of human MSCs derived from the placenta and their conditioned medium concentrate on skin regenerative processes in laboratory animals with LRI.

2. Materials and Methods

2.1. Study Groups

There were 80 laboratory animals used in the study (Wistar male rats aged 8–12 weeks and weighing 210.0 ± 30.0 g). The animals were obtained from the specialized laboratory animal nursery Pushchino, had the appropriate veterinary certificate, and were quarantined for 14 days. The study was approved by the section of the Academic Council (extract No. 43A, dated 25/9/2017) and at the meeting of the local bioethical committee (Protocol No. 8b dated 10/11/2012) of the State Research Center—Burnasyan Federal Medical Biophysical Center of Federal Medical Biological Agency.

Laboratory animals were randomized and divided into four groups (20 animals each) depending on the type of therapy performed:

Group 1—control (C)-irradiated rats without subsequent therapy;

Group 2—control (CM)-irradiated rats that received intradermal administration of culture medium (MesenCult) concentrate around the affected area three times, on days 1, 14, and 21;

Group 3—irradiated rats that received intradermal administration of human MSCs derived from the placenta (PL) three times, on days 1, 14, and 21;

Group 4—irradiated rats that received intradermal administration of concentrate of conditioned medium (CMPL) MSCs derived from the placenta three times on days 1, 14, and 21.

Each laboratory animal was observed 17 times: 1, 7, 14, 21, 28, 35, 42, 49, 56, 63, 70, 77, 84, 91, 98, 105, and 112 days after LRI modeling. During the examination, the condition of the laboratory animal was monitored with an assessment of its behavior, movement, cardiovascular and/or respiratory functions, changes in appetite and weight, body temperature, etc. On set days, the skin surface was examined and the course of the wound process was evaluated (the depth of damage to the skin, their size (length, width), the total area of the changed skin, the area of the open wound surface, the presence of discharge, blisters, scab, exfoliated epidermis, the color of the exposed dermis, fibrin plaque).

Animals were removed from the experiment 28, 42, 56, 70, 91, and 112 days from the beginning of the experiment.

2.2. Modeling of LRI

Modeling of relatively "soft" X-ray radiation of LRI was carried out on an LNK-268 (RAP100-10) X-ray unit (Diagnostika-M LLC, Moscow, Russia) with a radiation exposure mode at a dose of 110 Gy with a 0.1 mm aluminum filter, voltage 30 kV, beam current 6.1 mA, dose rate 21.1 Gy/min for 312 s (dose accuracy ± 5%, dose measurement uncertainty ± 6%) according to the proposed earlier method [10–12], leading to a short latency period and chronic skin ulcer in laboratory animals. After irradiation, the animals were seated in individual sterile boxes with an autonomous Smart Flow ventilation system (Tecniplast Group, Buguggiate, Italy), with free access to water and food.

2.3. Cultivation of MSCs

The experiment used nonpersonalized human MSCs derived from the placenta samples that are under long-term cryopreservation in a Cryobank. MSCs were cultured in a medium without xenogenic components (Stem Cell, Vancouver, Canada) with the addition of 100 U/mL of penicillin, 100 U/mL of streptomycin, and 2 mm of glutamine from the 3rd to the 5th passage. The resulting MSCs were administered to laboratory animals at a calculated dose of 2 million cells per 1 kg.

2.4. Immunological Characteristics and Viability of MSCs

The MSCs derived from the placenta immunophenotype was determined by flow cytometry. The expression of surface markers was evaluated using fluorochrome-labeled antibodies against CD34, CD45, CD90, CD105, CD73, and HLA-DR (BD Biosciences and Becman Coulter, Brea, CA, USA) on a FACSCanto II flow cytometer (Becton Dickinson, Durham, CA, USA) in accordance with the manufacturer's instructions.

Cell viability was evaluated using a 7-ADD dye that penetrates the cell's cytoplasmic membrane and binds to its DNA. The number of CD45-negative/7-ADD-positive cells was determined on a FACS Canto II flow cytometer (Becton Dickinson) in accordance with the manufacturer's instructions.

2.5. Obtaining a Conditioned Medium

The conditioned medium of MSCs derived from the placenta (CMPL) was taken into sterile tubes at 3–5 passages when the cells reached 80–90% confluence. A laboratory tangential flow filtration system, LabScale, developed for concentration, diafiltration, and microfiltration, was used to obtain CMPL. The CMPL was placed in a tangential flow filtration system and concentrated 8.08 times with an inlet pressure of 40–52 psi and an outlet pressure of 8–12 psi. The resulting volume was passed through a nylon syringe filter with a pore size of 0.22 μm (Corning, New York, NY, USA). The protein concentration was 648 μg/mL, IL-6: 853 pg/mL, IL-8: 8730 pg/mL, IL-10: 17.7 pg/mL, TGF-β: 1.0 pg/mL. The volume concentrate of conditioned medium for rats of the CMPL group for each injection was 0.4 mL. The introduction was carried out intradermally up to 12 points around the irradiation zone, retreating 2–3 mm from the edge.

2.6. Histological and Immunohistochemical Study

A skin flap was excised from the affected area (the area of the wound defect, with the adjacent skin and underlying muscles) and fixed in a 10% solution of neutral formalin. Further processing of excised samples was performed using standard histological methods. Preparations stained with hematoxylin and eosin were used for general assessment of the condition of the studied tissues.

Immunohistochemical examination of tissue samples on days 28 and 112 of the experiment was performed using an automated method, a Ventana BenchMark Ultra immunostainer with dewaxing and unmasking in an apparatus using antibodies to VEGF (Novocastra, Milton Keynes, UK, skin blood vessel endothelium marker), CD31 (Novocastra, endothelial cell marker), CD68 (Novocastra, macrophage marker), PGP9.5 (Novocastra, marker of differentiating neurons in the skin), Ki67 (Novocastra, marker of cell proliferation), FVIII (Novocastra, marker for platelet adhesion factor), collagens of types I and III (Novocastra), tissue inhibitor of metalloproteinases TIMP2 (Novocastra), and metalloproteinases of types 2 and 9 (Novocastra). Marker expression was evaluated semiquantitatively, assigning scores from 0 to 3, where 0 means no expression and 3 means fully expressed.

Statistical analysis of the results was performed using Microsoft Office Excel 2007 (Redmond, WA, USA), Statistica 6 (Round Rock, TX, USA) and ImageTool software (San Antonio, TX, USA. The Mann-Whitney test was used to assess the significance of the differences.

3. Results

3.1. Immunological Characteristics and Viability of MSCs

When analyzing the MSCs immunophenotype using flow cytometry, a high expression of MSCs markers (CD73, CD90, CD105) was detected in all cell cultures; markers of hematopoietic and lymphocytic origin were absent (CD34, CD45, HLA-DR). The immunophenotype met the requirements of the International organization of cell therapy for human MSCs [1]. MSCs maintained high activity and viability (98.21 ± 1.72%, 7-ADD) throughout the entire culture period (Figure 1).

3.2. Planimetric Analysis

The majority of animals in the study groups on day 7 showed a clearly visible area of altered skin, outlined by a demarcation line, with signs of dry or wet dermatitis. The average area of the changed skin was significantly lower in the C and CM groups compared to the PL and CMPL groups ($p \leq 0.05$) (Figure 2a). However, by day 14, the total changed skin area in the C, PL, and CMPL groups did not differ, whereas in the CM group it was less until the end of the experiment (Table 1).

From the 14th day of the experiment, the open wound surface of the skin was recorded in all groups of animals. The dynamics of reducing the area of the open wound surface was the same for all groups up to day 42 of the study. After that, we observed wave dynamics of increase and decrease of the open wound skin surface in all groups except CM (Figure 2b, Table 2).

On day 112, the area of the open wound surface in the CMPL group was 6.7 times smaller than in the control group. Complete healing of the open wound surface of the skin in the CM groups was observed in 40% and the CMPL in 60%, in the PL group in 20%, and in the C group there were no animals with a prolonged wound defect (Figure 2c).

3.3. Histological Examination

In all groups, an open wound defect covered with a purulent-necrotic crust was formed on the 28th day after irradiation. Weak, mainly perivascular lymphocytic-plasmocytic infiltration with an admixture of single neutrophilic granulocytes and moderate vascular proliferation of the microcirculatory bed was detected in the underlying dermis in the area of the defect bottom. Moderate thickening of the adjacent epidermis was noted along the edges of the wound defect, and few intraepidermal lymphocytes were recorded. At the same time, only in the CMPL group was the "creeping" of regenerating epithelium from one of the wound edges in the form of a strip 3–4 epithelial cells thick noted (Figure 3).

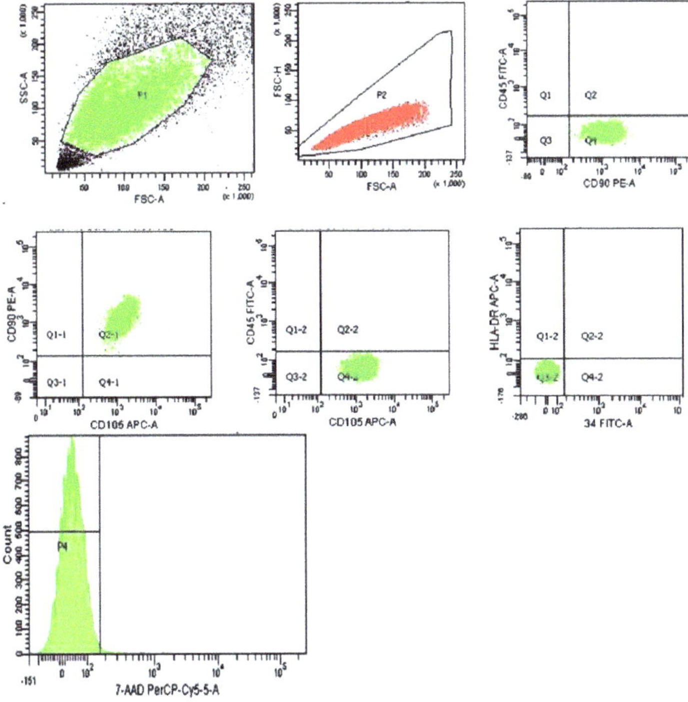

Figure 1. MSCs derived from the placenta immunophenotype: CD90+/CD105+73+/CD45−/CD34−/HLA-DR−, 7−ADD (99.5%).

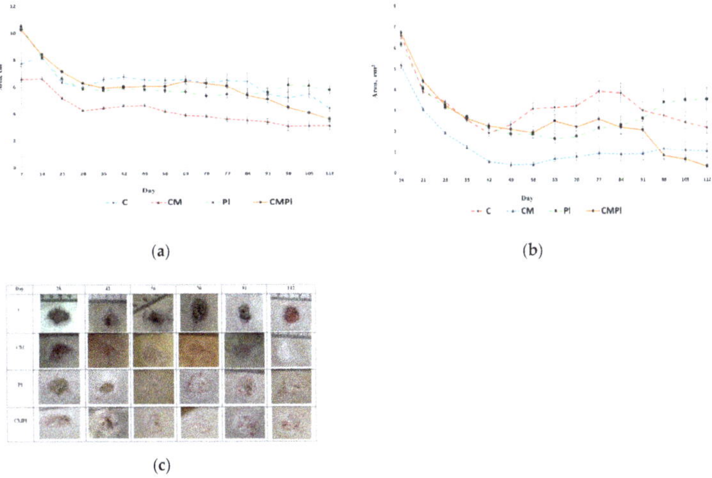

Figure 2. Dynamics of the LRI of animals: (**a**) dynamics of development of visible region of rat skin changes that occurred under X-ray radiation; (**b**) dynamics of development of an open wound surface of rat skin after exposure to X-ray radiation; (**c**) dynamics of healing of an open wound surface of the skin in animals (C: irradiated rats without subsequent therapy, CM: irradiated rats that received intradermal administration of culture medium (MesenCult) concentrate, PL: therapy of LRI using MSC derived from the placenta, CMPL: therapy of LRI using a concentrate of conditioned medium collected from a culture of the MSCs).

Table 1. The area of the total changed surface of animal skin in LRI (cm^2).

Day	C	CM	Pl	CMPL
7	7.76 ± 0.47	6.60 ± 0.42	10.5 ± 0.31 [1,2]	10.27 ± 0.19 [1,2]
14	8.26 ± 0.33	6.63 ± 0.26	8.16 ± 0.25	8.41 ± 0.21 [2]
21	6.36 ± 0.23	5.19 ± 0.18 [1]	6.63 ± 0.36 [2]	7.15 ± 0.2 [1,2]
28	5.93 ± 0.27	4.25 ± 0.13 [1]	5.88 ± 0.22 [2]	6.28 ± 0.18 [2]
35	6.56 ± 033	4.42 ± 0.0.17	5.74 ± 0.21 [2]	5.95 ± 0.19 [2]
42	6.76 ± 0.34	4.59 ± 0.21	5.92 ± 0.19	6.04 ± 0.21
49	6.54 ± 0.35	4.62 ± 0.24	5.80 ± 0.21	6.06 ± 0.39
56	6.50 ± 0.41	4.18 ± 0.26 [1]	5.71 ± 0.24 [2]	6.05 ± 0.41 [2]
63	6.56 ± 0.60	3.88 ± 0.23	5.64 ± 0.37 [2]	6.43 ± 0.54 [2]
70	6.33 ± 0.72	3.81 ± 0.26 [1]	5.33 ± 0.29	6.26 ± 0.58
77	6.49 ± 1.04	3.59 ± 0.43	5.44 ± 0.53	6.06 ± 0.41 [2]
84	6.41 ± 1.21	3.50 ± 0.54	5.45 ± 0.58	5.36 ± 0.47 [2]
91	5.50 ± 0.88	3.38 ± 0.47	5.60 ± 0.52 [2]	5.07 ± 0.58 [2]
98	5.19 ± 0.79	3.05 ± 0.71	6.13 ± 0.95	4.43 ± 0.31
105	5.47 ± 0.69	3.08 ± 0.67	6.06 ± 0.81	4.04 ± 0.23
112	4.35 ± 0.42	3.07 ± 0.60	5.75 ± 0.61	3.57 ± 0.30

Notes: [1] Significant differences in all groups compared to the control (C) ($p \leq 0.05$). [2] Significant differences between the PL and CMPL groups compared to the CM group ($p \leq 0.05$).

Table 2. Area of open wound surface of animal skin in LRI (cm^2).

Day	C	CM	Pl	CMPL
14	6.59 ± 0.28	5.19 ± 0.33 [1]	6.19 ± 0.23 [2]	6.73 ± 0.18 [2]
21	3.91 ± 0.2	3.05 ± 0.14 [1]	4.07 ± 0.22 [2]	4.42 ± 0.19 [1,2]
28	3.40 ± 0.19	1.92 ± 0.14 [1]	3.14 ± 0.2 [2]	3.25 ± 0.18 [2]
35	2.54 ± 0.19	1.23 ± 0.15 [1]	2.68 ± 0.21 [2]	2.60 ± 0.21 [2]
42	1.90 ± 0.28	0.55 ± 0.16 [1]	2.17 ± 0.27 [2]	2.25 ± 0.27 [2]
49	2.32 ± 0.34	0.39 ± 0.19 [1]	1.86 ± 0.29 [2]	2.09 ± 0.42 [2]
56	3.07 ± 0.53	0.41 ± 0.17 [1]	1.85 ± 0.37 [2]	1.94 ± 0.59 [2]
63	3.12 ± 0.70	0.68 ± 0.32 [1]	1.63 ± 0.46	2.48 ± 0.96
70	3.20 ± 0.71	0.78 ± 0.37 [1]	1.74 ± 0.55	2.19 ± 0.86
77	3.91 ± 0.97	0.94 ± 0.62 [1]	2.12 ± 0.86	2.56 ± 0.80
84	3.82 ± 0.95	0.90 ± 0.69 [1]	2.28 ± 0.87	2.17 ± 0.69
91	2.98 ± 0.83	0.93 ± 0.62 [1]	2.60 ± 0.81	2.05 ± 0.80
98	2.74 ± 0.69	1.16 ± 0.97 [1]	3.37 ± 1.32	0.82 ± 0.40 [1,2]
105	2.40 ± 0.62	1.09 ± 0.79 [1]	3.49 ± 1.21 [2]	0.65 ± 0.32 [1,2,3]
112	2.15 ± 0.57	1.04 ± 0.68	3.52 ± 1.14 [2]	0.32 ± 0.18 [1,2,3]

Notes: [1] Significant differences in all groups compared to the control (C) ($p \leq 0.05$). [2] Significant differences between the PL and CMPL groups compared to the CM group ($p \leq 0.05$). [3] Significant differences between the PL and CMPL groups ($p \leq 0.05$).

In group C, on day 56, the bottom of the skin defect reached the large subcutaneous muscle, and, in some cases, the subcutaneous fat. Pronounced edema and lymphocytic-plasmocytic infiltration of muscle tissue were determined. In the connective tissue of the dermis in the area of the edges of the wound defect, lymph-histiocytic infiltration with an admixture of neutrophils, granulation, and proliferation of microvessels were detected. The epidermis adjacent to the wound defect was thickened to 10–12 layers of cells, and focal hyperkeratosis, acanthosis, and degenerative changes in keratinocytes were noted. Areas of epithelial regeneration with a thickness of 4–8 epithelial cells were determined in the area of the defect edges (Figure 3(1b)); by the 112th day, pronounced purulent-necrotic changes in soft tissues appeared in the area of the defect bottom. Underlying connective tissue and moderate lymphocytic-plasmocytic infiltration were observed, with an admixture of neutrophilic granulocytes, moderate vascular proliferation of the microcirculatory bed, focal edema, and pronounced fibrotic changes. In most cases, areas of fibrosis and weakly expressed lymphoplasmocytic infiltration in the area of the large subcutaneous muscle and Hypoderma were detected. At the edges of the wound defect, there were large areas of regeneration of the integumentary epithelium in the form of a layer of cells 1–2 epithelial cells thick. The adjacent epidermis was thickened (up to 6–11 layers of cells), with signs of vacuole dystrophy and acanthosis (Figure 3(1c)).

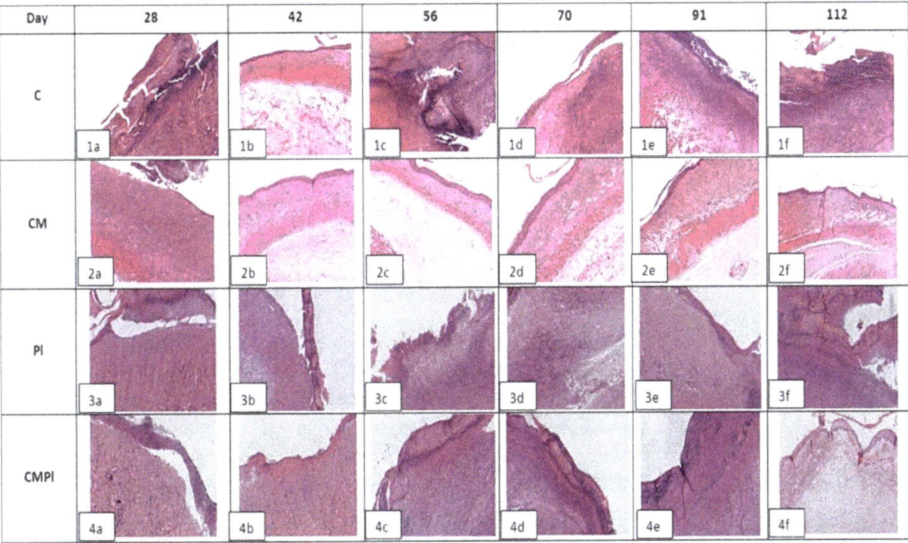

Figure 3. Histological examination of animal skin defects. Hemotoxylin-eosin staining, ×40 magnification for C and CM groups, ×20 for PL and CMPl groups (C: irradiated rats without subsequent therapy, CM: irradiated rats that received intradermal administration of culture medium (MesenCult) concentrate, PL: therapy of LRI using MSCs derived from the placenta, CMPL: therapy of LRI using a concentrate of conditioned medium collected from the culture of the MSCs).

In the CM group, signs of epithelization along the edges of the defect were noted on day 56 in most samples; in some cases a deep skin defect remained, which reached the subcutaneous fat and was covered with a purulent-necrotic crust. In all preparations, the large subcutaneous muscle was of the usual histological structure with moderate edema, subcutaneous fat with moderate edema, or moderate lymph with plasmocytic infiltration. In the underlying dermis, there were areas of fibrosis and weak perivascular lymphocytic infiltration (Figure 3(2b)). By the 112th day, only one observation revealed a large skin defect covered with a purulent-necrotic crust, which reached the subcutaneous fat with necrosis in the superficial parts and clusters of hemosiderophages in the deeper parts. In other preparations, the skin defect was partially or completely epithelized. In the dermis, there were areas of fibrosis and mild perivascular lymphocytic infiltration. In all cases, the adjacent epidermis was thickened (up to 10–11 layers of cells), with signs of severe dystrophy (Figure 3(2c)).

In the PL group on day 56, the bottom of the skin defect showed necrotic dermis, striated muscle, and underlying adipose tissue with pronounced neutrophilic infiltration. At the edges of the defect, there was moderate lymph-plasmocytic infiltration with an admixture of neutrophilic granulocytes, moderate proliferation of microcirculatory vessels, granulation, and fibrosis of striated muscle tissue (Figure 3(3b)). By the 112th day, the extensive skin defect was covered with a purulent-necrotic crust in all cases. Its bottom is represented by fibrotic connective tissue with angiomatosis, granulations, moderate lymph-plasmocytic infiltration with an admixture of neutrophilic granulocytes, and microvessel proliferation. The striated tissue at the bottom of the defect was not detected. There were extensive areas of fibrosis of the underlying adipose tissue. The epidermis at the edge of the wound defect was thickened (up to 8–10 layers of cells), with signs of vacuole dystrophy and proliferation of hair follicles (Figure 3(3c)).

In the CMPL group, on day 56, the open wound skin defect was covered with a purulent-necrotic crust, marginal epithelization was recorded over a longer length in most samples, and the thickness of the epithelial layer was 5–8 cells. The underlying dermis was moderately fibrotic with focal subepithelial edema and the presence of hair follicle rudiments in the amount of 1–3 in the field of vision. In the area of the defect bottom, muscle and adipose tissue were completely replaced by fibrous tissue with granulations with moderate lymph-plasmocytic infiltration with an admixture of neutrophilic granulocytes and pronounced microvessel proliferation (Figure 3(4b)). By the 112th day in all samples, the skin defect was completely epithelized and the thickness of the epithelial layer was 5–7 cells. The underlying dermis was focally fibrotic. There were rudiments of hair follicles (1–3) in the field of vision with focal proliferation of microcirculatory vessels. Large subcutaneous muscle was not detected in the central parts; it was replaced by connective tissue. There were no inflammatory changes (Figure 3(4b)).

3.4. Immunohistochemical Study

As a result of an immunohistochemical study, it was found that the number of newly formed vessels in whose endothelial cells the expression of CD31 was determined increased from day 28 to day 112 in the PL and CMPL groups (from 2.6 ± 1.0 to 10.97 ± 1.6 and from 4.1 ± 0.6 to 8.2 ± 1.8, respectively, $p \leq 0.05$), which indicated an increase in neoangiogenesis by the end of the experiment. Such changes were not detected in groups C and CM (Figure 4), nor for the vascular endothelial growth factor (VEGF) in endothelial cells and stroma cells in all the studied groups, except for CMPL (Figures 4d and 5b). In groups C, CM, and PL, an increase in FVIII expression in vascular endothelial cells was observed by day 112 of the experiment ($p \leq 0.05$) (Figures 4b and 5a).

In the course of the experiment, we noted an increase in the number of CD68-positive macrophages in the tissues surrounding the wound defect in groups C and PL (from 11.7 ± 1.4 and 12.9 ± 3.6 at 28 days to 24.73 ± 2.4 and 29.3 ± 3.5 at 112 days, respectively, $p \leq 0.05$), while in the CM group it was determined by the decrease in the number of these cells (22.1 ± 1.6 and 13.07 ± 1.8, $p \leq 0.05$), and in the CMPL group their number did not change (Figures 4a and 5a).

The number of regenerating nerve fibers expressing PGP9.5 increased by the end of the experiment in the C, CM, and PL groups ($p \leq 0.05$), and remained unchanged in the CMPL group (Figures 4a and 5a).

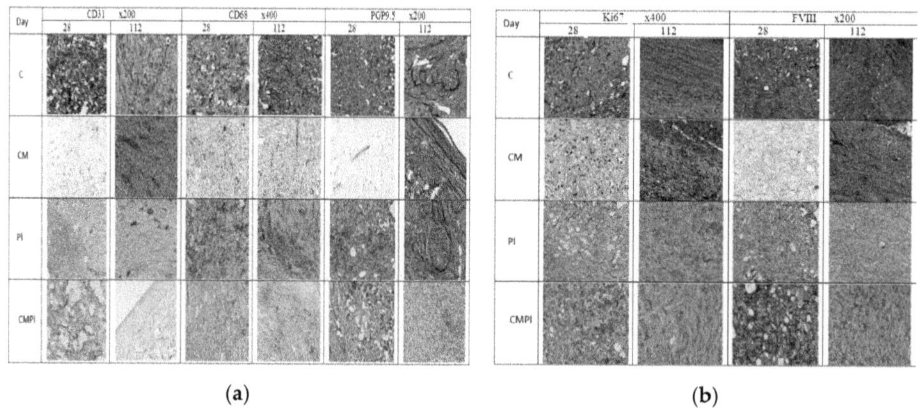

(a) (b)

Figure 4. *Cont.*

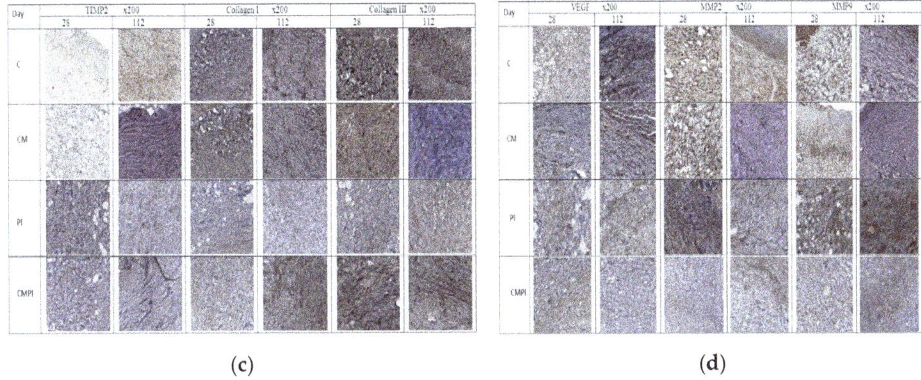

Figure 4. Immunohistochemical study of animal skin defects: (**a**) CD31, CD68, PGP9.5; (**b**) Ki67 and FVIII; (**c**) TIMP2, Collagen I, Collagen III, and FVIII; (**d**) VEGF, MMP2, and MMP9. (C: irradiated rats without subsequent therapy, CM: irradiated rats that received intradermal administration of culture medium (MesenCult) concentrate, PL: therapy of LRI using MSC derived from the placenta, CMPL: therapy of LRI using a concentrate of conditioned medium collected from the culture of the MSCs).

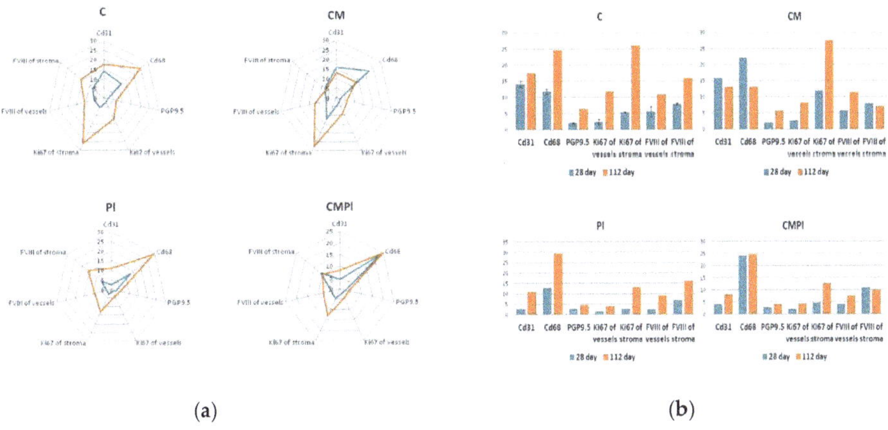

Figure 5. Immunohistochemical study of animal skin defects: (**a**) Absolute number of IHC markers in excised animal skin defect tissue samples per 10 visual fields; (**b**) semiquantitative assessment of the expression of IHC markers in excised animal skin defect samples (with a score from 0 to 3, where 0 is the absence of expression and 3 is full expression). (C: irradiated rats without subsequent therapy, CM: irradiated rats that received intradermal administration of culture medium (MesenCult) concentrate, PL: therapy of LRI using MSC derived from the placenta, CMPL: therapy of LRI using a concentrate of conditioned medium collected from culture of the MSCs. Blue: 28 days; orange: 112 days from the start of therapy. The score is from 0 to 3, where 0 is the absence of expression and 3 is full expression).

Expression of matrix metalloproteinases (MMP) 2 and 9, which led to the destruction of extracellular matrix proteins and stimulated cell migration and reproduction, decreased in all groups at the end of the experiment, with the exception of the CMPL group (Figure 4b,d), while expression of TIMP2, which is a tissue inhibitor of MMP, increased by day 112 in the C and CM groups, decreased in the PL group, and remained unchanged in the CMPL group (Figure 4b,c). The expression of "mature" type I collagen in the stroma increased in all groups from day 28 to day 112, with the exception of the PL group, while the expression of "immature" type III collagen in the stroma decreased by day 112 in the C, CM, and PL groups, but did not change in the CMPL group (Figure 4b,c).

4. Discussion

Skin LRIs pose a serious medical, social, and economic problem. It is known that early effects of skin damage from ionizing radiation (dry and wet dermatitis) are associated with damage to the epidermis, and late effects (skin atrophy, radiation necrosis, etc.) are the result of damage to the dermis [4]. Thus, the radiation target of the epidermis is the highly radiosensitive cells of the basal layer, and the vessels of the microcirculatory bed in the dermis. As a result, with deep radiation burns, necrotic and degenerative processes cover all layers of the skin, gradually spreading to the underlying tissues, up to the bone.

This study is aimed at preventing complications of late radiation injuries, improving the nature of the treatment course, and speeding up the healing time of skin LRIs. The introduction of MSCs, culture, and conditioned media was performed three times: in the absence of changes in the skin (day 1), after the formation of radiation dermatitis (day 14), and after the appearance of the wound surface (day 21) to activate reparative processes and angiogenesis in damaged skin, reduce the healing time of ulcerative defects, and prevent relapses.

In the course of the present study, the use of a conditioned medium concentrate and culture medium showed greater effectiveness in terms of reducing the time and area of the open wound surface. The greatest number of cases of complete healing of the affected skin of animals by the end of the observation period (112 days) was detected in the CMPL group, and healing was also noted in the CM group. In the same groups, no clinical signs of LRI recurrence were detected, but not in the C and PL groups (Figure 3).

In our previous study [7], the effectiveness of the use of MSCs of the gingival mucosa in LRI was shown. The absence of a significant effect when using MSCs derived from the placenta in this study is due to the tissue specificity of this cell source and the peculiarities of their production of paracrine factors.

According to the histological study, there was a decrease in inflammatory processes, the presence of hair follicle rudiments, and proliferation of microcirculatory vessels in the CMPL group, in contrast to other groups in which these changes were not so noticeable.

Given the nature of the radiation damage in all groups, the regenerative potential of cells in the affected area is significantly reduced, which was confirmed by data from immunohistochemical research. An increase in the expression of metalloproteinases (MMP 2 and 9), TIMP 2, collagen I and III, as well as the number of CD68 macrophages, was observed in the CMPL group on day 112, which probably indicates an increase in the rate of scarring and healing of the wound surface, which is confirmed by planimetric studies (Figure 4a,b).

5. Conclusions

Thus, the use of MSCs derived from the placenta conditioned medium concentrate (CMPL group) in severe LRI in laboratory animals accelerates the transition of the wound process to the stage of regeneration and epithelization. Interestingly, in one of the control groups, when using a culture medium concentrate (CM group), a significant decrease in the area of the wound surface was observed compared to other groups during the entire observation period. However, the analysis of histological and immunohistochemical studies does not allow us to unequivocally assert the effectiveness of this type of therapy.

Author Contributions: Conceptualization, T.A. and A.S.; investigation, V.B. (Vitaliy Brunchukov), D.U., A.R., I.K., V.N. and V.B. (Valentin Brumberg); resources, S.L., E.D. and K.P.; visualization, L.L.; writing—original draft preparation, T.A. and V.B. (Vitaliy Brunchukov); writing—review and editing, T.A., V.B. (Vitaliy Brunchukov) and M.B. All authors have read and agreed to the published version of the manuscript.

Funding: This work was supported by the Federal Target Program of the Russian Federation "Ensuring Nuclear and Radiation Safety for 2016–2020 and for the Period up to 2030".

Conflicts of Interest: The authors declare no conflict of interest.

References

1. Dominici, M.; Le Blanc, K.; Mueller, I.; Slaper-Cortenbach, I.; Marini, F.; Krause, D.; Deans, R.; Keating, A.; Prockop, D.; Horwitz, E. Minimal criteria for defining multipotent mesenchymal stromal cells. The International Society for Cellular Therapy position statement. *Cytotherapy* **2006**, *8*, 315–317. [CrossRef] [PubMed]
2. Hopewell, J.W.; E Coggle, J.; Wells, J.; Hamlet, R.; Williams, J.P.; Charles, M.W. The acute effects of different energy beta-emitters on pig and mouse skin. *Br. J. Radiol. Suppl.* **1986**, *19*, 47–51. [PubMed]
3. Kode, J.A.; Mukherjee, S.; Joglekar, M.V.; Hardikar, A.A. Mesenchymal stem cells: Immunobiology and role in immunomodulation and tissue regeneration. *Cytotherapy* **2009**, *11*, 377–391. [CrossRef] [PubMed]
4. Zheng, K.; Wu, W.; Yang, S.; Huang, L.; Chen, J.; Gong, C.; Fu, Z.; Zhang, L.; Tan, J. Bone marrow mesenchymal stem cell implantation for the treatment of radioactivity-induced acute skin damage in rats. *Mol. Med. Rep.* **2015**, *12*, 7065–7071. [CrossRef] [PubMed]
5. Akleev, A.V. *Chronic Radiation Syndrome in Residents of Coastal Villages of the Techa River*; Russian Federation: Chelyabinsk, Russian, 2012.
6. Astrelina, T.A.; Gomzikov, A.E.; Kobzeva, V.I. Assessment of the quality and safety of cryopreserved multipotent placental mesenchymal stromal cells in clinical practice. *Cell Transpl. Tissue Eng.* **2013**, *8*, 82–87.
7. Brunchukov, V.A.; Astrelina, T.A.; Nikitina, V.A.; Kobzeva, I.V.; Suchkova, Y.B.; Usupzhanova, D.Y.; Rastorgueva, A.A.; Karaseva, T.V.; Gordeev, A.V.; Maximova, O.A.; et al. Experimental Treatment of Radiation Skin Lesions with Mesenchymal Stem Cells and Their Conditioned Media. *Med Radiol. Radiat. Saf.* **2020**, *65*, 5–12. (In Russian) [CrossRef]
8. *Diagnosis, Treatment of Local Radiation Lesions and Their Long-Term Effects*; Federal clinical guidelines; FKG FMBA of Russia 2.6.7; Russian Federation: Moscow, Russia, 2015.
9. Onishchenko, G.G.; Popova, A.Y.; Romanovich, I.K.; Barkovskiy, I.K.; Kormanovskaya, T.A.; Shevkun, I.G. Radiation and hygienic certification and ESKID-information basis for management decisions to ensure radiation safety of the population of the Russian Federation. Message 2. Characteristics of sources and doses of radiation of the population of the Russian Federation. *Radiat. Hyg.* **2017**, *3*, 18–35.
10. Osanov, D.P. *Dosimetry and Radiation Biophysics of the Skin*; Energoatomizdat: Moscow, Russia, 1983; p. 152.
11. *Results of Radiation-Hygienic Certification in the Subjects of the Russian Federation for 2008: Radiation-Hygienic Passport of the Russian Federation*; Federal center of hygiene and epidemiology of Rospotrebnadzor, Federation: Moscow, Russia, 2009; p. 112.
12. Temnov, A.A.; Astrelina, T.A.; Rogov, K.A.; Lebedev, V.G.; Nasonova, T.A.; Lyrshchikova, A.V.; Deshevoy, Y.B.; Dobrynina, O.A.; Melerzanov, A.V.; Samoylov, A.S.; et al. Investigation of the influence of conditioned environment factors obtained in the cultivation of bone marrow mesenchymal stem cells on the course of severe local radiation skin lesions in rats. *Med. Radiol. Radiat. Saf.* **2018**, *63*, 35–39. [CrossRef]

Review

Rationale for the Use of Radiation-Activated Mesenchymal Stromal/Stem Cells in Acute Respiratory Distress Syndrome

Isabel Tovar [1,2], Rosa Guerrero [1,2], Jesús J. López-Peñalver [3], José Expósito [1,2,4] and José Mariano Ruiz de Almodóvar [5,*]

1 Departamento de Oncología Médica y Radioterapia, Servicio Andaluz de Salud (SAS), Avenida de las Fuerzas Armadas 2, 18014 Granada, Spain; mariai.tovar.sspa@juntadeandalucia.es (I.T.); rosa.guerrero.sspa@juntadeandalucia.es (R.G.); jose.exposito.sspa@juntadeandalucia.es (J.E.)
2 Instituto de Investigación Biosanitaria, Ibis Granada, Hospital Universitario Virgen de las Nieves, Avenida de las Fuerzas Armadas 2, 18014 Granada, Spain
3 Unidad de Radiología Experimental, Centro de Investigación Biomédica, Universidad de Granada, PTS Granada, 18016 Granada, Spain; jjpenalver@ugr.es
4 Departamento de Radiología y Medicina Física, Facultad de Medicina, Universidad de Granada, PTS Granada, 18016 Granada, Spain
5 Centro de Investigación Biomédica, Universidad de Granad, PTS Granada, 18016 Granada, Spain
* Correspondence: jmrdar@ugr.es

Received: 31 July 2020; Accepted: 31 August 2020; Published: 2 September 2020

Abstract: We have previously shown that the combination of radiotherapy with human umbilical-cord-derived mesenchymal stromal/stem cells (MSCs) cell therapy significantly reduces the size of the xenotumors in mice, both in the directly irradiated tumor and in the distant nonirradiated tumor or its metastasis. We have also shown that exosomes secreted from MSCs preirradiated with 2 Gy are quantitatively, functionally and qualitatively different from the exosomes secreted from nonirradiated mesenchymal cells, and also that proteins, exosomes and microvesicles secreted by MSCs suffer a significant change when the cells are activated or nonactivated, with the amount of protein present in the exosomes of the preirradiated cells being 1.5 times greater compared to those from nonirradiated cells. This finding correlates with a dramatic increase in the antitumor activity of the radiotherapy when is combined with MSCs or with preirradiated mesenchymal stromal/stem cells (MSCs*). After the proteomic analysis of the load of the exosomes released from both irradiated and nonirradiated cells, we conclude that annexin A1 is the most important and significant difference between the exosomes released by the cells in either status. Knowing the role of annexin A1 in the control of hypoxia and inflammation that is characteristic of acute respiratory-distress syndrome (ARDS), we designed a hypothetical therapeutic strategy, based on the transplantation of mesenchymal stromal/stem cells stimulated with radiation, to alleviate the symptoms of patients who, due to pneumonia caused by SARS-CoV-2, require to be admitted to an intensive care unit for patients with life-threatening conditions. With this hypothesis, we seek to improve the patients' respiratory capacity and increase the expectations of their cure.

Keywords: experimental radiotherapy; radiobiology; mesenchymal stem cells; cell therapy; exosome; annexin A1; acute respiratory-distress syndrome; COVID-19

1. Introduction

The investigation into mesenchymal stromal/stem cells (MSCs) has been of outstanding interest in the recent years [1]. Stromal cells are heterogeneous and contain several populations, including stem cells with different multipotential properties, committed progenitors and differentiated cells [2]. In all our experiments, we have used MSCs obtained from the human umbilical cord perivascular area of Wharton's jelly [3]. We have described these cells and assessed their phenotype [3], self-renewal potential, contractibility, differentiation [3–5], clonogenicity, radiosensitivity [6], secretion [5] and antitumoral activity both in basal conditions and after stimulation with X-rays [7,8].

We have recently shown that the combination of human umbilical-cord-derived MSCs cell therapy plus radiotherapy significantly reduces the size of established tumors in mice, both in the directly irradiated tumor and in the distant nonirradiated tumor [7] or in its metastasis [8]. These results support the hypothesis that human mesenchymal stromal/stem cells are radiosensitizers for local tumor radiotherapy, and simultaneously, they represent an effective tool for amplifying the systemic effects of radiotherapy. These out-of-target radiotherapy effects [9–11], promoted by MSCs are, in our view, of major interest [9,12].

We have also proved [7,8], that the preirradiation of MSCs triggers an important cellular change that transforms the MSCs into a source of molecules with very interesting pharmacologic proprieties. Amongst these actively secreted molecules, we have identified TRAIL and Dkk3 with very well-known antitumor activities, and annexin A1, whose activities we have previously reviewed [12] and now update here to include new data that demonstrate its anti-inflammatory and antiviral activity and its role in the regulation of hypoxia.

This secretion activity suggests a mechanistic explanation of how activated cells may positively spread their effect far from the place where they are applied. On this basis, we believe that exosomes, heavily loaded with annexin A1, will be liberated in the lungs after cell therapy with irradiated-MSCs cells, and this action would ameliorate symptoms in patients with sepsis in the lungs and in any other organs affected by septic shock.

A significant number of scientific reports are available demonstrating that gap junction, paracrine pathways and exocrine effects can transmit radiation-induced biological effects far from the place where the radiation is applied. These effects are frequently referred to as radiation-induced out-of-target effects. Multiple molecular signaling mechanisms [13] involving oxidative stress [14,15], kinases, inflammatory molecules [16,17], exosomes [8], microvesicles are postulated to contribute to bystander short- and long-range effects [12]. The anticancer immune response may also be activated by ionizing radiation, and a combination of different treatment strategies is promising in this field [11,18]. The activation of the immune system by the irradiated tumor to trigger the beneficial abscopal effect is decisively improving radiotherapy applications and their outcomes [19–22].

2. Mesenchymal Stromal/Stem Cells and Radiotherapy

It is generally acknowledged that MSCs can be found ubiquitously in many tissues and are not limited to those of mesodermal origin, such as bone marrow, adipose, muscle and bone.

Previous reports suggested a protective role for MSCs when combined with radiotherapy (RT) [23,24]. In effect, due to their properties, MSCs may be recognized as a therapeutic tool for treating radiation-induced tissue damage [25–27]. Several reports have shown that MSCs skillfully home onto neoplastics tissues [28,29] and together with tissue recovery functions MSCs prepare the microenvironment by controlling inflammatory processes to reduce the inflammation grade [30,31], where they might have the greatest therapeutic impact in vivo [32]. However, the amount of mesenchymal stromal/stem cells that are up-taken into injured tissues may not be sufficient to explain their strong overall protective effect.

The bioactivation of MSCs may be obtained indifferent ways [33] and the molecules secreted by the activated MSCs (MSCs*) might have an impact on several immune-cell lineages, establishing an advantageous sphere far away from its original location. We have proposed that exosomes liberated from radiation-activated MSCs* perform important intratumoral and systemic actions [8,12].

We are aware that cellular therapy with MSCs can be problematic in cancer therapy [12]. Therefore, it is important to emphasize that following irradiation MSCs become senescent and the senescence is associated with production of a senescent-associated secretory phenotype (SASP). The SASP has antitumor activity since it may induce senescence of neighboring cells by paracrine action [34]. Nevertheless, SASP can modify its composition and become much richer in proinflammatory factors and it has become evident that tumor-associated MSCs have a positive effect on tumor growth and the spread of metastasis [35] through the acquisition of a chemo- and radiotherapy resistance mechanisms [36], and it has been suggested that tumor cells can misuse SASP for their own growth [37]. On the other hand, it has been communicated that, in an inflammatory condition, the exosomes contained in the cancer cell secretome might have a role in the change of the normal MSCs cell phenotype toward a malignant one [36], which could be an impediment to MSCs therapeutic use.

Nevertheless, whether this innate tropism of MSCs toward the tumors and metastatic foci is related with cancer promotion or suppression remains controversial [35,37–40], and further studies on the interactions between cancerous cells and stromal components of tumor microenvironments have been proposed, which is imperative to allow the progress of more suitable treatments for cancer.

It is generally accepted that MSCs-based therapies are of major importance in regenerative medicine and, perhaps, in the future a solution for many other medical problems. However, the success of MSCs therapy relies on the efficiency of its administration and the biodistribution, engraftment, differentiation and secreting paracrine factors at the target sites [41]. Until now, there has been no universal delivery route for mesenchymal stromal/stem cells (MSCs) for different diseases [42]. In fact, efficient homing and migration toward lesion sites play an important role and the local transplantation of MSCs in spatial proximity to the lesion, as well as the systemic administration routes are being carefully explored [43]. There is growing evidence that mesenchymal stromal/stem cells based immunosuppression was mainly attributed to the effects of MSCs-derived extracellular vesicles [44] although it seems clear that transplanted MSCs can indeed leave the blood flow and transmigrate through the endothelial barrier, and reach the lesion site [45]. So, both mechanisms must be accepted until the underlying processes are better understood.

We know that in an uninjured mouse, exogenous intravenously injected MSCs rapidly accumulate within the lungs and are cleared from this site to other organs, such as the liver, within days [46]. As far as up-take in the lungs is concerned, the MSCs are able to release a wide variety of soluble mediators, including anti-inflammatory cytokines [47], antimicrobial and angiogenic macromolecules, and exosomes and microvesicles that are secreted to extravascular spaces [48]. All of the above leads us to believe that the amount of MSCs cells that engraft onto injured tissues may not be sufficient to account for their robust overall protective effects.

On the other hand, the induction of the mechanisms of epithelial and endothelial wound healing and the angiogenesis promotion has been attributed to macromolecules included in the exosomes released by the MSCs cells, which act as tools for defending the intestines from the damage produced by necrotizing enterocolitis experimentally induced in animal models [29]. This has been highly promising [23,49], and MSCs may be a well-thought-out therapeutic tool to treat radiation-induced tissue damage [30]. It is essential to highlight that the group of Chapel et al. has started a phase 2 clinical trial (ClinicalTrials.gov Identifier: NCT02814864) for the handling of severe collateral healthy tissue damage after radiation therapy in patients with prostate cancer, and this clinical trial is sustained by numerous reports focused on the use of MSCs for improving the damage severity on normal tissues after radiation treatment [46,50,51]. However, the damage severity and the mechanisms involved in the control of side effects after radiotherapy [52], as well as the role of MSCs in healthy-tissue radio-protection, are quite unknown.

We have included in Figure 1 a graphic summary of the widespread actions done by MSCs and MSCs*.

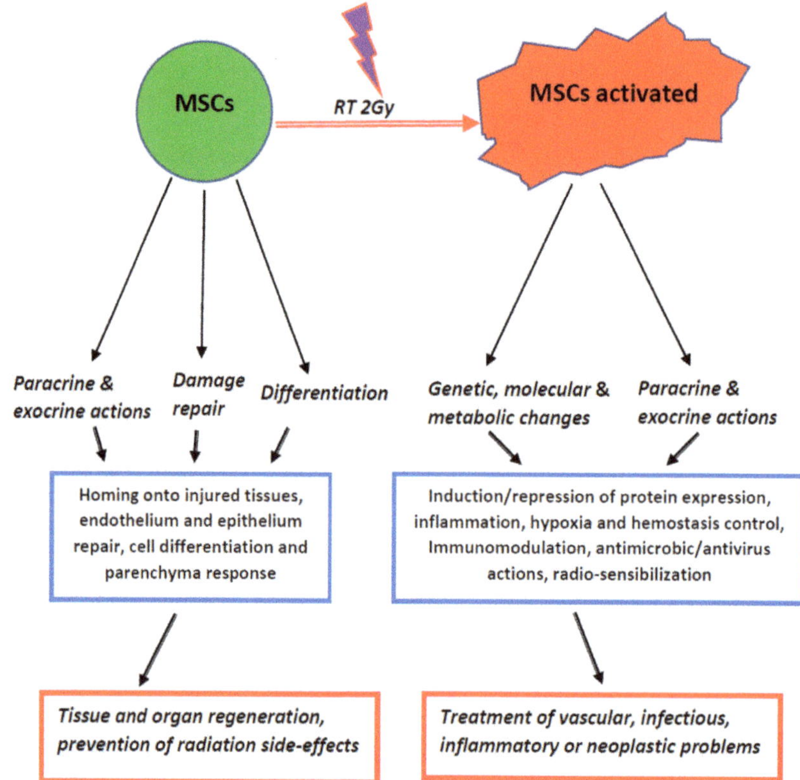

Figure 1. Graphic and schematic summary of cell actions, tissue response and possible therapeutic application of mesenchymal stromal/stem cells (MSCs) and activated MSCs.

3. Radiation-Activated Mesenchymal Stromal/Stem Cells

When we studied the exosome cargo before and after the activation of MSCs with RT, we discovered significant disparities in the results of the proteomic assessment of both samples. We described that there are qualitative, quantitative and functional differences amongst the proteins contained in the exosomes obtained from basal MSCs and activated MSCs* [8]. For more information in [8] see Supplementary Materials, additional file 1.

These findings demonstrate the profound metabolic change that these activated cell exosomes have undergone and the consequences after activation with radiation. Amongst the proteins representatives in exosomes released from MSCs*, we highlight the key components of cell–cell or cell–matrix adhesion and include annexin and integrins [8]. Between them, the presence of annexin A1 (ANXA1) is very noteworthy because it is always present in the exosomes released from MSCs* and constantly absent in MSCs. We verified these results using quantitative mRNA–PCR to measure the mRNA of this protein in MSCs and MSCs* and confirmed that mRNA is spectacularly induced in MSCs after irradiation [8]. After measuring quantitatively the mRNAs of the proteins of TRAIL, Dkk3 and ANXA1 in umbilical cord stromal stem-cells, before and after cell stimulation with 2 Gy low-energy transfer ionizing radiation, our previously published results [8] show a clear increase in their intracellular levels, compared with the levels found in basal situations (see these results in [8] supplementary material, Figure S2) and notice that the levels of mRNA of TRAIL and Dkk3 at 48 are strongly increased in treated cells compared to the basal levels ($p < 0.001$), whereas the levels of mRNA of ANXA1 are strongly increased at 24 h, and dramatically at 48 h of cell treatment, with the statistical differences found 24 and 48 h being very significant ($p < 0.0001$), which supports the massive presence of ANXA1 in the exosomes released by the radiation-stimulated MSCs.

4. Annexin A1 in the Inflammation and Hypoxia Processes Control

We stated that the existence of ANXA 1 in the exosomes separated from the culture medium of activated MSCs* and the absence of this protein in the medium withdrawn from the nonirradiated MSCs is a relevant outcome in our previous studies [8].

In relation with this protein, we would like to emphasize that after more than 30 years of research, annexins have been clearly recognized as key elements to control immune responses. The prototype component of this family, ANXA1, has been highly recognized as an anti-inflammatory factor involving cell mobility and the response of several components of the innate immune system [53]. However, it has now been recognized that ANXA1 also has important implications in maintaining homeostasis, fetal development, aging processes and in the evolution of several diseases such as cancer [54,55]. Inflammation is a tightly regulated mechanism, initiated following tissue damage or infection. If unrestrained or unsolved, the inflammation may lead to further tissue damage and give rise to persistent inflammatory diseases and autoimmunity with eventual loss of organ function. It is now evident that the outcome of inflammation is an active process that occurs during an intense inflammatory incident [56]. After MSCs activation, the released ANXA1 might diminish the gathering of neutrophils in the tissue injured in several ways. Additionally, ANXA1 promotes neutrophil apoptosis and acts on macrophages to stimulate the phagocytosis and the removal of dead neutrophils [56,57], and leads to the rapid reconstruction of tissue homeostasis. Inflammation resolve is controlled by several endogenous factors involving macromolecules and proteins, such as ANXA1, and their presence is relevant in many diseases [58]. The study of ANXA1 in relationship with the innate immune system has focused mainly on the anti-inflammatory and proresolving actions through its binding to the formyl-peptide receptor 2 (FPR2)/ALX receptor. There is much evidence that ANXA1, and its mimetic peptides [58], may have an important role in alleviating complications associated with ischemia–reperfusion injury [59]. Moreover, the presence of chronic inflammation in tumors is common and facilitates tumor growth, metastatic dissemination and treatment resistance [60]. Physical abnormality of tumor vasculature, including its chaotic structure, enlarged interstitial pressure, increased stiffness and hypoxia, are physical barriers in tumor treatment [61] are inspiring new anticancer strategies aimed at targeting the tumoral tissue to normalize these physical irregularities [61,62].

ANXA1 is an endogenous inhibitor of NF-κB that can be induced in cancer cells and experimental tumors by potent anti-inflammatory glucocorticoids and modified nonsteroidal anti-inflammatory drugs [49]. In this context, ANXA1 has long been classified as an anti-inflammatory protein due to its actions on leukocyte-mediated immune responses. However, it is now well known that ANXA1 has extensive effects further from the immune system, with consequences in maintaining the homeostatic atmosphere within the whole body due to its capacity to influence cellular signaling, hormonal secretion and diseases [63]. Upon an injury, epithelial wound shutting is a excellently adjusted process that re-establishes homeostasis, but in chronic diseases it is related with nonhealing vascular lesions; in this processes ANXA1 is involved as a preresolving mediator [64].

Moreover, new studies indicating an intracellular function of ANXA1 have now been published. In effect, using AnxA1 knockout mice, it has been noted that ANSA1 is essential for IL-1β release both in vivo as in vitro [65]. Furthermore, we know that ANXA1 colocalize and exactly connect with NLRP3, suggesting the activity of ANXA1 in inflammasome initiation is independent of its anti-inflammatory role via FPR2 [65]. These mechanisms, which could be of major importance in the resolution of lung inflammation and in septic shock through cytokine storm control, deserve more research.

5. Annexin A1 in the Treatment of Inflammation

The significance of annexin A1 (ANXA1), a 37 kDa monomeric protein, to stress response is that its synthesis and release are controlled by glucocorticoids (GCs). After release, it has been shown that

ANXA1 could strongly downregulate polymorphonuclear leukocyte migration into inflammatory sites and accelerate their apoptosis, upregulating the monocyte migration into the inflammatory sites [66].

Recently, the role of ANXA1 in the treatment of acute radiation-induced lung damage has been studied and the causes of its action examined [67]. Neuroinflammation initiated by damage-associated molecular patterns has been implicated in adverse neurological outcomes following lethal hemorrhagic shock and polytrauma [68]. Results obtained by Ma Q. et al. [68] show that attractive proresolving pharmacological approaches, such as annexin-A1 biomimetic peptides, can efficiently attenuate neuroinflammation and reveal a novel complex role for ANXA1 as a therapeutic and a prophylactic drug due to its ability to strengthen endogenous proresolving, anti-thrombo-inflammatory mechanisms in cerebral ischemia–reperfusion injury. Finally, it has been announced that recombinant human ANXA1 may represent a novel candidate for the treatment of diabetes type 2 and/or its complications [69,70].

6. Annexin A1 and Lung Diseases

Endogenous glucocorticoids are proresolving intermediaries, a model of which is the endogenous glucocorticoid-regulated protein annexin A1. Because silicosis is an occupational lung disease typified by persistent inflammation and fibrosis, models regarding this illness have been studied to test the therapeutic properties of the ANXA1 on experimental silicosis [66]. The authors have demonstrated that the therapeutic administration of N-terminal peptide of ANXA1 (Ac2-26) in ischemia–reperfusion-provoked lung injury might substantially attenuate the lung edema and proinflammatory cytokine production, thus reducing oxidative stress, apoptosis, neutrophil infiltration and lung tissue injury, perhaps via the activation of the N-formyl peptide receptor [66].

A similar result was published in an experimental study made with animals affected by bleomycin-induced lung fibrosis that were treated with an ANXA1 peptido-mimetic, administrated prophylactically (from day 0 to 21) or therapeutically (from day 14 onward), which improved signs of both inflammation and fibrosis [71]. Together these data show a pathophysiological relevance for ANXA1 in lung inflammation and in fibrosis, and may open up a new approach for the pharmacological handling of pneumonia and lung fibrosis. Currently, the resolution of inflammation, once considered to be a passive process, has recently been revealed to be an active and precisely controlled process. In the resolution stage of acute inflammation, new mediators, including lipoxins and resolvins, which are members of the specific proresolving mediators of inflammation, are released [72].

Acute lung injury and the more severe forms of acute respiratory distress syndrome, ALI/ARDS, are relatively common syndromes in seriously ill patients and are related with a high rate of morbidity and mortality. Recently, new evidence has shown that the resolution of inflammation might be an active and highly regulated process. Specific proresolving mediators (SPMs), have been proved to produce strong immune-resolving effects, such as cell proliferation, migration and the clearance of apoptotic cells and microorganisms. Therefore, the effective and timely control of inflammation could be the key step to maintain effective host defense and the restoration of homeostasis. Therefore, this reveals a new mechanism for pulmonary edema fluid reabsorption in which SPMs, amongst them annexin A1, might offer new chances to design "reabsorption-targeted" treatments with high levels of precision in controlling acute lung injury [73]. It is also widely acknowledged that to survive, edema fluid should be removed for patients with ALI/ARDS [74].

Moreover, lung endotoxemia is characterized by neutrophil accumulation, enlarged vascular permeability and parenchymal damage. In relation with toxic problems, it has been proposed that the molecular reactions stimulated by ANXA1 peptidomimetic Ac2-26 lead to the control of leukocyte activation/migration and both cytokine production and lung injury that are generated by lipopolysaccharides [75]. It was also published that ANXA1 may accelerate the resolution of inflammation in acute radiation-induced lung damage through the inhibition of IL-6 and myeloperoxidase inflammatory cytokines, demonstrating that ANXA1 may have a therapeutic role as treatment target for acute-radiation lung damage [76].

Moreover, it is well known that pattern recognition receptors (PRRs) are key elements in the innate immune response. FPR2/ALXR, a receptor modulated for specialized proresolving mediators of inflammation, amongst them annexin A1, has been shown to be one of the receptors implicated in inflammation process control. This has encouraged the research community to search for and develop new anti-inflammatory/proresolution small molecules to control inflammation through the activation of FPR2/ALXR [44].

We believe that the protective function of the ANXA1-FPR2 signaling axis recently described in viral infections it is very important [60]. The formyl peptide receptor (FPR) 2 is a pattern recognition receptor that, in addition to proinflammatory, pathogen-derived compounds, also recognizes the anti-inflammatory endogenous ligand annexin A1 (ANXA1), and it has been shown that ANXA1, via FPR2, controls inflammation and bacterial dissemination during pneumococcal pneumonia by promoting host defenses, suggesting ANXA1-based peptides as a novel therapeutic strategy to control pneumococcal pneumonia [77].

In this context, it has been described that mice with the influenza A virus (IAV) infection in the murine model treated with ANXA1 displayed significantly attenuated pathology upon a subsequent IAV infection with significantly improved survival, impaired viral replication in the respiratory tract and less severe lung damage.

7. COVID-19: The Magnitude of the Problem

Most countries in the world are suffering a significant spread of SARS-CoV-2, causing pandemic effects. The clinical presentation of the SARS-CoV-2 infection varies from asymptomatic or with light symptoms to clinical situations characterized by respiratory insufficiency requiring mechanical ventilation and intensive care, to multiorgan dysfunction syndrome with signs and symptoms such as sepsis, septic shock and multisystem failure. It also is true, unfortunately, that all the countries in the world do not have the capacity to solve this problem due to the lack of therapeutic measures that could have the appropriate impact. The problem is massive. Therefore, there is a great need to contemplate new methods to improve patients' biological resistance to SARS-CoV-2 by using mesenchymal stromal/stem cells [78]. We know that SARS-CoV-2 invade cells through the ACE2 receptor widely expressed in human cells, including the alveolar epithelium and the capillary endothelium. The MSCs are ACE2 negative. So, the transplanted cells are unable to participate in the spread of the infection.

For the healthcare services, the two key imperative necessities in the SARS-CoV-2 infection are to hinder and reduce infection rates, and to decrease the death rate of those infected. The accumulating epidemiological analyses, connected with country-based mitigation strategies, and with estimations that about 80% COVID-19 patients have mild or asymptomatic disease, 14% severe disease, and 6% are critically ill, support a permanent need for the treatment of SARS-CoV-2 infection and COVID-19 pneumonia in the long term.

According to preliminary estimates of severity that were based on a recent analysis of data from EU/EEA countries and the UK available in the *European Surveillance System TESSy* and online country reports (for countries whose data were incomplete or missing in *TESSy*) and summarized by the *European Centre for Disease Prevention and Control* (*ECDC*), we know that amongst all the cases of patients affected, hospitalization has occurred in 32% of cases reported from 26 countries, and cases with severe illness (requiring ICU and/or respiratory support) have accounted for 2.4% cases reported from 16 countries. Moreover, amongst hospitalized cases, severe illness was reported in 9.2% of hospitalized cases in 19 countries and death occurred in the 11% of the hospitalized cases in 21 countries. The age-specific hospitalization rates amongst all cases showed elevated risk amongst those aged 60 years and over. Finally, a strong estimate for the COVID-19 case death rate is still lacking and theoretically biased by partial outcome data and differences in testing policies and procedures.

The number of people affected worldwide is progressive and continuously growing, and SARS-CoV-2 has infected more than 24.5 million people and killed more than 830,000 people in different countries, areas or territories with cases (ECDC on 28 August 2020). The worldwide lethality (average) is ≈3.38% with a range of 0.1% to 14.0% depending on the country.

The magnitude of the problem is enormous and terrifying.

8. Clinical Trials of MSCs Transplantation in Patients with COVID-19 Pneumonia

MSC products are quickly arising as promising treatment candidates for the COVID-19 pandemic. It is well known that septic shock is associated with a considerable viral load in terms of both mortality and morbidity for survivors of this illness. Preclinical sepsis studies advise that mesenchymal stromal/stem cells (MSCs) may moderate inflammation, improve pathogen clearance and tissue repair and reduce death. Because MSCs have not been assessed in humans with septic shock, a clinical trial that examines safety and tolerability of MSCs is mandatory before proceeding to a randomized controlled trial to study patient outcomes. This has been performed by L.A. McIntyre et al. [79] and their results show that the infusion of freshly cultured allogenic bone-marrow-derived MSCs, up to a dose of 3 million cells/kg, into patients with septic shock seems safe and, consequently, the results of the phase I dose escalation and safety trial provide researchers with the rationale and argument to now conduct larger trials to study the efficacy of MSCs in a clinical trial in patients with septic shock [80]; the clinical trial is registered with the www.clinicaltrials.gov (NCT02421484) reference.

Preclinical and early clinical data suggest that human umbilical cord stromal MCSs, because of their anti-inflammatory and immunomodulatory actions, are able to heal tissues affected and thus improve recovery rates [81]. Additionally, this treatment also seems to be antimicrobial. Two recent studies from China [78,82] have examined whether MSCs could be useful for treating SARS-CoV-2/COVID pneumonia, based on known immune modulatory and reparative abilities of stem cells. Both studies show an outstanding reversal of symptoms, even in severe to critical circumstances. These clinical studies not only recognize a novel therapeutic approach, but also the reality of natural processes able to reduce acute inflammatory pneumonia.

Following the intravenous transplantation of MSCs, a noteworthy population of cells accumulates in the lung, which together with their immunomodulatory effect, could protect alveolar epithelial cells, recover the pulmonary microenvironment, avoid pulmonary fibrosis and cure lung dysfunction. It has been suggested that MSCs have cured or significantly improved the functional outcomes of seven patients without any detected side effects. The pulmonary function and symptoms of these seven patients were significantly improved in two days after MSCs transplantation. Furthermore, the gene expression profile revealed MSCs were $ACE2^-$ and TMPRSS2, which showed that the MSCs were free from the SARS-CoV-2 infection. Thus, the intravenous cellular transplantation was safe and efficient for handling in patients with COVID pneumonia, particularly for the patients in a seriously severe condition [78].

Given the uncertainties in this area, Golchin et al. [83] have reviewed published clinical trials and hypotheses to offer useful information to researchers and those involved in stem-cell therapy. In their study, they considered a new approach to enhance patients' immunological responses to COVID-19 pneumonia using MSCs and debating the aspects of this proposed treatment. However, currently, there are no approved MSC-based approaches for the prevention and/or treatment of COVID-19 patients; nevertheless, clinical trials are ongoing.

The immunomodulatory and anti-inflammatory properties of MSCs in the treatment of respiratory diseases have been confirmed by 17 completed clinical studies, and also more than 70 trials have been registered in this regard (https://clinicaltrials.gov).

Many of the critically ill COVID-19 patients are in a hypercoagulable or procoagulant situation and with a high probability for disseminated intravascular coagulation, thromboembolism and thrombotic multiorgan catastrophe, another cause of the high death rate. Therefore, it is mandatory to only use well-characterized and safe MSCs in the most urgent and experimental treatments [84]. Moreover, in order to alleviate patients with SARS-CoV-2 infection, the obvious risk of adverse thrombotic reactions after the transplant of high doses of poorly typified cell product, an obligatory a set of significant procedures for combining innate immune hemocompatibility examination into the usual patients' characterization and clinical procedures, before applying MSCs cell therapies has been proposed [84].

Of course, cost effectiveness and the speed of medicinal formulation and transport are topics to be considered for MSCs-based therapy for COVID-19, but without a doubt, whatever the cost the life of a human being is priceless. Nevertheless, the clinical use of MSCs therapy to treat COVID-19 seems promising. Therefore, bearing in mind that MSCs therapy could become an important contribution to terminate the high COVID-19 death rates and prevent long-term functional side effects in those who survive disease, it is essential that the funding agencies invest more into the development of MSCs suitable for safe clinical applications [71].

However, it is very important to underline that scientists are tirelessly trying to obtain a vaccine for SARS-CoV-2 infection and COVID-19 pneumonia, as well as therapeutics to treat this disease [83], and that now a vaccine to protect against SARS-CoV-2 infection has been assessed for safety, tolerability and immunogenicity of a recombinant adenovirus type-5 (Ad5) vectored vaccine expressing the spike glycoprotein of a grave acute respiratory syndrome coronavirus 2 (SARS-CoV-2) variety [85]. These recently published results show that the vaccine is safe and immunogenic at 28 days postvaccination. Humoral responses against SARS-CoV-2 hit the highest point at day 28 postvaccination in healthy adults, and quick specific T-cell responses were observed from day 14 postvaccination. These findings imply that the Ad5 vectored SARS-CoV-2/COVID vaccine deserves more research [85] and an ongoing phase 2 trial in China (NCT04341389) will offer more data on the safety and immunogenicity of the Ad5 vectored SARS-CoV-2/COVID-19 vaccine. The progress in this field is extremely fast, and an excellent update on the subject can be found in [86].

9. Conclusions and Perspectives

The present global health crisis involving the appearance and rapid spread of a new coronavirus has encouraged the worldwide scientific community to consider how it can help to combat this mounting viral pandemic.

Amongst all the different mesenchymal stromal/stem cells that might be used, umbilical cord stem cells seem to be the most desirable for a series of reasons that have been very well explained by S. Atluri et al. [81]. Considering together both the previous reports and our own knowledge, and research on the exceptional abilities of proliferation [5,7], secretion [4] and differentiation [17,71] of the umbilical cord mesenchymal stromal/stem cells that we have investigated [7,8], we have also decided to recommend umbilical cord mesenchymal stromal/stem cells as a vehicle for annexin A1 for septic shock treatment.

The activation of these MSCs with a 2 Gy low-LET radiation dose produces an important increase in the cell-released exosomes and these nanovesicles, which can reach all the tissues and organs affected, contain a very specific load of proteins, including annexin A1 [8,12], whose activity in situations of infection, inflammation and hypoxia has been intensively discussed in the previous sections of this paper. This protein together with the endothelium-repair functions characteristic of MSCs must play a major role in the treatment of the septic shock and pneumonia related with SARS-CoV-2 infection.

Moreover, it is generally accepted that the efficacy of transplanted MSCs actually seems to be independent of the physical proximity of the transplanted cells to damaged tissue. Supposedly a vectorized signaling system, we now believe that the exosomes released from radiation-activated-MSCs cells can reach other organs different from the lungs, where they will be up-taken after intravenous injection and thus extend the anti-inflammatory and antimicrobiological effects of the treatment, to cover systemic problems such as the treatment of patients with septic shock in general and for COVID-19 at this particular time.

This hypothesis provides a rationale for the therapeutic efficacy of MSCs and their secreted exosomes in patients with clinical conditions characterized by respiratory failure necessitating mechanical ventilation and medical assistance in the intensive care unit, for multiorgan insufficiency and systemic manifestations such as sepsis, septic shock and multiple organ dysfunction cases.

Lastly, a scheme for our hypothetical cellular therapy in patients with acute respiratory distress syndrome would be an intravenous infusion of 6 million/kg of patient-weight divided into two parts: (a) 3 million nonirradiated-MSCs/kg of patient-weight, to take advantage of the protective, regenerative and repair MSCs-effects at the lung–vasculature and (b) 3 million preirradiated-MSCs*/kg of patient-weight, to achieve, as soon as possible within the patients, the loaded-exosomes with ANXA1 that clinical-grade umbilical cord MSCs* are able to produce after radiation stimulation and thus, take advantage of the extensive range of anti-thrombo-inflammatory, antiviral and immunomodulatory actions associated with this protein.

Finally, we want to clarify that this paper only presents a hypothesis and that the possibility of treating patients is still far off because we lack the necessary experimental data, which would prove the applicability, efficiency and security necessary to further the hypothesis in its transition from the laboratory bench to the patient's bed. Therefore, more work is necessary to promote this idea and use activated MSCs* as a therapy for patients with COVID-19, but that is our challenge and we are optimistic of a positive outcome.

Author Contributions: I.T., R.G., J.J.L.-P., J.E. and J.M.R.d.A. contributed to the study conception and design. The first draft of the manuscript was written by J.M.R.d.A. and all authors commented on previous versions of the manuscript. All authors have read and agreed to the published version of the manuscript.

Funding: This work was supported by Ministerio de Economía y Competividad, MINECO: SAF2012-40011-C02-02 and SAF2015-70520-R, RTICC RD12/0036/0026 to J.M.R.d.A.

Acknowledgments: The authors thank A.L. Tate for revising their English text.

Conflicts of Interest: The authors declare no conflict of interest.

References

1. Keating, A. Mesenchymal stromal cells: New directions. *Cell Stem Cell* **2012**, *10*, 709–716. [CrossRef]
2. Squillaro, T.; Peluso, G.; Galderisi, U. Clinical Trials With Mesenchymal Stem Cells: An Update. *Cell Transplant.* **2016**, *25*, 829–848. [CrossRef]
3. Farias, V.A.; Linares-Fernandez, J.L.; Penalver, J.L.; Paya Colmenero, J.A.; Ferron, G.O.; Duran, E.L.; Fernandez, R.M.; Olivares, E.G.; O'Valle, F.; Puertas, A.; et al. Human umbilical cord stromal stem cell express CD10 and exert contractile properties. *Placenta* **2011**, *32*, 86–95. [CrossRef]
4. Peñalver, J.L.; Linares-Fernández, J.L.; de Araujo Farías, V.; López-Ramón, M.V.; Tassi, M.; Oliver, F.J.; Moreno-Castilla, C.; deAlmodóvar, J.M.R. Activated carbon cloth as support for mesenchymal stem cell growth and differentiation to osteocyte. *Carbon* **2009**, *47*, 3574–3577. [CrossRef]
5. de Araújo Farias, V.; López-Peñalver, J.J.; Sirés-Campos, J.; López-Ramón, M.V.; Moreno-Castilla, C.; Oliver, F.J.; de Almodóvar, J.M.R. Growth and spontaneous differentiation of umbilical-cord stromal stem cells on activated carbon cloth. *J. Mater. Chem. B* **2013**, *1*, 3359–3368. [CrossRef]
6. Gomez-Millan, J.; Katz, I.S.; Farias Vde, A.; Linares-Fernandez, J.L.; Lopez-Penalver, J.; Ortiz-Ferron, G.; Ruiz-Ruiz, C.; Oliver, F.J.; Ruiz de Almodovar, J.M. The importance of bystander effects in radiation therapy in melanoma skin-cancer cells and umbilical-cord stromal stem cells. *Radiother. Oncol.* **2012**, *102*, 450–458. [CrossRef]
7. De Araujo Farias, V.; O'Valle, F.; Lerma, B.A.; Ruiz de Almodovar, C.; Lopez-Penalver, J.J.; Nieto, A.; Santos, A.; Fernandez, B.I.; Guerra-Librero, A.; Ruiz-Ruiz, M.C.; et al. Human mesenchymal stem cells enhance the systemic effects of radiotherapy. *Oncotarget* **2015**, *6*, 31164–31180. [CrossRef]
8. de Araujo Farias, V.; O'Valle, F.; Serrano-Saenz, S.; Anderson, P.; Andres, E.; Lopez-Penalver, J.; Tovar, I.; Nieto, A.; Santos, A.; Martin, F.; et al. Exosomes derived from mesenchymal stem cells enhance radiotherapy-induced cell death in tumor and metastatic tumor foci. *Mol. Cancer* **2018**, *17*, 122. [CrossRef] [PubMed]
9. Lara, P.C.; Lopez-Penalver, J.J.; Farias Vde, A.; Ruiz-Ruiz, M.C.; Oliver, F.J.; Ruiz de Almodovar, J.M. Direct and bystander radiation effects: A biophysical model and clinical perspectives. *Cancer Lett.* **2015**, *356*, 5–16. [CrossRef]
10. Mothersill, C.E.; Moriarty, M.J.; Seymour, C.B. Radiotherapy and the potential exploitation of bystander effects. *Int. J. Radiat. Oncol. Biol. Phys.* **2004**, *58*, 575–579. [CrossRef]

11. Formenti, S.C.; Demaria, S. Systemic effects of local radiotherapy. *Lancet Oncol.* **2009**, *10*, 718–726. [CrossRef]
12. Farias, V.A.; Tovar, I.; Del Moral, R.; O'Valle, F.; Exposito, J.; Oliver, F.J.; Ruiz de Almodovar, J.M. Enhancing the Bystander and Abscopal Effects to Improve Radiotherapy Outcomes. *Front. Oncol.* **2019**, *9*, 1381. [CrossRef] [PubMed]
13. Decrock, E.; Hoorelbeke, D.; Ramadan, R.; Delvaeye, T.; De Bock, M.; Wang, N.; Krysko, D.V.; Baatout, S.; Bultynck, G.; Aerts, A.; et al. Calcium, oxidative stress and connexin channels, a harmonious orchestra directing the response to radiotherapy treatment? *BBA Mol. Cell. Res.* **2017**, *1864*, 1099–1120. [CrossRef] [PubMed]
14. Azzam, E.I.; de Toledo, S.M.; Little, J.B. Oxidative metabolism, gap junctions and the ionizing radiation-induced bystander effect. *Oncogene* **2003**, *22*, 7050–7057. [CrossRef] [PubMed]
15. Kim, S.M.; Oh, J.H.; Park, S.A.; Ryu, C.H.; Lim, J.Y.; Kim, D.S.; Chang, J.W.; Oh, W.; Jeun, S.S. Irradiation enhances the tumor tropism and therapeutic potential of tumor necrosis factor-related apoptosis-inducing ligand-secreting human umbilical cord blood-derived mesenchymal stem cells in glioma therapy. *Stem cells* **2010**, *28*, 2217–2228. [CrossRef]
16. Mothersill, C.; Seymour, C.B. Radiation-induced bystander effects–implications for cancer. *Nat. Rev. Cancer* **2004**, *4*, 158–164. [CrossRef]
17. Prise, K.M.; Schettino, G.; Folkard, M.; Held, K.D. New insights on cell death from radiation exposure. *Lancet Oncol.* **2005**, *6*, 520–528. [CrossRef]
18. Formenti, S.C.; Demaria, S.; Barcellos-Hoff, M.H.; McBride, W.H. Subverting misconceptions about radiation therapy. *Nat. Immunol.* **2016**, *17*, 345. [CrossRef]
19. Golden, E.B.; Apetoh, L. Radiotherapy and immunogenic cell death. *Semin. Radiat. Oncol.* **2015**, *25*, 11–17. [CrossRef]
20. Golden, E.B.; Formenti, S.C. Radiation therapy and immunotherapy: Growing pains. *Int. J. Radiat. Oncol. Biol. Phys.* **2015**, *91*, 252–254. [CrossRef]
21. Petroni, G.; Formenti, S.C.; Chen-Kiang, S.; Galluzzi, L. Immunomodulation by anticancer cell cycle inhibitors. *Nat. Rev. Immunol.* **2020**. [CrossRef]
22. Demaria, S.; Formenti, S.C. The abscopal effect 67 years later: From a side story to center stage. *Br. J. Radiol.* **2020**, *93*. [CrossRef] [PubMed]
23. Chang, P.Y.; Qu, Y.Q.; Wang, J.; Dong, L.H. The potential of mesenchymal stem cells in the management of radiation enteropathy. *Cell Death Dis.* **2015**, *6*, e1840. [CrossRef]
24. Maziarz, R.T.; Devos, T.; Bachier, C.R.; Goldstein, S.C.; Leis, J.F.; Devine, S.M.; Meyers, G.; Gajewski, J.L.; Maertens, J.; Deans, R.J.; et al. Single and multiple dose MultiStem (multipotent adult progenitor cell) therapy prophylaxis of acute graft-versus-host disease in myeloablative allogeneic hematopoietic cell transplantation: A phase 1 trial. *Biol. Blood Marrow Transplant.* **2015**, *21*, 720–728. [CrossRef]
25. Klein, D.; Schmetter, A.; Imsak, R.; Wirsdorfer, F.; Unger, K.; Jastrow, H.; Stuschke, M.; Jendrossek, V. Therapy with Multipotent Mesenchymal Stromal Cells Protects Lungs from Radiation-Induced Injury and Reduces the Risk of Lung Metastasis. *Antioxid. Redox Signal* **2016**, *24*, 53–69. [CrossRef]
26. Bernardo, M.E.; Cometa, A.M.; Locatelli, F. Mesenchymal stromal cells: A novel and effective strategy for facilitating engraftment and accelerating hematopoietic recovery after transplantation? *Bone Marrow Transplant.* **2012**, *47*, 323–329. [CrossRef]
27. Nicolay, N.H.; Liang, Y.; Lopez Perez, R.; Bostel, T.; Trinh, T.; Sisombath, S.; Weber, K.J.; Ho, A.D.; Debus, J.; Saffrich, R.; et al. Mesenchymal stem cells are resistant to carbon ion radiotherapy. *Oncotarget* **2015**, *6*, 2076–2087. [CrossRef]
28. Loebinger, M.R.; Sage, E.K.; Davies, D.; Janes, S.M. TRAIL-expressing mesenchymal stem cells kill the putative cancer stem cell population. *Br. J. Cancer* **2010**, *103*, 1692–1697. [CrossRef]
29. Loebinger, M.R.; Janes, S.M. Stem cells as vectors for antitumour therapy. *Thorax* **2010**, *65*, 362–369. [CrossRef]
30. Nicolay, N.H.; Lopez Perez, R.; Debus, J.; Huber, P.E. Mesenchymal stem cells—A new hope for radiotherapy-induced tissue damage? *Cancer Lett.* **2015**, *366*, 133–140. [CrossRef]
31. Skripcak, T.; Belka, C.; Bosch, W.; Brink, C.; Brunner, T.; Budach, V.; Buttner, D.; Debus, J.; Dekker, A.; Grau, C.; et al. Creating a data exchange strategy for radiotherapy research: Towards federated databases and anonymised public datasets. *Radiother. Oncol.* **2014**, *113*, 303–309. [CrossRef] [PubMed]
32. Murphy, M.B.; Moncivais, K.; Caplan, A.I. Mesenchymal stem cells: Environmentally responsive therapeutics for regenerative medicine. *Exp. Mol. Med.* **2013**, *45*, e54. [CrossRef] [PubMed]

33. Lee, R.H.; Yoon, N.; Reneau, J.C.; Prockop, D.J. Preactivation of human MSCs with TNF-alpha enhances tumor-suppressive activity. *Cell Stem Cell* **2012**, *11*, 825–835. [CrossRef]
34. Ozcan, S.; Alessio, N.; Acar, M.B.; Mert, E.; Omerli, F.; Peluso, G.; Galderisi, U. Unbiased analysis of senescence associated secretory phenotype (SASP) to identify common components following different genotoxic stresses. *Aging* **2016**, *8*, 1316–1329. [CrossRef]
35. O'Malley, G.; Heijltjes, M.; Houston, A.M.; Rani, S.; Ritter, T.; Egan, L.J.; Ryan, A.E. Mesenchymal stromal cells (MSCs) and colorectal cancer: A troublesome twosome for the anti-tumour immune response? *Oncotarget* **2016**, *7*, 60752–60774. [CrossRef]
36. Hass, R.; von der Ohe, J.; Ungefroren, H. Potential Role of MSC/Cancer Cell Fusion and EMT for Breast Cancer Stem Cell Formation. *Cancers* **2019**, *11*, 1432. [CrossRef]
37. Shi, Y.; Du, L.; Lin, L.; Wang, Y. Tumour-associated mesenchymal stem/stromal cells: Emerging therapeutic targets. *Nat. Rev. Drug Discov.* **2017**, *16*, 35–52. [CrossRef]
38. Chowdhury, R.; Webber, J.P.; Gurney, M.; Mason, M.D.; Tabi, Z.; Clayton, A. Cancer exosomes trigger mesenchymal stem cell differentiation into pro-angiogenic and pro-invasive myofibroblasts. *Oncotarget* **2015**, *6*, 715–731. [CrossRef]
39. Wu, S.; Ju, G.Q.; Du, T.; Zhu, Y.J.; Liu, G.H. Microvesicles derived from human umbilical cord Wharton's jelly mesenchymal stem cells attenuate bladder tumor cell growth in vitro and in vivo. *PLoS ONE* **2013**, *8*, e61366. [CrossRef]
40. Vieira de Castro, J.; Gomes, E.D.; Granja, S.; Anjo, S.I.; Baltazar, F.; Manadas, B.; Salgado, A.J.; Costa, B.M. Impact of mesenchymal stem cells' secretome on glioblastoma pathophysiology. *J. Transl. Med.* **2017**, *15*. [CrossRef]
41. Yun, W.S.; Aryal, S.; Ahn, Y.J.; Seo, Y.J.; Key, J. Engineered iron oxide nanoparticles to improve regenerative effects of mesenchymal stem cells. *Biomed. Eng. Lett.* **2020**, *10*, 259–273. [CrossRef]
42. Liu, Z.; Mikrani, R.; Zubair, H.M.; Taleb, A.; Naveed, M.; Baig, M.; Zhang, Q.; Li, C.; Habib, M.; Cui, X.; et al. Systemic and local delivery of mesenchymal stem cells for heart renovation: Challenges and innovations. *Eur. J. Pharmacol.* **2020**, *876*. [CrossRef]
43. Nitzsche, F.; Muller, C.; Lukomska, B.; Jolkkonen, J.; Deten, A.; Boltze, J. Concise Review: MSC Adhesion Cascade-Insights into Homing and Transendothelial Migration. *Stem Cells* **2017**, *35*, 1446–1460. [CrossRef]
44. Harrell, C.R.; Jovicic, N.; Djonov, V.; Arsenijevic, N.; Volarevic, V. Mesenchymal Stem Cell-Derived Exosomes and Other Extracellular Vesicles as New Remedies in the Therapy of Inflammatory Diseases. *Cells* **2019**, *8*, 1605. [CrossRef]
45. Carreras-Planella, L.; Monguio-Tortajada, M.; Borras, F.E.; Franquesa, M. Immunomodulatory Effect of MSC on B Cells Is Independent of Secreted Extracellular Vesicles. *Front. Immunol.* **2019**, *10*. [CrossRef]
46. Moussa, L.; Pattappa, G.; Doix, B.; Benselama, S.L.; Demarquay, C.; Benderitter, M.; Semont, A.; Tamarat, R.; Guicheux, J.; Weiss, P.; et al. A biomaterial-assisted mesenchymal stromal cell therapy alleviates colonic radiation-induced damage. *Biomaterials* **2017**, *115*, 40–52. [CrossRef]
47. Lee, R.H.; Pulin, A.A.; Seo, M.J.; Kota, D.J.; Ylostalo, J.; Larson, B.L.; Semprun-Prieto, L.; Delafontaine, P.; Prockop, D.J. Intravenous hMSCs improve myocardial infarction in mice because cells embolized in lung are activated to secrete the anti-inflammatory protein TSG-6. *Cell Stem Cell* **2009**, *5*, 54–63. [CrossRef]
48. Hu, S.; Park, J.; Liu, A.; Lee, J.; Zhang, X.; Hao, Q.; Lee, J.W. Mesenchymal Stem Cell Microvesicles Restore Protein Permeability Across Primary Cultures of Injured Human Lung Microvascular Endothelial Cells. *Stem Cells Transl. Med.* **2018**, *7*, 615–624. [CrossRef]
49. Rager, T.M.; Olson, J.K.; Zhou, Y.; Wang, Y.; Besner, G.E. Exosomes secreted from bone marrow-derived mesenchymal stem cells protect the intestines from experimental necrotizing enterocolitis. *J. Pediatr. Surg.* **2016**, *51*, 942–947. [CrossRef]
50. Bessout, R.; Demarquay, C.; Moussa, L.; Rene, A.; Doix, B.; Benderitter, M.; Semont, A.; Mathieu, N. TH17 predominant T-cell responses in radiation-induced bowel disease are modulated by treatment with adipose-derived mesenchymal stromal cells. *J. Pathol.* **2015**, *237*, 435–446. [CrossRef]
51. Van de Putte, D.; Demarquay, C.; Van Daele, E.; Moussa, L.; Vanhove, C.; Benderitter, M.; Ceelen, W.; Pattyn, P.; Mathieu, N. Adipose-Derived Mesenchymal Stromal Cells Improve the Healing of Colonic Anastomoses Following High Dose of Irradiation Through Anti-Inflammatory and Angiogenic Processes. *Cell Transplant.* **2017**, *26*, 1919–1930. [CrossRef] [PubMed]

52. Lopez, E.; Guerrero, R.; Nunez, M.I.; del Moral, R.; Villalobos, M.; Martinez-Galan, J.; Valenzuela, M.T.; Munoz-Gamez, J.A.; Oliver, F.J.; Martin-Oliva, D.; et al. Early and late skin reactions to radiotherapy for breast cancer and their correlation with radiation-induced DNA damage in lymphocytes. *Breast. Cancer Res.* **2005**, *7*, R690–R698. [CrossRef] [PubMed]
53. Weyd, H. More than just innate affairs—On the role of annexins in adaptive immunity. *J. Biol. Chem.* **2016**, *397*, 1017–1029. [CrossRef] [PubMed]
54. Guo, C.; Liu, S.; Sun, M.Z. Potential role of Anxa1 in cancer. *Future Oncol.* **2013**, *9*, 1773–1793. [CrossRef] [PubMed]
55. Boudhraa, Z.; Bouchon, B.; Viallard, C.; D'Incan, M.; Degoul, F. Annexin A1 localization and its relevance to cancer. *Clin. Sci.* **2016**, *130*, 205–220. [CrossRef]
56. Alessandri, A.L.; Sousa, L.P.; Lucas, C.D.; Rossi, A.G.; Pinho, V.; Teixeira, M.M. Resolution of inflammation: Mechanisms and opportunity for drug development. *Pharmacol. Ther.* **2013**, *139*, 189–212. [CrossRef]
57. Soehnlein, O.; Lindbom, L. Phagocyte partnership during the onset and resolution of inflammation. *Nat. Rev. Immunol.* **2010**, *10*, 427–439. [CrossRef]
58. Fredman, G.; Tabas, I. Boosting Inflammation Resolution in Atherosclerosis: The Next Frontier for Therapy. *Am. J. Pathol.* **2017**, *187*, 1211–1221. [CrossRef]
59. Ansari, J.; Kaur, G.; Gavins, F.N.E. Therapeutic Potential of Annexin A1 in Ischemia Reperfusion Injury. *Int. J. Mol. Sci.* **2018**, *19*, 1211. [CrossRef]
60. Shalapour, S.; Karin, M. Immunity, inflammation, and cancer: An eternal fight between good and evil. *J. Clin. Investig.* **2015**, *125*, 3347–3355. [CrossRef]
61. Hanahan, D.; Weinberg, R.A. Hallmarks of Cancer: The Next Generation. *Cell* **2011**, *144*, 646–674. [CrossRef] [PubMed]
62. Shen, Y.; Wang, X.; Lu, J.; Salfenmoser, M.; Wirsik, N.M.; Schleussner, N.; Imle, A.; Freire Valls, A.; Radhakrishnan, P.; Liang, J.; et al. Reduction of Liver Metastasis Stiffness Improves Response to Bevacizumab in Metastatic Colorectal Cancer. *Cancer Cell* **2020**, *37*, 800–817. [CrossRef] [PubMed]
63. Sheikh, M.H.; Solito, E. Annexin A1: Uncovering the Many Talents of an Old Protein. *Int. J. Mol. Sci.* **2018**, *19*, 1045. [CrossRef] [PubMed]
64. Leoni, G.; Nusrat, A. Annexin A1: Shifting the balance towards resolution and repair. *J. Biol. Chem.* **2016**, *397*, 971–979. [CrossRef] [PubMed]
65. Galvao, I.; de Carvalho, R.V.H.; Vago, J.P.; Silva, A.L.N.; Carvalho, T.G.; Antunes, M.M.; Ribeiro, F.M.; Menezes, G.B.; Zamboni, D.S.; Sousa, L.P.; et al. The role of annexin A1 in the modulation of the NLRP3 inflammasome. *Immunology* **2020**, *160*, 78–89. [CrossRef] [PubMed]
66. Trentin, P.G.; Ferreira, T.P.; Arantes, A.C.; Ciambarella, B.T.; Cordeiro, R.S.; Flower, R.J.; Perretti, M.; Martins, M.A.; Silva, P.M. Annexin A1 mimetic peptide controls the inflammatory and fibrotic effects of silica particles in mice. *Br. J. Pharmacol.* **2015**, *172*, 3058–3071. [CrossRef]
67. Sun, Z.; Shi, K.; Yang, S.; Liu, J.; Zhou, Q.; Wang, G.; Song, J.; Li, Z.; Zhang, Z.; Yuan, W. Effect of exosomal miRNA on cancer biology and clinical applications. *Mol. Cancer* **2018**, *17*. [CrossRef]
68. Ma, Q.; Zhang, Z.; Shim, J.K.; Venkatraman, T.N.; Lascola, C.D.; Quinones, Q.J.; Mathew, J.P.; Terrando, N.; Podgoreanu, M.V. Annexin A1 Bioactive Peptide Promotes Resolution of Neuroinflammation in a Rat Model of Exsanguinating Cardiac Arrest Treated by Emergency Preservation and Resuscitation. *Front. Neurosci.* **2019**, *13*. [CrossRef]
69. Purvis, G.S.D.; Solito, E.; Thiemermann, C. Annexin-A1: Therapeutic Potential in Microvascular Disease. *Front. Immunol.* **2019**, *10*. [CrossRef]
70. Purvis, G.S.D.; Collino, M.; Loiola, R.A.; Baragetti, A.; Chiazza, F.; Brovelli, M.; Sheikh, M.H.; Collotta, D.; Cento, A.; Mastrocola, R.; et al. Identification of AnnexinA1 as an Endogenous Regulator of RhoA, and Its Role in the Pathophysiology and Experimental Therapy of Type-2 Diabetes. *Front. Immunol.* **2019**, *10*. [CrossRef]
71. Damazo, A.S.; Sampaio, A.L.; Nakata, C.M.; Flower, R.J.; Perretti, M.; Oliani, S.M. Endogenous annexin A1 counter-regulates bleomycin-induced lung fibrosis. *BMC Immunol.* **2011**, *12*. [CrossRef] [PubMed]
72. Corminboeuf, O.; Leroy, X. FPR2/ALXR agonists and the resolution of inflammation. *J. Med. Chem.* **2015**, *58*, 537–559. [CrossRef] [PubMed]
73. Wang, Q.; Yan, S.F.; Hao, Y.; Jin, S.W. Specialized Pro-resolving Mediators Regulate Alveolar Fluid Clearance during Acute Respiratory Distress Syndrome. *Chin. Med. J.* **2018**, *131*, 982–989. [CrossRef] [PubMed]

74. Sznajder, J.I. Alveolar edema must be cleared for the acute respiratory distress syndrome patient to survive. *Am. J. Respir. Crit. Care Med.* **2001**, *163*, 1293–1294. [CrossRef] [PubMed]
75. Da Cunha, E.E.; Oliani, S.M.; Damazo, A.S. Effect of annexin-A1 peptide treatment during lung inflammation induced by lipopolysaccharide. *Pulm Pharmacol. Ther.* **2012**, *25*, 303–311. [CrossRef]
76. Han, G.; Lu, K.; Xu, W.; Zhang, S.; Huang, J.; Dai, C.; Sun, G.; Ye, J. Annexin A1-mediated inhibition of inflammatory cytokines may facilitate the resolution of inflammation in acute radiation-induced lung injury. *Oncol. Lett.* **2019**, *18*, 321–329. [CrossRef]
77. Machado, M.G.; Tavares, L.P.; Souza, G.V.S.; Queiroz-Junior, C.M.; Ascencao, F.R.; Lopes, M.E.; Garcia, C.C.; Menezes, G.B.; Perretti, M.; Russo, R.C.; et al. The Annexin A1/FPR2 pathway controls the inflammatory response and bacterial dissemination in experimental pneumococcal pneumonia. *FASEB J.* **2020**, *34*, 2749–2764. [CrossRef]
78. Leng, Z.; Zhu, R.; Hou, W.; Feng, Y.; Yang, Y.; Han, Q.; Shan, G.; Meng, F.; Du, D.; Wang, S.; et al. Transplantation of ACE2(-) Mesenchymal Stem Cells Improves the Outcome of Patients with COVID-19 Pneumonia. *Aging Dis.* **2020**, *11*, 216–228. [CrossRef]
79. McIntyre, L.A.; Stewart, D.J.; Mei, S.H.J.; Courtman, D.; Watpool, I.; Granton, J.; Marshall, J.; Dos Santos, C.; Walley, K.R.; Winston, B.W.; et al. Cellular Immunotherapy for Septic Shock. A Phase I Clinical Trial. *Am. J. Respir. Crit. Care Med.* **2018**, *197*, 337–347. [CrossRef]
80. Nolta, J.A.; Galipeau, J.; Phinney, D.G. Improving mesenchymal stem/stromal cell potency and survival: Proceedings from the International Society of Cell Therapy (ISCT) MSC preconference held in May 2018, Palais des Congres de Montreal, Organized by the ISCT MSC Scientific Committee. *Cytotherapy* **2020**, *22*, 123–126. [CrossRef]
81. Atluri, S.; Manchikanti, L.; Hirsch, J.A. Expanded Umbilical Cord Mesenchymal Stem Cells (UC-MSCs) as a Therapeutic Strategy in Managing Critically Ill COVID-19 Patients: The Case for Compassionate Use. *Pain Physician* **2020**, *23*, E71–E83. [PubMed]
82. Wu, Z.; McGoogan, J.M. Characteristics of and Important Lessons From the Coronavirus Disease 2019 (COVID-19) Outbreak in China: Summary of a Report of 72314 Cases From the Chinese Center for Disease Control and Prevention. *JAMA* **2020**. [CrossRef] [PubMed]
83. Golchin, A.; Seyedjafari, E.; Ardeshirylajimi, A. Mesenchymal Stem Cell Therapy for COVID-19: Present or Future. *Stem Cell Rev.* **2020**. [CrossRef]
84. Moll, G.; Drzeniek, N.; Kamhieh-Milz, J.; Geissler, S.; Volk, H.D.; Reinke, P. MSC Therapies for COVID-19: Importance of Patient Coagulopathy, Thromboprophylaxis, Cell Product Quality and Mode of Delivery for Treatment Safety and Efficacy. *Front. Immunol.* **2020**, *11*. [CrossRef]
85. Zhu, F.C.; Li, Y.H.; Guan, X.H.; Hou, L.H.; Wang, W.J.; Li, J.X.; Wu, S.P.; Wang, B.S.; Wang, Z.; Wang, L.; et al. Safety, tolerability, and immunogenicity of a recombinant adenovirus type-5 vectored COVID-19 vaccine: A dose-escalation, open-label, non-randomised, first-in-human trial. *Lancet* **2020**, *395*, 1845–1854. [CrossRef]
86. Corey, L.; Mascola, J.R.; Fauci, A.S.; Collins, F.S. A strategic approach to COVID-19 vaccine R&D. *Science* **2020**, *368*, 948–950. [CrossRef]

Article

Exposure of Human Skin Organoids to Low Genotoxic Stress Can Promote Epithelial-to-Mesenchymal Transition in Regenerating Keratinocyte Precursor Cells

Sophie Cavallero [1,2,3,4,†,‡], Renata Neves Granito [1,2,3,4,†,§], Daniel Stockholm [5], Peggy Azzolin [1,2,3,4], Michèle T. Martin [1,2,3,4,*] and Nicolas O. Fortunel [1,2,3,4,*]

1 Laboratoire de Génomique et Radiobiologie de la Kératinopoïèse, Institut de Biologie François Jacob, CEA/DRF/IRCM, 91000 Evry, France; socavallero@gmail.com (S.C.); re_neves@yahoo.com.br (R.N.G.); peggy.azzolin@cea.fr (P.A.)
2 INSERM U967, 92260 Fontenay-aux-Roses, France
3 Université Paris-Saclay, 75013 Paris 11, France
4 Université Paris-Diderot, 78140 Paris 7, France
5 Ecole Pratique des Hautes Etudes, PSL Research University, UMRS 951, Genethon, 91002 Evry, France; stockho@genethon.fr
* Correspondence: michele.martin@cea.fr (M.T.M.); nicolas.fortunel@cea.fr (N.O.F.); Tel.: +33-1-60-87-34-91 (M.T.M.); +33-1-60-87-34-92 (N.O.F.); Fax: +33-1-60-87-34-98 (M.T.M. & N.O.F.)
† These authors contributed equally.
‡ Present address: DEBR/Unité Rad/Institut de Recherche Biomédicale des Armées, 1 Place du Général Valérie André, 91223 Brétigny sur Orge, France.
§ Present address: Federal University of São Paulo (UNIFESP), Rua Silva Jardim, 136, Santos 11015020, SP, Brazil.

Received: 18 July 2020; Accepted: 14 August 2020; Published: 18 August 2020

Abstract: For the general population, medical diagnosis is a major cause of exposure to low genotoxic stress, as various imaging techniques deliver low doses of ionizing radiation. Our study investigated the consequences of low genotoxic stress on a keratinocyte precursor fraction that includes stem and progenitor cells, which are at risk for carcinoma development. Human skin organoids were bioengineered according to a clinically-relevant model, exposed to a single 50 mGy dose of γ rays, and then xeno-transplanted in nude mice to follow full epidermis generation in an in vivo context. Twenty days post-xenografting, mature skin grafts were sampled and analyzed by semi-quantitative immuno-histochemical methods. Pre-transplantation exposure to 50 mGy of immature human skin organoids did not compromise engraftment, but half of xenografts generated from irradiated precursors exhibited areas displaying focal dysplasia, originating from the basal layer of the epidermis. Characteristics of epithelial-to-mesenchymal transition (EMT) were documented in these dysplastic areas, including loss of basal cell polarity and cohesiveness, epithelial marker decreases, ectopic expression of the mesenchymal marker α-SMA and expression of the EMT promoter ZEB1. Taken together, these data show that a very low level of radiative stress in regenerating keratinocyte stem and precursor cells can induce a micro-environment that may constitute a favorable context for long-term carcinogenesis.

Keywords: human epidermis; keratinocytes; stem cells; precursor cells; low-dose γ irradiation; regeneration; dysplasia; epithelial-to-mesenchymal transition (EMT); ZEB1

1. Introduction

The development of human three-dimensional (3D) organoids by bioengineering now permits these models to make an increasing contribution to deciphering developmental processes, tissue and organ physiology and pathophysiological contexts. Notably, achievements have been reported with heart organoids in the domain of myocardial infarction and drug cardiotoxicity [1], with brain organoids for studies of hypoxic brain injury and prematurity [2] and medulloblastoma modeling [3], with liver organoids for studies of normal [4] and cancer [5] development, and with organoids modeling normal and cancer contexts in the digestive tract [6,7]. In skin, human 3D organoids have demonstrated efficiency in the modeling of pathophysiological contexts, such as defects in the epidermal barrier associated with atopic dermatitis [8], or epidermal cancer proneness in xeroderma pigmentosum [9]. Notably, bioengineered 3D epidermises have contributed to the knowledge of keratinocyte stem and progenitor cells [10–13].

Today, deciphering the adverse impacts of normal tissue exposure to low radiation doses constitutes a biomedical research field of growing interest, due to their increasing use in medical diagnosis technologies such as computed tomography (CT) scans, for both adult and pediatric patients [14,15]. Radiotherapy (RT) is also a source of low-dose exposure for normal tissues and organs surrounding the targeted tumor. As the number of cancer survivors and their lifespans increase thanks to constant improvement of diagnostic methods and medical management, the problem of RT complications is becoming a medical issue of growing importance. Skin is of particular concern regarding RT adverse reactions, as this organ can develop different types of short- and long-term radio-pathologies [16]. Our group and I. Turesson's group in Sweden have shown that human skin, and notably the epidermis, can develop different types of complications after exposure to high [17] or low [18,19] radiation doses, complications such as erythema, epidermitis, dysplasia, as well as acanthosis and carcinoma in the long-term. However, the contributions of specific target cell populations in these pathophysiological processes still require in-depth studies.

Here, we have investigated the effects of low radiation doses on the capacity of keratinocyte stem and progenitor cells to ensure epidermis regeneration, and have explored the cellular perturbations at the origin of radio-induced disorders that can affect this tissue. We show that pre-transplantation exposure of keratinocyte precursor cells to a single dose of 50 mGy, at the initial stage of 3D epidermis generation, induced focal dysplasia in xenografted epidermises, exhibiting characteristics of epithelial-to-mesenchymal transition (EMT).

2. Materials and Methods

2.1. Human Tissue and Cell Materials

The present study was approved by the review board of the iRCM (Institut de Radiobiologie Cellulaire et Moléculaire, CEA (French Alternative Energies and Atomic Energy Commission), Fontenay-aux-Roses, France), and is in accordance with the scientific, ethical, safety and publication policy of CEA (CODECO number DC-2008-228, reviewed by the ethical research committee IDF-3). Human skin tissue from adult healthy donors was collected in the context of breast reduction surgery, after informed consent. Epidermal keratinocytes and dermal fibroblasts were extracted after enzymatic treatment. Frozen banked samples of human epidermal holoclone keratinocytes, generated and characterized in [11], were studied as a model of immature skin keratinocyte precursor cells.

2.2. Bi-Dimensional Culture of Keratinocytes

Holoclone keratinocyte samples were thawed and amplified in bi-dimensional mass conditions one week before use for skin substitute bioengineering. Cultures were performed in a serum-containing medium, in the presence of a feeder layer of human dermal fibroblasts growth-arrested by γ irradiation (60 Gy), as described in [11]. Plastic devices coated with type I collagen were used (Biocoat, BD Biosciences, Le Pont de Claix, France). Composition of the serum-containing medium included DMEM and Ham's F12 media (Gibco, ThermoFisher, Les Ulis, France) (*v/v*, 3/1 mixture), 10% fetal calf serum

(Hyclone, Fisher Scientific, Illkirch, France), 10 ng/mL epidermal growth factor (EGF) (Chemicon, Fisher Scientific, Illkirch, France), 5 μg/mL transferrin (Sigma, Saint-Quentin Fallavier, France), 5 μg/mL insulin (Sigma, Saint-Quentin Fallavier, France), 0.4 μg/mL hydrocortisone (Sigma, Saint-Quentin Fallavier, France), 180 μM adenine (Sigma, Saint-Quentin Fallavier, France), 2 mM tri-iodothyronine (Sigma, Saint-Quentin Fallavier, France), 2 mM L glutamine (Gibco, ThermoFisher, Les Ulis, France) and 100 U/mL penicillin/streptomycin (Gibco, ThermoFisher, Les Ulis, France).

2.3. *Three-Dimensional Skin Substitute Bioengineering*

Plasma-based human skin substitutes were reconstructed according to [12,20,21]. Human plasma (generous gift from Pr Lataillade, Biomedical Research Institute of French Armies (IRBA), INSERM U1197 Clamart, France), was mixed on ice with 4.68 mg/mL sodium chloride (Fresenius Medical Care, Savigny, France), 0.8 mg/mL $CaCl_2$ (Laboratoire Renaudin, Itxassou, France), 9.7 μg/mL Exacyl (tranexamic acid) (Sanofi, France) and human dermal fibroblasts. The mixture was spread in 9.6 cm^2 Petri dishes (BD Biosciences, Le Pont de Claix, France) and plasma fibrin was allowed to polymerize for 30 min at 37 °C. Fibrin gels were then covered with keratinocyte growth medium (same composition as that used for 2D cultures). The next day, keratinocytes were seeded onto these dermal substrates, at the density of 2400 cells/cm^2. After 2 weeks of culture (medium changed every 2 days), skin substitutes were ready for xenografting.

2.4. *Skin Substitute Irradiation*

Exposition of skin substitutes to ionizing radiation was performed at the initial step of epidermis regeneration by keratinocyte precursor cells. Accordingly, samples were irradiated 6 h after keratinocyte seeding onto dermal reconstructs, a time sufficient for their recovery from trypsinization and attachment to their new environment. A 137Cs source was used (γ rays, IBL637 irradiator, Cis-Bio international, Saclay, France). Low dose irradiations (50 mGy) were performed at the dose rate of 50 mGy/min, and irradiations at higher dose (2 Gy) at the dose rate of 850 mGy/min. Control samples were sham-irradiated (same processing except γ ray delivery). The control, 50 mGy and 2 Gy cohorts comprise respectively 14, 14 and 13 skin substitutes.

2.5. *Skin Substitute Xenografting*

Experimental procedures [12,21] were approved by the ethical committee CEEA-51 from the Center for Exploration and Experimental Functional Research (CERFE) (Genople®, Evry, France). Experiments and housing were managed at CERFE under appropriate aseptic conditions. Immuno-deficient athymic Nude *Foxn1^{nu}* mice (ENVIGO, Gannat, France) were used as recipients for the xenografting of human skin substitutes. Mice were anesthetized via intraperitoneal injection of ketamine (Centravet, Maisons-Alfort, France) and xylasine (Centravet, Maisons-Alfort, France), and maintained onto a heated surface to avoid hypothermia. A full-thickness disk of dorsal skin (~1 cm^2) was removed. This mouse skin piece was then devitalized by serial freezing in liquid nitrogen and thawing, and kept for use as a bio-bandage. The wound bed was covered with an equivalent surface of human bioengineered skin substitute. The devitalized piece of mouse skin was then sutured to the mouse skin border, to cover and transiently protect the xenograft site. This bio-bandage was removed 1 week later under isoflurane (Axience, Pantin, France) anesthesia (anesthetic unit from Minerve, Esternay, France). Mice were euthanized 20 days post-xenografting for analysis of human regenerated skin characteristics using the cervical dislocation method, under anesthesia. Notably, previous characterization of the present xenograft model by live imaging performed on grafts generated with [GFP^+] transduced keratinocytes has shown no diffuse mixing between the human and mouse epidermises, indicating the absence of recruitment of mouse epithelial cells within human xenografts (Figure S1 and [12]).

2.6. Preparation of Skin Sections for Histological Characterization

After dissection from euthanized recipient mice, human regenerated skin samples were washed in PBS, and then fixed for 1 day in a buffered solution of 10% formalin (Labonord, Villeneuve D'ascq, France). Fixed tissue samples were dehydrated by graded successive ethanol treatment and then paraffin-embedded. Paraffin sections of 5 μm thickness were prepared (Novaxia histology laboratory, Saint Laurent Nouan, France). For histological characterization, sections were colored with hematoxylin-eosin-saffron (HES), and then scanned and converted into high resolution digital slides using the Axio Scan.Z1 system (Zeiss, Marly le Roi, France) at the imagery platform of the Genethon institute (Evry, France), or using the Pannoramic scan II system (3D Histech, Budapest, Hungary) at the histology platform of INRA-CEA (Jouy-en-Josas, France).

2.7. Quantitative Histology

2.7.1. Percentage of Section Length Displaying Abnormal Epidermis Histology

For the determination of section length corresponding to defective regeneration, 6 whole-length HES-stained sections were systematically considered for each individual xenograft. Epidermis areas displaying abnormal organization were identified by the experimenter and their length was measured using a digital scale. Percentages reported on whole section lengths were calculated. Data were expressed as dot plots cumulating the analysis of multiple regions for each section to ensure representativity (numbers are indicated in figure legends).

2.7.2. Detection of Non-Cohesive Spaces within Regenerated Epidermises

For quantification of abnormally regenerated epidermis areas, semi-quantitative estimation of non-cohesive spaces was performed on 6 whole-length HES-stained sections for each xenograft. Digital images were converted into binary pictures in greyscale using Fiji software. After establishing a threshold based on tissue-free areas of slides, expanded intercellular spaces present within regenerated epidermises were converted into red pixels, which were automatically quantified using an in-house routine (imagery platform of the Genethon institute, Evry, France). Dot plots cumulated the analyses of multiple regions for each section (numbers are indicated in figure legends).

2.7.3. Assessment of Keratinocyte Polarity

For each individual xenograft, analysis was performed on 2 sections stained with DAPI (Fluoroshield™, Sigma-Aldrich, Saint-Quentin Fallavier, France), in which 3 regions of interest (ROIs) were defined, corresponding to ~ 800 μm section length. For the analysis focused on areas displaying abnormal organization, these regions were identified by the experimenter and selected as ROIs. A mask was defined by the experimenter to extract the basal keratinocyte layer and characterize nuclei orientation versus the dermo-epidermal junction (JDE) plane. Angle measurements were performed automatically using a routine developed with Fiji software, and data were plotted into 18 angle categories (from 0° to 90°) using R software (Genethon imagery platform, Evry, France).

2.8. Section Processing for Immunofluorescence Analyses

Paraffin-embedded sections (Novaxia histology laboratory, Saint Laurent Nouan, France) were deparaffinized in xylene and rehydrated in ethanol-H_2O. Antigen retrieval was then performed by immersion of paraffin sections for 20 min in Target Retrieval Solution (Dako, Glostrup, Denmark) at 95 °C. Non-specific antibody binding was blocked either by incubation in a 2% BSA (bovine serum albumin) solution or in serum. Staining was performed using non-conjugated primary antibodies, revealed using fluorochrome-conjugated secondary antibodies. Negative controls were performed, corresponding to the staining procedure without primary antibody, and showed no signal. Antibodies and blocking condition are listed in Table 1. Nuclei were stained with DAPI (Fluoroshield™, Sigma-Aldrich, Saint-Quentin Fallavier,

France). Image acquisition was performed using a Leica SP8 fluorescence imaging system. For fluorescence semi-quantitative analysis, stained sections were converted into high resolution digital slides using the Axio Scan.Z1 (Zeiss, Marly le Roi, France) at the Genethon imagery platform (Evry, France).

2.9. Semi-Quantitative Immunofluorescence Analyses

All marker analyses were performed on 2 stained sections for each xenograft. For analysis of keratinocyte polarity, 3 ROIs were defined per section, corresponding to ~ 800 μm length. Areas displaying abnormal organization were identified by the experimenter and considered as ROIs for their specific characterization. For analysis of VANGL2 expression, a mask was defined to extract the basal keratinocyte layer. VANGL2 signal level (arbitrary units, a.u.) was determined using a routine developed with Fiji software (Genethon imagery platform, Evry, France). DAPI staining of nuclei was used for signal normalization. For analysis of ZEB1 expression, positive keratinocytes were counted by the experimenter, and percentage of $ZEB1^+$ cells was calculated. DAPI staining (Fluoroshield™, Sigma-Aldrich, Saint-Quentin Fallavier, France) was used to identify and count all nuclei within epidermises. For analysis of E-cadherin and α-smooth muscle actin (α-SMA) expression, fluorescence levels (a.u.) were determined within epidermises using a routine developed with Fiji software (Genethon imagery platform, Evry, France). Keratin-14 (K14) and β-catenin were used to specifically mark keratinocytes.

2.10. Apoptosis Assay

Search for genomic DNA fragmentation associated with late apoptosis was performed in sections of skin substitutes pre-xenografting and regenerated skin 20 days post-xenograting, using the TUNEL (terminal deoxynucleotidyl transferase dUTP nick-end labeling) principle. The In Situ Cell Death Detection Kit was used (Roche Molecular Biochemicals, Mannheim, Germany), according to the manufacturer's instructions. Technical positive controls corresponded to sections treated for 10 min with 1500 U/mL recombinant DNase I (Roche Molecular Biochemicals, Mannheim, Germany). Nuclei were colored with DAPI (Fluoroshield™, Sigma-Aldrich, Saint-Quentin Fallavier, France). Image acquisition was performed using a Leica SP8 fluorescence imaging system (Leica microsystems, Nanterre, France).

2.11. Statistics

Statistical analyses were achieved using GraphPad Prism software (GraphPad Software, San Diego, CA, USA). Statistical significance of the data was determined using the Mann–Whitney U test.

Table 1. Antibodies.

Primary Antibodies	Blocking Reagents
Rabbit polyclonal anti-ZEB1 [H-102] (sc-25388, Santa Cruz, Heidelberg, Germany)	BSA
Rabbit polyclonal anti-involucrin (ab53112, Abcam, Paris, France)	BSA
Rabbit polyclonal anti-αSMA (ab5694, Abcam, Paris, France)	Serum
Mouse monoclonal anti-lamin 5 (ab78286, Abcam, Paris, France)	Diagomics, Blagnac, France
Mouse monoclonal anti-cytokeratin 14 [LL002] (Leica Biosystems, Nanterre, France)	BSA
Mouse monoclonal anti-βcatenin [15B8] (ab6301, Abcam, Paris, France)	Serum
Mouse monoclonal anti-E-cadherin [M168] (ab76055, Abcam, Paris, France)	Serum
Secondary Antibodies	
Goat anti-Mouse, Alexa Fluor®594 conjugate (A-11032, ThermoFisher scientific, Les Ulis, France)	
Goat anti-Mouse, Alexa Fluor®488 conjugate (A-11001, ThermoFisher scientific, Les Ulis, France)	
Goat anti-Rabbit, Alexa Fluor®594 conjugate (A-11037, ThermoFisher scientific, Les Ulis, France)	
Donkey anti-Rabbit, FITC conjugate (ab97063, Abcam, Paris, France)	

3. Results

3.1. Experimental Design

A functional approach was designed to model the impact of ionizing radiation (IR) on the regenerative capacity of human epidermal keratinocyte precursor cells (Figure 1). The cellular material used in this study is defined as holoclone keratinocytes (Figure 1A), which correspond to the progeny of single keratinocyte stem cells [11]. These cells exhibit extensive long-term growth potential in bidimensional (2D) culture, as well as genomic stability and efficient epidermal regeneration, as assessed by in vitro epidermis reconstruction and in vivo xenografting [11,12], thus showing functional characteristics of a highly immature population of cultured precursors. Three-dimensional (3D) skin substitutes were bioengineered (Figure 1B) according to a preclinical process [20]. Skin reconstructs were either exposed or not exposed to ionizing radiation (IR) (Figure 1C) at an early stage of epidermis development, corresponding to a non-stratified keratinocyte basal monolayer (see Figure 1B). Samples were submitted to a single exposure to IR at a low dose (50 mGy) or at a higher dose (2 Gy). The next day, irradiated and non-irradiated tissue samples were xenografted onto recipient nude mice (Figure 1D) [21]. This experimental process enabled the study of human epidermis regeneration in an in vivo context up to complete differentiation. Fully differentiated human epidermis substitutes were then characterized (Figure 1E). The model repeatedly gave rise to normally organized and differentiated epidermises.

3.2. *Irradiation Did Not Compromise Epidermis Regeneration but Induced Local Dysplasia*

Wound re-epithelialization was macroscopically observed in all xenografted mice, indicating that keratinocyte precursors were globally functional in the three experimental conditions. Moreover, in xenografted skin sections colored with hematoxylin-eosin-saffron (HES), epidermis stratification and the presence of a horny layer could be observed in all conditions. Quantitative histological analysis was then performed on HES-colored sections to characterize epidermal regeneration (Figure 2). In xenografts performed with non-irradiated keratinocyte precursors (controls), epidermal development was similar to that of a native epidermis in a majority of the observed areas, exhibiting a regular basal layer and correctly stratified supra-basal layers. Notably, differentiated granular and horny layers were clearly identifiable (Figure 2A). In some areas, a discrete disorganization affecting the basal and spinous layers was observed (Figure 2B), which represented an average of 3% of the section length, and corresponded to the background rate of epidermal irregularity of the experimental model. A maximum of 12% irregular areas was observed in two out of 14 control xenografts (Figure 2B). Accordingly, the 12% value was considered a threshold for categorizing abnormal epidermis regeneration. Epidermises regenerated by irradiated keratinocyte precursor cells exhibited local marked disorganization of the basal and spinous layers, alteration of the dermo-epidermis junction and infiltration of keratinocytes in the dermis, which characterized abnormalities termed dysplastic areas (DAs) (Figure 2A). In xenografts performed with keratinocyte precursors that received the IR dose of 2 Gy, DAs represented an average of 10% of the section length, with 10 out of 13 xenografts above the control cohort threshold value, and a maximum DA extent of 42% ($p < 0.0001$ versus non-irradiated) (Figure 2B,C). A quite unexpected observation was the marked impact of the low-dose IR of 50 mGy on regenerated epidermis characteristics—an average of 12% of the section length corresponded to DAs in this condition. Notably, seven out of 14 xenografts were above the control cohort threshold value, with a maximum observed DA extent of 93% ($p < 0.0001$ versus non-irradiated) (Figure 2B,C). Taken together, these observations showed that a single exposure to IR can perturb epidermis regeneration by human keratinocyte precursors, even at a dose of 50 mGy. Importantly, a search for DNA fragmentation (TUNEL assay), performed either on 3D skin substitutes, in vitro 24 h post-irradiation or in vivo 20 days post-xenografting, did not detect any positive signal (Figure 2D,E), documenting an absence of apoptosis. In addition, the presence or absence of p16^{INK4a}, which exerts the function of stress-induced senescence promotion in keratinocytes [22], was assessed by immunofluorescence. No p16^{INK4a} staining was observed in keratinocytes, either in normal epidermis areas or in dysplastic regions of xenografts (Figure S2 and Table S1), suggesting that senescence is not a major mechanism in dysplasia and EMT development. The next parts of the study were then focused on samples in the low-dose conditions.

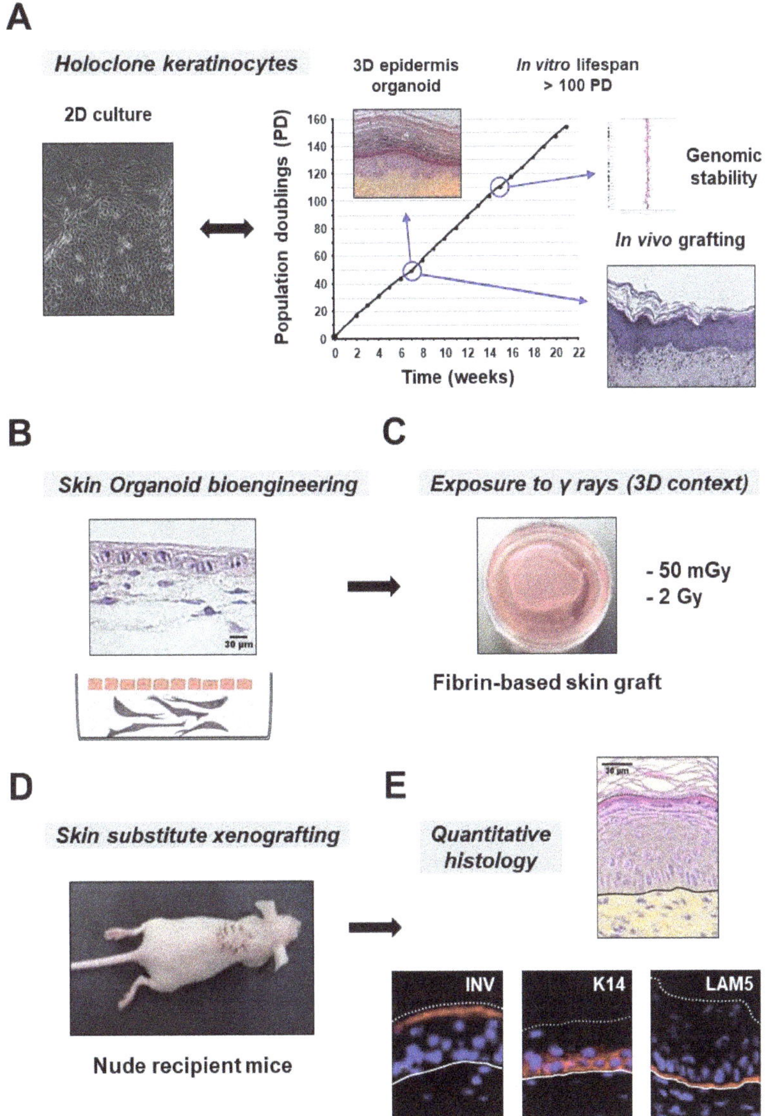

Figure 1. Study architecture. (**A**) Holoclone keratinocytes were used as a cellular model of cultured human epidermal precursor cells. (**B**) Bioengineering of an immature three-dimensional (3D) human skin substitute using holoclone keratinocytes for epidermis regeneration. A typical section colored with hematoxylin-eosin-saffron (HES) is shown. (**C**) Single-exposure of 3D human skin substitutes to ionizing radiation (IR): 50 mGy (dose rate: 50 mGy/min), 2 Gy (dose rate: 850 mGy/min) or sham irradiation. (**D**) The next day, xenografting of irradiated and non-irradiated 3D human skin substitutes in recipient nude mice, which enables full maturation of human epidermises in an in vivo context. (**E**) Removal and sampling of human grafts 20 days post-xenografting for quantitative histology and analysis of marker expression patterns. Pictures from a typical section with normal histology (HES coloration), and normal expression pattern of epidermal markers are shown: keratin-14 (K14), laminin-5 (LAM5) and involucrin (INV).

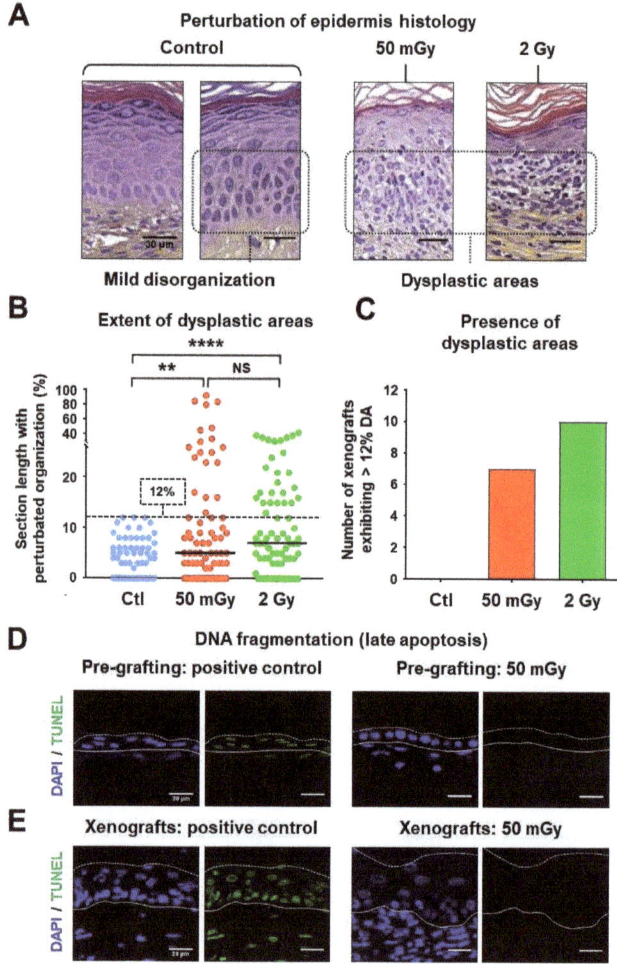

Figure 2. Local dysplastic areas developed in xenografted epidermises originating from irradiated keratinocyte precursors. (**A**) HES coloration of human skin samples 20 days post-xenografting. Representative pictures were selected for the visualization of the normal histology of control epidermises, the mild disorganization considered as the background of the xenograft model, and examples of dysplastic areas (DA) that were characteristic of irradiated conditions. (**B**) Estimation of the percentage of tissue section length displaying mild disorganization or DA. A total of 14 xenografts were performed for the control (Ctl) and 50 mGy conditions, and 13 were performed for the 2 Gy condition. Dot plots cumulated the analyses of 6 sections for each xenograft. Bars correspond to median values (NS $p > 0.05$; ** $p < 0.01$; **** $p < 0.0001$, Mann–Whitney U test). (**C**) The histogram shows the numbers of xenografts that displayed DAs corresponding to at least 12% of epidermis section length: $n = 7$ out of 14 xenografts for the 50 mGy condition; $n = 10$ out of 13 xenografts for the 2 Gy condition. (**D**,**E**) Search for genomic DNA fragmentation associated with late apoptosis using the terminal deoxynucleotidyl transferase dUTP nick-end labeling (TUNEL) assay in pre-grafting skin substitutes (24 h post-irradiation) (**D**) and 20 days post-xenografting (**E**). Technical positive controls corresponded to sections treated with DNase. Nuclei were colored with DAPI. No signal was detected either in the sham-irradiated or in the irradiated conditions at both experimental stages. Photographs shown are representative of $n = 14$ xenografts for both the control and 50 mGy conditions.

3.3. Dysplastic Areas Exhibited Impaired Polarity of Basal Keratinocytes

As an abnormal organization of the keratinocyte basal layer was systematically observed in dysplastic areas, a particular focus was made on this epidermal compartment. In healthy skin, basal keratinocytes are oriented perpendicularly to the dermo-epidermal junction (JDE), whereas loss of polarity occurs in various pathophysiological processes including epithelial-to-mesenchymal transition (EMT). Measurements of basal nuclei orientation were performed on the xenograft sections by image analysis of stained nuclei (Figure 3). Nuclei orientations versus the JDE plane were determined (Figure 3A) and classified according to three categories: nearly perpendicular (angles between 60° and 90°), nearly parallel (angles between 0° and 30°) and oblique (angles between 30° and 90°) (Figure 3B). In control xenografts, a majority of nuclei had a nearly perpendicular orientation, and similar data were obtained in normal areas of xenografts from irradiated keratinocyte precursors. In contrast, a marked increase of oblique and nearly horizontal nuclei orientations was detected within DAs, demonstrating a significant loss of epithelial polarity ($p < 0.0001$ versus repartition in normal areas) (Figure 3B). To further document this observation, the expression pattern of VANGL2, a membrane protein involved in the regulation of cell polarity and migration [23], was then analyzed in xenograft sections by immunofluorescence (Figure 3C,D). This protein was expressed in all basal keratinocytes in normal epidermis, whereas it was reduced or absent in dysplastic cells (Figure 3C). Semi-quantitative image analysis confirmed that the VANGL2 level was significantly reduced in DAs ($p < 0.0001$ versus control areas) (Figure 3D). In summary, a loss of basal keratinocyte polarity that spatially correlated with a perturbated expression of VANGL2 was identified as a marked characteristic of dysplastic areas.

3.4. Defective Cell–Cell Interactions Were Observed in Dysplastic Areas

In a non-pathological context, the interfollicular epidermis forms a cohesive structure devoid of large intercellular spaces. Microscopic observation of xenograft sections pointed out the presence of visible spaces within DAs, the extent of which were then estimated by semi-quantitative image analysis using an in-house algorithm (Figure 4A,B). Analysis of the whole section length showed that empty spaces were globally augmented in xenografts generated with irradiated keratinocyte precursors ($p < 0.005$ versus control xenografts) (Figure 4B). When the analysis was focused on DAs, this parameter was even more significantly increased, due to the presence of large non-cohesive zones ($p < 0.0001$ versus control xenografts and versus normal areas from the 50 mGy conditions) (Figure 4A,B). Considering the regulatory link between VANGL2 and E-cadherin [24], the expression patterns of the latter cell–cell adhesion molecule were analyzed in xenograft sections by immunofluorescence (Figure 4C,D). In xenografts generated with non-irradiated keratinocyte precursors (Figure 4C), as well as outside DAs in xenografts from irradiated cells, a typical expression pattern of E-cadherin on keratinocyte membranes was observed, thinly contouring cells within the basal and supra-basal layers. In DAs, a more blurred pattern, associated with a globally lower expression level, was observed (Figure 4C). Semi-quantitative analysis of fluorescence signals confirmed the marked reduction of E-cadherin level in DAs ($p < 0.0001$ versus control xenografts; $p < 0.01$ versus whole-length 50 mGy xenograft sections) (Figure 4D). In summary, these observations identified defective connectivity as a characteristic of dysplastic areas.

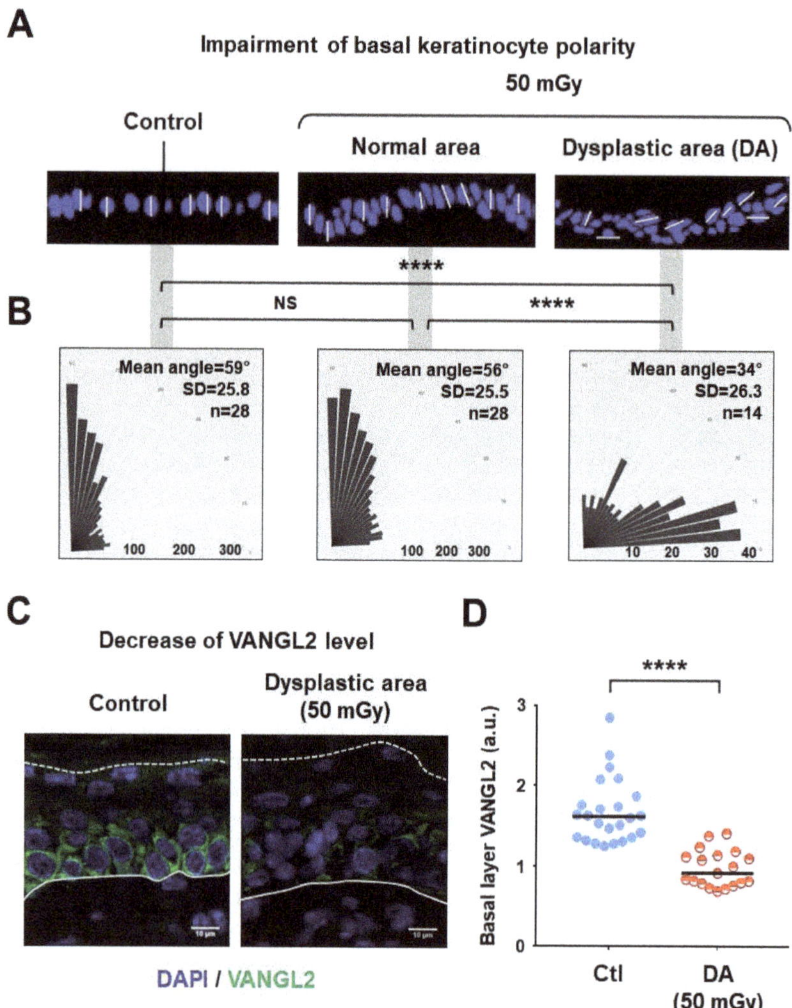

Figure 3. Perturbations of basal keratinocyte polarity in dysplastic epidermis areas of xenografts. (**A**) Imaging of basal keratinocyte nuclei orientation versus the dermo-epidermal junction (JDE) plane, after coloration with DAPI. Typical zoomed pictures of basal layer sections are shown, with white bars added to illustrate some perpendicular, oblique and parallel nuclei orientations. (**B**) Distribution of basal keratinocyte nuclei according to angle versus the JDE plan into 18 angle categories from 0° to 90°, characterized by automated image analysis. The vertical axis represents angle values and the horizontal axis numbers of cells in the different angle categories. Analysis was performed on 14 different xenografts for all conditions, n indicates numbers of analyzed dysplastic areas (NS $p > 0.05$; **** $p < 0.0001$, Mann–Whitney U test). (**C**) Immunofluorescence detection of VANGL2 protein. Representative pictures were selected for the visualization of VANGL2 in basal keratinocytes from normal epidermis regions, showing its decrease or absence in basal keratinocytes from dysplastic areas. Nuclei were colored with DAPI. (**D**) Quantification of VANGL2 fluorescence in the epidermis basal layer (arbitrary units, a.u.), showing a reduced signal within DAs. Dot plots cumulated the analyses of 24 normal regions from 14 control (Ctl) xenografts and 19 DAs from the 14 xenografts of the 50 mGy conditions. Bars correspond to median values. (**** $p < 0.0001$, Mann–Whitney U test).

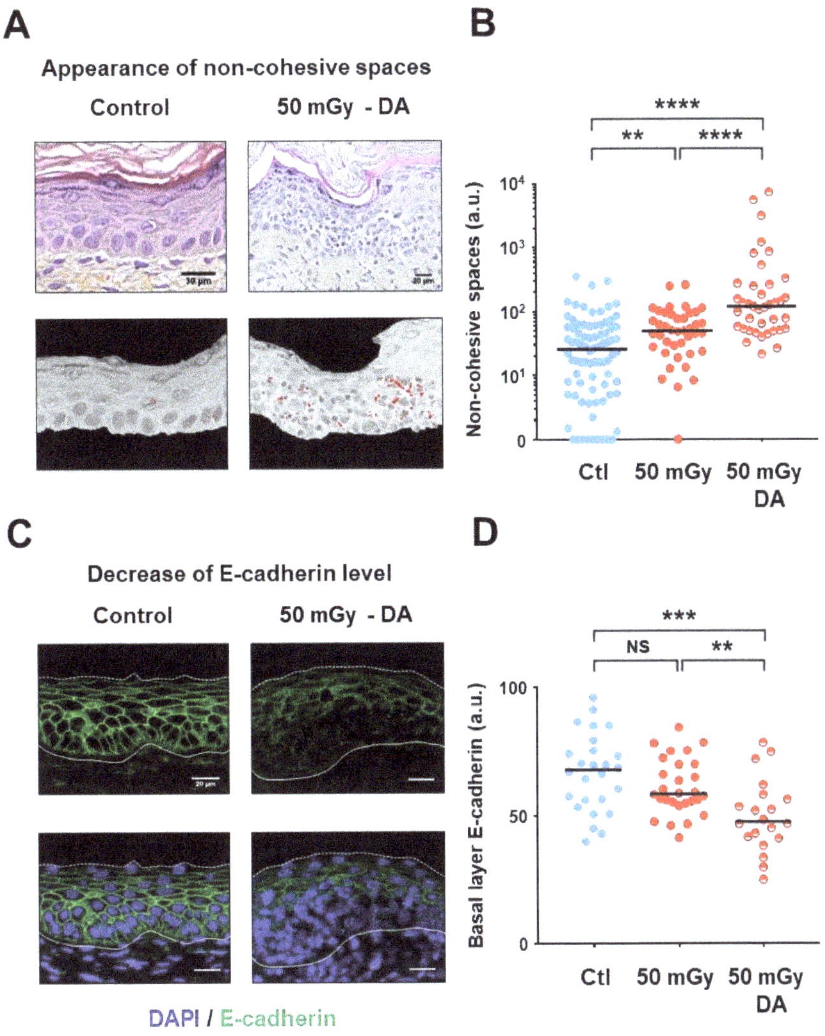

Figure 4. Defective cell-cell′ cohesiveness in dysplastic areas. (**A**) Typical pictures of HES sections illustrating cohesive epidermis in normal regions and the presence of visible non-cohesive spaces within DAs (top panel). Conversion of spaces into red pixels by automated image processing (bottom panel) for semi-quantitative analysis. (**B**) Semi-quantitative analysis of non-cohesive spaces based on red pixel conversion (arbitrary units, a.u.), revealing the presence of significant non-cohesive zones in DAs. Dot plots cumulated the analyses of 82 normal regions from control (Ctl) xenografts, 41 random regions from the 50 mGy xenografts and 39 selected regions corresponding to DAs in the 50 mGy xenografts. Bars correspond to median values (** $p < 0.01$; **** $p < 0.0001$, Mann–Whitney U test). (**C**) Immunofluorescence detection of E-cadherin. Representative pictures were selected for the visualization of E-cadherin expression in normal epidermis regions, showing its decrease or absence in basal keratinocytes from DAs. Nuclei were colored with DAPI. (**D**) Semi-quantitative analysis of E-cadherin (arbitrary units, a.u.), showing a lower signal within DAs. Dot plots cumulated the analyses of 26 normal regions from the 14 control (Ctl) xenografts and 20 DAs from the 14 xenografts of the 50 mGy condition. Bars correspond to median values (NS $p > 0.05$; ** $p < 0.01$; **** $p < 0.0001$, Mann–Whitney U test).

3.5. Epithelial-to-Mesenchymal Transition Markers Were Detected in Epidermal Dysplastic Areas

Considering the major importance of E-cadherin and stable cellular adherens junctions in the pathophysiological process of epithelial-to-mesenchymal transition (EMT) [25], expression of EMT effectors and markers was analyzed in xenografts. Firstly, expression of the transcription factor ZEB1 was investigated (Figure 5A,B). In control xenografts (Figure 5A), as well as in normal regions of xenografts from irradiated keratinocyte precursors (not shown), ZEB1 expression was almost exclusively restricted to the dermis, and rarely observed in epidermal keratinocytes (Figure 5A). Quantification performed on whole-length sections indicated that ZEB1-positive cells represented about 3% or less of most sections (Figure 5B). In contrast, observation of xenografts from irradiated keratinocyte precursors showed that ZEB1-positive keratinocytes were abundantly present in DAs (Figure 5A), reaching up to 18% of keratinocytes ($p < 0.0001$ versus control xenografts, and versus whole-length 50 mGy xenograft sections) (Figure 5B). Of note, this increase was not significant when considering whole-length 50 mGy xenograft sections, compared to controls (Figure 5B), suggesting that activation of ZEB1 expression did not concern all keratinocytes. Of note, β-catenin expression on keratinocyte membranes was also impaired in DAs. Expression of α-smooth muscle actin (α-SMA), a typical mesenchymal marker, was then analyzed in association with keratin-14 (K14), used as a specific counterstaining of basal keratinocytes (Figure 5C,D). In control xenografts (Figure 5C), as well as in normal regions of xenografts from irradiated keratinocyte precursors (not shown), α-SMA was exclusively distributed within the dermis. This distribution was strongly modified in DAs, with the detection of α-SMA signaling in cells that co-expressed K14 (Figure 5C), showing an ectopic expression in keratinocytes. α-SMA signal quantification focused on DAs showed a detectable signal above control values ($p < 0.0001$ versus control xenografts, and versus whole-length 50 mGy xenograft sections) (Figure 5D). As for ZEB1, the level of α-SMA did not appear to be significantly affected when considering whole-length 50 Gy xenograft sections, compared to the background signal of controls (Figure 5D). Taken together, these results identified a significant link between the pathophysiological process of EMT and characteristics of dysplastic areas.

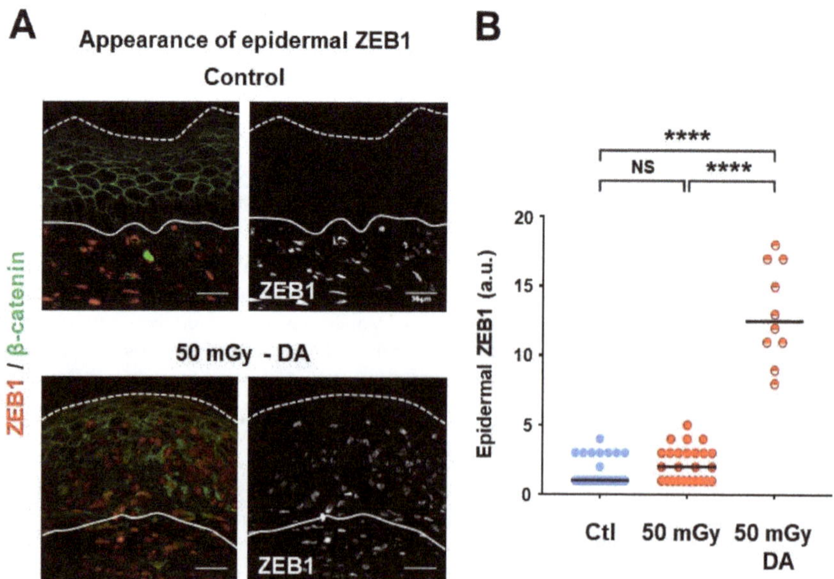

Figure 5. *Cont.*

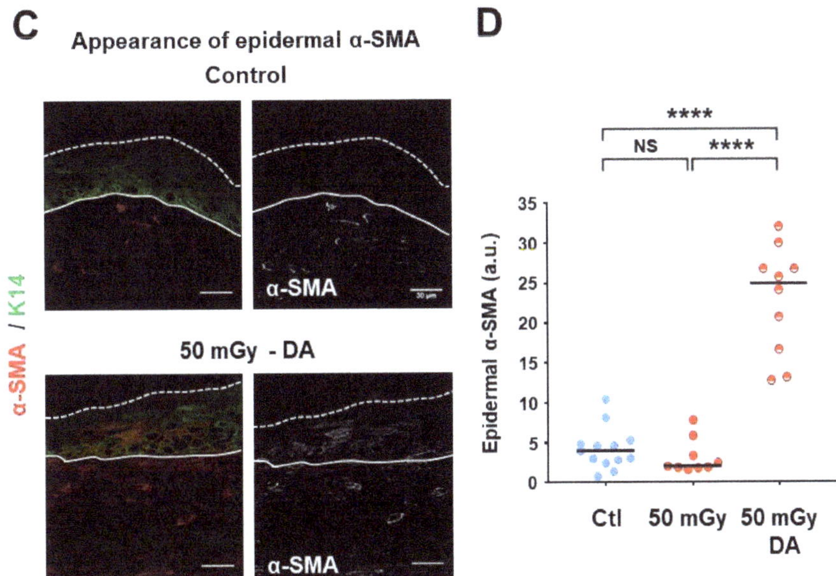

Figure 5. Detection of epithelial-to-mesenchymal transition markers in dysplastic areas. (**A**) Immunofluorescence detection of ZEB1. Representative pictures were selected for the visualization of dermal ZEB1 expression in normal skin regions and its ectopic presence in epidermal DAs. Gray-tone visualization of ZEB1 signal is also shown. (**B**) Quantitative analysis showed rare ZEB1-positive keratinocytes in normal epidermis regions, and abundant ZEB1-positive keratinocytes in DAs. Dot plots cumulated the analyses of 26 normal regions from the 14 control (Ctl) xenografts, 25 random regions from the 14 xenografts of the 50 mGy condition and 10 selected regions corresponding to DAs in 50 mGy xenografts. Bars correspond to median values (NS $p > 0.05$; **** $p < 0.0001$). β-catenin staining, which marked keratinocyte contours, revealed impaired epidermal organization and its reduced expression. (**C**) Immunofluorescence detection of α-smooth muscle actin (α-SMA). Representative pictures were selected for the visualization of dermal α-SMA expression in normal skin regions and its abnormal presence in epidermal DAs. Sections were stained for keratin 14 to mark basal keratinocytes. Gray-tone visualization of α-SMA signal is also shown. (**D**) Dot plots cumulated the analyses of 13 normal regions from control (Ctl) xenografts, 9 random regions from the 50 mGy xenografts and 10 selected regions corresponding to DAs in 50 mGy xenografts. Bars correspond to median values (NS $p > 0.05$; **** $p < 0.0001$).

4. Discussion

High radiation doses such as those delivered during radiation therapy (RT) produce pathological changes in mesenchymal tissues with long-term alterations of fibroblast phenotypic and functional characteristics that may impair the quality of life of treated cancer patients [16,26,27], whereas the pathophysiology of epithelial complications of RT has been investigated less [17]. Furthermore, the effects of very low doses of genotoxic stress are poorly documented [19,28], although dysplasia has been reported in the skin of radiotherapy patients [18]. We have here addressed the question of possible adverse reactions subsequent to the exposition of keratinocyte stem and progenitor cells to low radiation doses. The dose analyzed here (50 mGy) is in the range of those delivered to normal tissues adjacent to the target tumor volume during radiotherapy, and is relevant for biomedical diagnostic procedures, notably scanner imaging. It is much lower than the dose limit accepted for the induction of carcinoma, which has been proposed to be around 500 mGy, notably based on the Japanese atomic bomb survivor study [29]. The key contribution of the present study was the demonstration that dysplasia and epithelial-to-mesenchymal transition (EMT) develop in epidermises generated by keratinocyte stem and precursor cells exposed to a single low dose of γ irradiation, thus documenting a micro-environment favoring the development of skin cancer.

Epidermal holoclone keratinocytes provided a model representative of an immature population of cultured precursors containing functional stem and progenitor cells. These cells correspond to the clonal progeny of single keratinocyte stem cells that were functionally characterized according to an extensive growth potential exceeding 100 population doublings through successive subcultures, and the capacity for epidermis reconstruction in vitro and regeneration in vivo [11,12]. Importantly, the stem cell status attributed to holoclones has been demonstrated in vivo by cellular tracing in the entirely regenerated epidermis of an epidermolysis bullosa patient engrafted with an autologous, genetically corrected skin substitute [30]. Moreover, the fibrin-based epidermis organoids that were used in this study corresponded to an adaption from a clinically relevant model of bioengineered skin substitute [20,31]. The present epidermis regeneration approach permits modeling of skin stem and progenitor cell properties and potentialities in conditions characterized by a higher level of cell proliferation and metabolic activities than in the context of healthy skin homeostasis. These conditions probably exacerbate radiosensitivity and thus allow observation of cell and tissue responses to low stress levels.

The semi-quantitative imaging approaches that were set-up to characterize the epidermal distribution patterns of molecular markers provided clues for the understanding of genotoxic stress-induced epidermal dysplasia and EMT development. Firstly, the marked alteration of VANGL2 patterns that was detected in association with the impaired orientation of basal keratinocyte nuclei constituted a relevant parameter, due to the involvement of this membrane protein in the regulation of cell polarity and migration [23]. VANGL2 is a central component of the planar cell polarity signaling pathway (PCP), which is essential for correct epidermal development and morphogenesis [32,33], as well as epidermal wound repair [34]. The loss of epithelial polarity is a key process in the early steps of EMT. Weakening and disruption of cell–cell contacts, which were documented here by the presence of non-cohesive spaces and a local decrease in E-cadherin level, are typical characteristics of the pathophysiological process of EMT [25]. Finally, detection of an ectopic expression of the mesenchymal marker α-SMA in keratinocytes consolidated the EMT-like phenotypic switch occurring in radiation-induced dysplastic areas (DAs) [35]. Among the various transcription factors involved in EMT, ZEB1 has been described as a major early player in its development, later favoring epithelial tumor progression in association with E-cadherin suppression [36]. We show here the appearance of ectopic ZEB1 protein in keratinocytes of dysplastic epidermis, thus providing a link between low-dose irradiation of holoclone keratinocyte precursor cells and potential initiation sites of the carcinoma development cascade. Notably, perturbated functions of the 'wingless' (WNT) signaling pathways, as suggested here by the loss of the β-catenin protein, associated with focal dysplasia, might constitute a promoter event of the pathophysiological processes described here.

In conclusion, we have developed an approach based on in vitro bioengineering of human skin organoids, coupled with in vivo xenografting in immune-deficient mice, to explore the pathophysiological consequences of low-dose γ-irradiation exposure of epidermal stem and progenitor cells on their subsequent regenerative capacity. We have observed that a single 50 mGy radiation dose was sufficient to promote local perturbations in regenerated epidermises with cellular and molecular characteristics of dysplasia and EMT, which may constitute an initial risk for the future development of carcinomas. Interestingly, this approach is directly applicable to other biomedical research domains, for example characterization of the skin's defenses and responses to pollution [37].

Supplementary Materials: The following are available online at http://www.mdpi.com/2073-4409/9/8/1912/s1, Figure S1: Absence of epidermis mixing between recipient mice skin and regenerated human epidermis grafts. Figure S2 and Table S1: Absence of p16^{INK4a} in human epidermises regenerated by irradiated and non-irradiated keratinocyte precursors. Supplementary Methods.

Author Contributions: S.C.: investigation, methodology, resources, formal analysis, validation and original draft preparation; R.N.G.: investigation, methodology, resources, formal analysis and validation; D.S.: investigation, methodology, resources and formal analysis; P.A.: investigation, methodology and resources; M.T.M.: conceptualization, methodology, formal analysis, visualization, validation, writing—original draft preparation and writing—review and editing, supervision, funding acquisition; N.O.F.: conceptualization, methodology, formal analysis, visualization, validation, writing—original draft preparation and writing, review and editing, supervision, funding acquisition. All authors have read and agreed to the published version of the manuscript.

Funding: This work was supported by grants from: CEA and INSERM (UMR967); "Délégation Générale de l'Armement" (DGA) grants; FUI-AAP13 and the "Conseil Général de l'Essonne" within the STEMSAFE grant; EURATOM (RISK-IR, FP7, grant 323267); and "Electricité de France" (EDF). Genopole® (Evry, France) also provided support for equipment and infrastructures.

Acknowledgments: We wish to thank S. Bouet and A. Boukadiri (histology platform, UMR 1313 GABI, INRA/CEA, Jouy en Josas) for technical assistance. Our thanks also go to H. Serhal and Y. Diaw of the Clinique de l'Essonne (Evry), who kindly provided human skin samples from healthy donors. We thank J.-J. Lataillade and M. Trouillas (IRBA, INSERM U1197, Clamart) for helpful discussions on the skin regeneration model.

Conflicts of Interest: The authors declare no conflict of interest.

References

1. Richards, D.J.; Li, Y.; Kerr, C.M.; Yao, J.; Beeson, G.C.; Coyle, R.C.; Chen, X.; Jia, J.; Damon, B.; Wilson, R.; et al. Human cardiac organoids for the modelling of myocardial infarction and drug cardiotoxicity. *Nat. Biomed. Eng.* **2020**, *4*, 446–462. [CrossRef] [PubMed]
2. Pașca, A.M.; Park, J.Y.; Shin, H.W.; Qi, Q.; Revah, O.; Krasnoff, R.; O'Hara, R.; Willsey, A.J.; Palmer, T.D.; Pașca, S.P. Human 3D cellular model of hypoxic brain injury of prematurity. *Nat. Med.* **2019**, *25*, 784–791. [CrossRef] [PubMed]
3. Ballabio, C.; Anderle, M.; Gianesello, M.; Lago, C.; Miele, E.; Cardano, M.; Aiello, G.; Piazza, S.; Caron, D.; Gianno, F.; et al. Modeling medulloblastoma in vivo and with human cerebellar organoids. *Nat. Commun.* **2020**, *11*, 583. [CrossRef] [PubMed]
4. Huch, M.; Gehart, H.; van Boxtel, R.; Hamer, K.; Blokzijl, F.; Verstegen, M.M.; Ellis, E.; van Wenum, M.; Fuchs, S.A.; de Ligt, J.; et al. Long-term culture of genome-stable bipotent stem cells from adult human liver. *Cell* **2015**, *160*, 299–312. [CrossRef]
5. Broutier, L.; Mastrogiovanni, G.; Verstegen, M.M.; Francies, H.E.; Gavarró, L.M.; Bradshaw, C.R.; Allen, G.E.; Arnes-Benito, R.; Sidorova, O.; Gaspersz, M.P.; et al. Human primary liver cancer-derived organoid cultures for disease modeling and drug screening. *Nat. Med.* **2017**, *23*, 1424–1435. [CrossRef]
6. Clevers, H. Modeling Development and Disease with Organoids. *Cell* **2016**, *165*, 1586–1597. [CrossRef]
7. Drost, J.; van Boxtel, R.; Blokzijl, F.; Mizutani, T.; Sasaki, N.; Sasselli, V.; de Ligt, J.; Behjati, S.; Grolleman, J.E.; van Wezel, T.; et al. Use of CRISPR-modified human stem cell organoids to study the origin of mutational signatures in cancer. *Science* **2017**, *358*, 234–238. [CrossRef]
8. Pendaries, V.; Malaisse, J.; Pellerin, L.; Le Lamer, M.; Nachat, R.; Kezic, S.; Schmitt, A.M.; Paul, C.; Poumay, Y.; Serre, G.; et al. Knockdown of filaggrin in a three-dimensional reconstructed human epidermis impairs keratinocyte differentiation. *J. Investig. Dermatol.* **2014**, *134*, 2938–2946. [CrossRef]
9. Bernerd, F.; Asselineau, D.; Vioux, C.; Chevallier-Lagente, O.; Bouadjar, B.; Sarasin, A.; Magnaldo, T. Clues to epidermal cancer proneness revealed by reconstruction of DNA repair-deficient xeroderma pigmentosum skin in vitro. *Proc. Natl. Acad. Sci. USA* **2001**, *98*, 7817–7822. [CrossRef]
10. Fortunel, N.O.; Hatzfeld, J.A.; Rosemary, P.-A.; Ferraris, C.; Monier, M.-N.; Haydont, V.; Longuet, J.; Brethon, B.; Lim, B.; Castiel, I.; et al. Long-term expansion of human functional epidermal precursor cells: Promotion of extensive amplification by low TGF-beta1 concentrations. *J. Cell Sci.* **2003**, *116*, 4043–4052. [CrossRef]
11. Fortunel, N.O.; Cadio, E.; Vaigot, P.; Chadli, L.; Moratille, S.; Bouet, S.; Roméo, P.-H.; Martin, M.T. Exploration of the functional hierarchy of the basal layer of human epidermis at the single-cell level using parallel clonal microcultures of keratinocytes. *Exp. Dermatol.* **2010**, *19*, 387–392. [CrossRef] [PubMed]
12. Fortunel, N.O.; Chadli, L.; Coutier, J.; Lemaître, G.; Auvré, F.; Domingues, S.; Bouissou-Cadio, E.; Vaigot, P.; Cavallero, S.; Deleuze, J.-F.; et al. KLF4 inhibition promotes the expansion of keratinocyte precursors from adult human skin and of embryonic-stem-cell-derived keratinocytes. *Nat. Biomed. Eng.* **2019**, *3*, 985–997. [CrossRef] [PubMed]

13. Larderet, G.; Fortunel, N.O.; Vaigot, P.; Cegalerba, M.; Maltère, P.; Zobiri, O.; Gidrol, X.; Waksman, G.; Martin, M.T. Human side population keratinocytes exhibit long-term proliferative potential and a specific gene expression profile and can form a pluristratified epidermis. *Stem Cells* **2006**, *24*, 965–974. [CrossRef] [PubMed]
14. Kopp, M.; Loewe, T.; Wuest, W.; Brand, M.; Wetzl, M.; Nitsch, W.; Schmidt, D.; Beck, M.; Schmidt, B.; Uder, M.; et al. Individual calculation of effective dose and risk of malignancy based on Monte Carlo simulations after whole body computed tomography. *Sci. Rep.* **2020**, *10*, 9475. [CrossRef] [PubMed]
15. Bernier, M.O.; Baysson, H.; Pearce, M.S.; Moissonnier, M.; Cardis, E.; Hauptmann, M.; Struelens, L.; Dabin, J.; Johansen, C.; Journy, N.; et al. Cohort profile: The EPI-CT study: A european pooled epidemiological study to quantify the risk of radiation-induced cancer from paediatric CT. *Int. J. Epidemiol.* **2019**, *48*, 379–381. [CrossRef]
16. Martin, M.T.; Vulin, A.; Hendry, J.H. Human epidermal stem cells: Role in adverse skin reactions and carcinogenesis from radiation. *Mutat. Res.* **2016**, *770*, 349–368. [CrossRef]
17. Sivan, V.; Vozenin-Brotons, M.-C.; Tricaud, Y.; Lefaix, J.-L.; Cosset, J.-M.; Dubray, B.; Martin, M.T. Altered proliferation and differentiation of human epidermis in cases of skin fibrosis after radiotherapy. *Int. J. Radiat. Oncol. Biol. Phys.* **2002**, *53*, 385–393. [CrossRef]
18. Turesson, I.; Nyman, J.; Qvarnström, F.; Simonsson, M.; Book, M.; Hermansson, I.; Sigurdardottir, S.; Johansson, K.A. A low-dose hypersensitive keratinocyte loss in response to fractionated radiotherapy is associated with growth arrest and apoptosis. *Radiother. Oncol.* **2010**, *94*, 90–101. [CrossRef]
19. Qvarnström, F.; Simonsson, M.; Nyman, J.; Hermansson, I.; Book, M.; Johansson, K.A.; Turesson, I. Double strand break induction and kinetics indicate preserved hypersensitivity in keratinocytes to subtherapeutic doses for 7weeks of radiotherapy. *Radiother. Oncol.* **2017**, *122*, 163–169.
20. Alexaline, M.M.; Trouillas, M.; Nivet, M.; Bourreau, E.; Leclerc, T.; Duhamel, P.; Martin, M.T.; Doucet, C.; Fortunel, N.O.; Lataillade, J.-J. Bioengineering a human plasma-based epidermal substitute with efficient grafting capacity and high content in clonogenic cells. *Stem Cells Transl. Med.* **2015**, *4*, 643–654. [CrossRef]
21. Fortunel, N.O.; Bouissou-Cadio, E.; Coutier, J.; Martin, M.T. Iterative three-dimensional epidermis bioengineering and xenografting to assess long-term regenerative potential in human keratinocyte precursor cells. *Methods Mol. Biol.* **2020**, *2109*, 155–167. [PubMed]
22. Sasaki, M.; Kajiya, H.; Ozeki, S.; Okabe, K.; Ikebe, T. Reactive oxygen species promotes cellular senescence in normal human epidermal keratinocytes through epigenetic regulation of p16(INK4a.). *Biochem. Biophys. Res. Commun.* **2014**, *452*, 622–628. [CrossRef] [PubMed]
23. Williams, B.B.; Cantrell, V.A.; Mundell, N.A.; Bennett, A.C.; Quick, R.E.; Jessen, J.R. VANGL2 regulates membrane trafficking of MMP14 to control cell polarity and migration. *J. Cell Sci.* **2012**, *125*, 2141–2147. [CrossRef] [PubMed]
24. Nagaoka, T.; Inutsuka, A.; Begum, K.; Bin Hafiz, K.M.; Kishi, M. Vangl2 regulates E-cadherin in epithelial cells. *Sci. Rep.* **2014**, *4*, 6940. [CrossRef] [PubMed]
25. Zhitnyak, I.Y.; Rubtsova, S.N.; Litovka, N.I.; Gloushankova, N.A. Early events in actin cytoskeleton dynamics and E-cadherin-mediated cell-cell adhesion during epithelial-mesenchymal transition. *Cells* **2020**, *9*, 578. [CrossRef]
26. Martin, M.; Lefaix, J.; Delanian, S. TGF-beta1 and radiation fibrosis: A master switch and a specific therapeutic target? *Int. J. Radiat. Oncol. Biol. Phys.* **2000**, *47*, 277–290. [CrossRef]
27. Reisdorf, P.; Lawrence, D.A.; Sivan, V.; Klising, E.; Martin, M.T. Alteration of transforming growth factor-beta1 response involves down-regulation of Smad3 signaling in myofibroblasts from skin fibrosis. *Am. J. Pathol.* **2001**, *159*, 263–272. [CrossRef]
28. Mezentsev, A.; Amundson, S.A. Global gene expression responses to low- or high-dose radiation in a human three-dimensional tissue model. *Radiat. Res.* **2011**, *175*, 677–688. [CrossRef]
29. Sugiyama, H.; Misumi, M.; Kishikawa, M.; Iseki, M.; Yonehara, S.; Hayashi, T.; Soda, M.; Tokuoka, S.; Shimizu, Y.; Sakata, R.; et al. Skin cancer incidence among atomic bomb survivors from 1958 to 1996. *Radiat. Res.* **2014**, *181*, 531–539. [CrossRef]
30. Hirsch, T.; Rothoeft, T.; Teig, N.; Bauer, J.W.; Pellegrini, G.; De Rosa, L.; Scaglione, D.; Reichelt, J.; Klausegger, A.; Kneisz, D.; et al. Regeneration of the entire human epidermis using transgenic stem cells. *Nature* **2017**, *551*, 327–332. [CrossRef]

31. Ronfard, V.; Rives, J.-M.; Neveux, Y.; Carsin, H.; Barrandon, Y. Long-term regeneration of human epidermis on third degree burns transplanted with autologous cultured epithelium grown on a fibrin matrix. *Transplantation* **2000**, *70*, 1588–1598. [CrossRef] [PubMed]
32. Box, K.; Joyce, B.W.; Devenport, D. Epithelial geometry regulates spindle orientation and progenitor fate during formation of the mammalian epidermis. *Elife* **2019**, *8*, e47102. [CrossRef] [PubMed]
33. Kimura-Yoshida, C.; Mochida, K.; Nakaya, M.A.; Mizutani, T.; Matsuo, I. Cytoplasmic localization of GRHL3 upon epidermal differentiation triggers cell shape change for epithelial morphogenesis. *Nat. Commun.* **2018**, *9*, 4059. [CrossRef] [PubMed]
34. Caddy, J.; Wilanowski, T.; Darido, C.; Dworkin, S.; Ting, S.B.; Zhao, Q.; Rank, G.; Auden, A.; Srivastava, S.; Papenfuss, T.A.; et al. Epidermal wound repair is regulated by the planar cell polarity signaling pathway. *Dev. Cell* **2010**, *19*, 138–147. [CrossRef] [PubMed]
35. Fontana, F.; Raimondi, M.; Marzagalli, M.; Sommariva, M.; Limonta, P.; Gagliano, N. Epithelial-to-mesenchymal transition markers and CD44 isoforms are differently expressed in 2D and 3D cell cultures of prostate cancer cells. *Cells* **2019**, *8*, 143. [CrossRef]
36. Miro, C.; Di Cicco, E.; Ambrosio, R.; Mancino, G.; Di Girolamo, D.; Cicatiello, A.G.; Sagliocchi, S.; Nappi, A.; De Stefano, M.A.; Luongo, C.; et al. Thyroid hormone induces progression and invasiveness of squamous cell carcinomas by promoting a ZEB-1/E-cadherin switch. *Nat. Commun.* **2019**, *10*, 5410. [CrossRef]
37. Svoboda, E. When skin's defence against pollution fails. *Nature* **2018**, *563*, S89–S90. [CrossRef]

MDPI
St. Alban-Anlage 66
4052 Basel
Switzerland
Tel. +41 61 683 77 34
Fax +41 61 302 89 18
www.mdpi.com

Cells Editorial Office
E-mail: cells@mdpi.com
www.mdpi.com/journal/cells

www.ingramcontent.com/pod-product-compliance
Lightning Source LLC
LaVergne TN
LVHW070918170726
843515LV00005B/816

* 9 7 8 3 0 3 6 5 6 9 0 1 7 *